Water
Distribution
Systems

ice
Institution of Civil Engineers

publishing

Water Distribution Systems

Edited by

Dragan A. Savić
University of Exeter, UK

John K. Banyard
Independent consultant, Warwick, UK

Published by ICE Publishing, 40 Marsh Wall, London E14 9TP.

Distributors for ICE Publishing books are
USA: Publishers Storage and Shipping Corp., 46 Development
Road, Fitchburg, MA 01420
Australia: DA Books and Journals, 648 Whitehorse Road, Mitcham
3132, Victoria

First published 2011

Also available from ICE Publishing

Basic Water Treatment, Fourth edition.
C. Binnie and M. Kimber. ISBN 978-0-7277-3608-6
*Pressure Transients in Water Engineering: A Guide to Analysis and
Interpretation of Behaviour.*
J. Ellis. ISBN 978-0-7277-3592-8
Whole Life Costing for Water Distribution Network Management.
P. Skipworth, M. Engelhardt, A. Cashman, D. Savić, A. Saul and
G. Walters. ISBN 978-0-7277-3166-1
Wastewater Treatment and Technology.
C. Forster (ed.) ISBN 978-0-7277-3229-3

www.icevirtuallibrary.com

A catalogue record for this book is available from the British Library

ISBN: 978-0-7277-4112-7

© Thomas Telford Limited 2011

ICE Publishing is a division of Thomas Telford Ltd, a wholly-owned
subsidiary of the Institution of Civil Engineers (ICE).

FSC
www.fsc.org
MIX
Paper from
responsible sources
FSC® C013604

Typeset by Academic + Technical, Bristol
Index created by Indexing Specialists (UK) Ltd, Hove, East Sussex
Printed and bound in Great Britain by CPI Antony Rowe,
Chippenham and Eastbourne

Contents

Preface

There are already a large number of textbooks covering hydraulics and water engineering, so why do we need yet another one to fill our bookshelves?

While the above is certainly a valid question, we, the editors, with long experience in both academia and with water service providers, were aware that there was a gap. With privatisation of the UK water industry in the late 1980s, which brought changes to the organisation of the industry, including a high level of regulation, a reduction in in-house expertise and a decline in research investment, there is a need to provide a useful reference book for practising engineers, particularly to provide information on up-to-date practice in today's increasingly complex water industry. This is important, as practitioners in the developed (and the less-developed) parts of the world face not only classical design and management problems but also ever-increasing environmental and sustainability requirements and concerns, and, at the same time, few engineers can hope to keep pace with the vast amount of material presented at conferences and seminars. We also felt that there needed to be a book which would provide a suitable guide for final-year undergraduates and MSc students of water and environmental engineering courses, but which at the same time could be a useful reference after graduation when they enter employment with the water industry, environmental protection agencies or consultancies.

We have not tried to cover the whole of the water industry but, rather, have focused on water distribution, where there have been major advances in the engineer's ability to optimise solutions, and obtain levels of understanding that have been denied to previous generations of practitioners.

To achieve this aim, we have assembled authors from academia, consultancy and the water service providers, to ensure that each chapter provides a balanced view of not only what is theoretically possible but also what is practical both in the design office and in the world of water distribution system operation.

We hope that our readers will find this book helpful in their working lives, and that it will not become yet another tome that gathers dust on the bookshelfs.

Dragan A. Savić FICE, FCIWEM
University of Exeter, UK

John K. Banyard OBE, F.R.Eng, FCGI, FICE, FCIWEM
Independent consultant, Warwick, UK

Abbreviations

ABC	activity-based costing
ACO	ant colony optimisation
AoS	appraisal of sustainability
ASR	aquifer storage and recovery
AM	area meter
AMP	asset management plan
AZNP	average zone night pressure
BPSO	best practicable sustainable option
BMV	burst main valve
BABE	bursts and background estimates
CAPEX	capital expenditure
CAMS	catchment management strategy
CFD	computational fluid dynamics
CARE-W	Computer Aided Rehabilitation of Water Networks
CBA	cost–benefit analysis
CEA	cost-effectiveness analysis
DDA	demand-driven analysis
DEM	digital elevation model
DG	Director General
DRM	discolouration risk model
DAF	dissolved air flotation
DMA	distribution management areas
DOMS	distribution operation and maintenance strategies
DMA	district metered area
DCM	domestic consumption monitor
DWI	Drinking Water Inspectorate
DWSP	drinking water safety plan
ELL	economic level of leakage
ECF	electro-coagulation–flotation
EGL	energy grade line
EA	Environment Agency
EIA	environmental impact assessment
EPA	Environmental Protection Agency
ET	evapotranspiration
EPR	evolutionary polynomial regression
EPS	extended period simulation
FOSM	first order second moment model
FDO	flexible design option
GA	genetic algorithm
GIS	geographic information system
GGA	global gradient algorithm
GAC	granular activated carbon
GUI	graphical user interface
GHG	greenhouse gas
GDP	gross domestic product
GSS	Guaranteed Standards Scheme
HDA	Head-driven analysis
HDN	heuristic derived from nature
HDPE	high-density polyethylene
HPPE	high-performance polyethylene

HGL	hydraulic grade line
IPCC	Intergovernmental Panel on Climate Change
IRR	internal rate of return
IMF	International Monetary Fund
ISO	International Organization for Standardization
IWA	International Water Association
lcd	litres per capita per day
MIEX	magnetic ion exchange
MDPE	medium-density polyethylene
MCS	Monte Carlo simulation
MCA	multi-criteria analysis
MOGA	multi-objective genetic algorithm
NOM	natural organic matter
NTU	nephelometric turbidity units
NPSH	net positive suction head
NPV	net present value
NHPP	non-homogeneous Poisson process
NRV	non-return valve
Ofwat	Office of Water Services
OSEC	on-site electrolytic chlorination
OPEX	operational expenditure
OPI	operational performance index
OECD	Organisation for Economic Co-operation and Development
OPA	overall performance assessment
PI	performance indicator
PE	polyethylene
PU	polyurethane
PVC	polyvinyl chloride
PPRA	pre- and post-rehabilitation assessment
PMA	pressure managed area
PRV	pressure-reducing valve
PSV	pressure-sustaining valve
PDF	probability density function
PFI	public finance initiative
PSBR	public sector borrowing requirement
RCP	rapid crack propagation
SOSM	second order second moment model
SoSI	security of supply index
SA	simulated annealing
SSF	slow sand filtration
SIC	Standard Industrial Classification
SEA	strategic environmental assessment
SCADA	supervisory control and data acquisition
DWI	The Drinking Water Inspectorate
WRAS	The Water Regulations Advisory Scheme
UKAS	UK Accreditation Service
UKWIR	UK Water Industry Research
UNCED	UN Conference on Environment and Development
uPVC	unplasticised polyvinyl chloride

WDS	water distribution system
WIS	Water Industry Standard
WIS	water into supply
WRc	Water Research Centre
WSP	water safety plan
WSP	water service provider
WLC	whole-life costing
WCED	World Commission on Environment and Development
WHO	World Health Organization

List of contributors

John K. Banyard *Independent consultant, Warwick, UK*
Joby Boxall *University of Sheffield, UK*
David Butler *Centre for Water Systems, University of Exeter, UK*
Rob Casey *Thames Water, Bourne End, UK*
Adrian Cashman *University of the West Indies, Barbados*
Neil Dewis *Yorkshire Water Services, Bradford, UK*
Ken Gedman *Consultant, UK*
Nigel J. D. Graham *Imperial College London, UK*
Frank Grimshaw *Severn Trent Plc., Birmingham, UK*
Barrie Holden *Anglian Water Services Ltd, Peterborough, UK*
Adrian Johnson *MWH, Edinburgh, UK*
Paul Jowitt *SISTech Ltd, Heriot-Watt University, Edinburgh, UK*
Zoran Kapelan *University of Exeter, UK*
Myles Key *South West Water Ltd, Exeter, UK*
John Machell *University of Sheffield, UK*
Adrian McDonald *Environment Faculty, University of Leeds, UK*
Harrison Mutikanga *National Water and Sewerage Corporation,
    Kampala, Uganda*
Michael Pocock *Veolia Water Central, Hatfield, UK*
Mark Randall-Smith *Water Engineer, Verwood, UK*
Clare Ridgewell *Essex & Suffolk Water, Chelmsford, UK*
Adrian Saul *University of Sheffield, UK*
Dragan A. Savić *University of Exeter, UK*
Tiku Tanyimboh *University of Strathclyde, UK*
Seneshaw Tsegaye *University of Birmingham, UK*
Kalanithy Vairavamoorthy *University of Birmingham, UK*
Howard S. Wheater *Imperial College London, UK*

Water Distribution Systems
ISBN: 978-0-7277-4112-7

ICE Publishing: All rights reserved
doi: 10.1680/wds.41127.001

ice

Institution of Civil Engineers

publishing

Chapter 1
Historical development of water distribution practice

Dragan A. Savić University of Exeter, UK
John K. Banyard Independent consultant, Warwick, UK

1.1   Introduction

One of the problems facing the modern water distribution engineer is the longevity of the infrastructure. Today, those fortunate enough to live in the developed nations of the world take the supply of potable water for granted. Yet, it is within living memory when many houses had a single cold water tap, with no bathroom, and certainly no washing machine or dishwasher. The very significant increase in the volume of water consumed, not to mention far higher quality standards, is often overlooked and the technology is viewed as long established, with the result that an impression is created that the industry has ossified and lacks 'innovation' (Council for Science and Technology, 2009). Even more sadly, it is possible to find practitioners who encourage these beliefs, by arguing that ideas that were appropriate for their predecessors must still be satisfactory today. For that reason alone it is worth taking a short space within this book to review the historical development of water treatment and distribution. However, it is also important to provide a basic foundation of how technology has evolved, to help better understand today's technology and good practice. This will hopefully assist in preparing the way for future developments to be introduced to further advance today's practice.

This chapter will consider the development of four distinct technologies, all of which are essential to the modern distribution engineer:

- water treatment
- pipe materials
- pipeline hydraulics
- analysis of pipe networks.

It will conclude with a brief look at some of the challenges that are emerging, which will ensure that innovation will continue to be an essential driving force for the industry.

1.2   History of water treatment and supply

It is impossible to say precisely when the first installations of artificial water supply were constructed. It is well understood that man must always have needed access to clean water for survival. For early nomadic peoples, this was a matter of finding a clean river or spring.

Early conurbations were sited near to water sources, and there is evidence of early civilisations going back to at least the fourth millennium BC. The earliest form of water engineering appears to have been the construction of irrigation canals, but at some stage wells must have been constructed. Rather than get involved in lengthy discussions as to where or when the first water supply system was constructed, it will meet our needs in understanding the historical developments of water supply if we rely on Roman sources, in particular the work of Frontinus (35–103 AD) (Frontinus, 1980). Frontinus was certainly not the first Roman to write about water supply: about 100 years earlier, the architect Vitruvius (*ca.* 75–15 BC) had produced a large work on architecture (Vitruvius, 1970) which included among many other topics the construction of aqueducts. However, Frontinus is a more helpful source because he was appointed the manager of Rome's water supply in 97 AD. Furthermore, he left a report on his work, which has survived as a textbook, which, well beyond his intentions, has been used to instruct (willingly or otherwise) generations of Latin students.

Frontinus was not an engineer: he was an extremely successful professional soldier who in 76 AD was appointed the governor of Britain. At the end of his term as governor he returned to Rome, having already written a book outlining military stratagems. He was faced by a new challenge, one for which he was not wholly prepared, and having tackled it he produced *De aquis urbis Romae* (Frontinus, 1980). In this, he sets out his understanding of the history of Rome's water supply, saying that for the first 441 years of Rome's existence, Rome was supplied by wells, springs and, of course, the River Tiber. However, around 312 BC the first aqueduct was brought into use, known as the Appian aqueduct, after Appius Claudius Crassus, who was also responsible for the Appian Way. The aqueduct was around 5 miles long, and much of it was constructed underground. Frontinus goes on to detail a further seven aqueducts, all constructed before he took office. Today's engineers might care to ponder the engineering feat of building a structure which Frontinus describes as still being in use some 400 years after its initial construction.

It is interesting that Frontinus condemns the construction of the Alsietian (or Augusta) aqueduct by the emperor Augustus, describing the water as unwholesome and not used for consumption by the people. This is evidence that there was at this time clear under-standing of the link between water and illness; indeed, it would be surprising if this were not the case.

The book reveals that, in modern parlance, Frontinus inherited 'a mess'. He explains how he had all of the aqueducts surveyed and drawings produced, so that he did not have to waste his time going out to view problems personally: his subordinates could explain the issue to him with the help of an appropriate drawing. He had the aqueducts relined with lead to prevent leakage, and pursued vigorously the owners of villas along the route of the aqueducts who had tapped into them illegally to provide a free water supply to their properties. There was no water treatment as it would be recognised today; instead, water discharged into tanks at the end of the aqueducts, where impurities could settle out. Water was generally distributed around Rome by water sellers, and Frontinus has some harsh words for them. Frontinus' military background enabled him to recognise the need for standards and discipline, and he applied these concepts very successfully to the management of Rome's water supply; in doing so he provided a good indication of the tasks of the asset manager, which would be recognised some 1900 years later.

Although Frontinus was only concerned with Rome's aqueducts, the provision of water supplies was extremely important to all of Rome's cities. Perhaps the most famous is the spectacular Pont du Gard near the French city of Nîmes, but it is by no means unique, and, for example, the ruins of the aqueduct that brought water into the City of Barcelona are still visible near to the gothic cathedral.

Unfortunately, with the decline and eventual fall of the Roman Empire, the aqueducts fell into disrepair, and the population returned to the methods that had served Rome for the first 440 years of its existence (if Frontinus is indeed correct). The same situation persisted across Europe for many hundreds of years. It appears that in medieval times, monasteries started to pipe water as an addition to the supplies from wells on which they were frequently founded, and it is possible that some of this water found its way to the local population (Barty-King, 1992).

In 1589 Sir Francis Drake was instrumental in providing the City of Plymouth with a new water supply known as Drake's Leat, dug by hand (although not in the single day ascribed by legend), and in 1613 the New River, promoted and largely financed by Myddelton, was completed, in order to supply fresh water to the ever-expanding City of London from the River Lea some 20 miles away. However, none of these schemes can really match the technical achievements of the Romans. Some 1400–1500 years after Frontinus, civilisation was only beginning to catch up with the Romans of his era.

The lack of water treatment and clean water supplies began to manifest themselves as the Industrial Revolution took place. Farm workers flocked to the new industrial cities to better their existence, but this placed huge strains on both water supplies and sanitation, both of which were extremely basic, and outbreaks of typhoid and cholera became commonplace, albeit by no means continuous.

In 1819, the poet Shelley reflected the state of cities in his poem *Peter Bell the Third* (Shelley, 1839), with the lines:

'Hell is a city rather like London,
A populous and smoky city'

The state of the working classes was exposed in a report published in 1842 (Chadwick, 1842), and the quotations in Box 1.1 adequately demonstrate the misery of those days.

Box 1.1 Extracts from *Report on the Sanitary Condition of the Labouring Population of Great Britain* (Chadwick, 1842)

'The various forms of epidemic, endemic and other disease caused, or aggravated or propagated chiefly among the labouring classes by atmospheric impurities produced by decomposing animal and vegetable substances, by damp and filth and close and overcrowded dwellings prevail among the population in every part of the kingdom . . .'

'That such disease wherever its attacks are frequent is always founding connection with the above circumstances . . .'

'The formation of all habits of cleanliness is obstructed by defective supplies of water.'

Although Chadwick's report did start the movement to provide better living conditions, particularly in terms of sanitation, the lack of scientific understanding about the cause of disease was a major hurdle.

It is difficult now to fully understand why the link between impure water and disease was not appreciated, particularly as there is ample evidence that there was a desire for clean, wholesome water. In 1852, the Metropolis Water Act required all water derived from the Thames and supplied in London within 5 miles of St Paul's Cathedral to be first filtered, but, even so, the accepted medical theory for much of the 19th century was that typhoid and cholera were airborne diseases, spread by the breathing in of miasmas (foul air) and had nothing to do with water, which was considered 'clean' as long as it was clear and not turbid.

Even after the now-celebrated Broad Street (Soho) pump incident in 1854, there was little change in the then-established view as to the cause of cholera and typhoid. Dr John Snow, a medical practitioner, had produced a paper as early as 1849 suggesting that cholera was not spread by miasmas (Snow, 1849), but the paper produced little interest. A careful reading of Chadwick's Report shows that he fully subscribed to the miasmic theory. In 1854 there was an outbreak of cholera in Soho, a district of London loosely bounded by what in modern London are Oxford Street, Regent Street, Leicester Square and Charing Cross Road. There was nothing exceptional about this: London in those days suffered regular outbreaks of both cholera and typhoid, as did most cities and large towns within the UK, and indeed in Europe and the USA. However, Dr Snow came to realise that all of his patients drew water from the same pump in Broad Street. He plotted the progress of the outbreak on a map and demonstrated to the public authorities that the outbreak in Soho was linked to the Broad Street pump, and persuaded them to remove the handle of the pump, after which the outbreak rapidly subsided. It is ironic that in the paper that he subsequently wrote in 1855 (Snow, 1855), concerning the outbreak and his work in stopping it, he identified a number of patients who had used the Broad Street pump in preference to alternative sources closer to their homes. Subsequent investigation showed that the well which supplied the Broad Street pump was located very close to a cess pit. It is significant that despite the success in stopping this particular outbreak of cholera, the pump handle was reinstated by the authorities once the epidemic had passed.

Snow's work was not accepted by the medical establishment; his case was not helped by the fact that he had subjected the water to chemical and microscopic examination and failed to identify any agent that he could categorically state was the cause of the cholera outbreak. Today this seems to us almost incredible, particularly as it was well known from microscopic examination that there were organisms in the water. However, it was to be another 30 years before the 'germ theory of disease' was to be established by the French scientist Louis Pasteur. It is impossible to put a precise date on Pasteur's work, as he published a number of papers from 1865, but by 1880 the case for the germ theory was established as correct. Pasteur did not invent the germ theory – it had existed as a hypothesis for many years – what he did was to demonstrate convincingly that the germ theory was correct and that the more widely accepted miasmic theory was invalid.

Looking back at historical actions is always difficult, and today it seems strange that the perceived wisdom of the medical establishment supported the miasmic theory. It was certainly well known that drinking dirty water led to illness, a fact recognised for centuries: this is one of the reasons given for medieval monasteries brewing beer, although there may also have been other incentives! Nonetheless, until Pasteur's work, the majority opinion was that serious illnesses such as typhoid, cholera and even the plague were transmitted by miasmas, and it is against that background that the development of the principal water treatment processes will be explored.

The first recorded instance of the use of a filtration system for water treatment was in Paisley, Scotland, in 1808. However, this did not reflect some farsighted public health concerns by the city fathers; rather, it was installed by John Gibb to improve his bleachery. The town is sited on the side of the River Cart, which was notorious for becoming turbid in times of storm. This variability in water quality affected the colouring of the cloth being bleached. It is reported that he was so successful that he was able to sell surplus water to those who wished to pay for it.

The first municipal installation of water filters (slow filters) was in Chelsea by James Simpson in 1829.

The first installation of filters in the USA was in Richmond, Virginia, in 1833, and there was a large installation of so-called 'English' filters in Poughkeepsie in New York, installed by James P. Kirkwood in 1872. Once Pasteur's work became recognised, the need to remove bacteria was understood, and thereafter the efficacy of water filtration as a means of avoiding outbreaks of disease became more readily demonstrable; however, although a well-designed filter made a tremendous improvement to the quality of water, it could not guarantee purity.

In 1895, George W. Fuller, working in Louisville in the USA, was attempting to find the most appropriate way of treating the waters of the Ohio river, and successfully combined the addition of chemical coagulants and water filtration to produce the now-classic two-stage process that is at the heart of many water treatment plants around the world. He formed his own consultancy practice subsequently, and was responsible for a much larger installation in Little Falls, New Jersey, in 1902.

Fuller also worked on the development of so-called rapid gravity filters, which are commonplace today, being cleaned mechanically as part of the operational cycle rather than depending on manual excavation of sand associated with the original slow sand filters. Although the basic concept of chemical coagulation and clarification, followed by filtration, has remained for the last 100 years, that does not mean that there has been no progress. Doubtless if Fuller were to visit a modern treatment works he would recognise his basic process flow sheet, but he would also be amazed at the sophistication now deployed. The coagulation process is now carefully controlled by computers, and the separation process is undertaken by a variety of clarifier designs which improve performance beyond anything that he was able to achieve. Even his dependence on gravity for separation has been replaced at times by the use of dissolved air flotation. More recently, the advent of membrane filtration has started to challenge the traditional flow sheet, but these devices, employing as they do filtration at a truly molecular level, still require protection in the form of roughing filters, or the use of chemical coagulants to break the molecular bonds before separation takes place across the membrane with the use of a partial vacuum.

Despite the huge advances in coagulation and filtration, there remains one further element in the process of modern water treatment that has had a major impact on the work of all water distribution engineers. This is disinfection. While Fuller and his contemporaries were pursuing the removal of dangerous bacteria from the water supplies, an alternative approach was also being developed, that of simply killing the bacteria. Of course, the chemicals used had to be harmless to man, and hence strong oxidising agents were adopted.

There are several claims for the first use of chlorine in treating drinking water, around the start of 20th century, particularly in Belgium (both Middelkerke and Antwerp are cited), and it is probable that it did not occur in the UK, although there are some references to trials in Maidstone in 1897. What is well established is that in 1903 there was an outbreak of typhoid at Fulborne Asylum in Cambridgeshire: the authorities sought permission from the House of Lords to introduce chlorine into the drinking water, but the request was refused on the grounds that it would be too dangerous. Twelve months later there was an outbreak of cholera in Lincoln, and, one assumes learning from the experience at Fulborne, the city fathers did not seek any approval but went ahead with chlorine dosing, following which the outbreak rapidly abated. At this point they made a fatal mistake by stopping the use of chlorine, and suffered a further outbreak several years later. However, Lincoln is generally credited with the first use of chlorine in water treatment in the UK. From that time on, the use of chlorine became increasingly commonplace in the control of both typhoid and cholera. In 1909 the city of New Jersey became the first municipality in the USA to use chlorine as a permanent element of its water purification process. Gradually, disinfection with chlorine (or compounds of chlorine such as hypochlorite) became part of the established water treatment process flow sheet.

In 1935 there was an outbreak of typhoid in Croydon, UK, where chlorine had not been used in the treatment process. Following that outbreak, the UK legislation was changed, and it became a requirement for potable water to contain a residual disinfectant when distributed through the public water supply network. Since that date there has not been a single reported case of water-borne cholera or typhoid within the UK. The reader will find a more detailed explanation of current water treatment practice in Chapter 4, 'Water supply systems'.

The requirement for a residual disinfectant provides a major challenge to the distribution engineer. Initially, it was sufficient for the water to leave the treatment works with a high chlorine residual and for this to simply decay as it passed through the distribution system. However, public opinion turned against the taste of highly chlorinated water, and despite the inaugural address of one president of the American Water Works Association in the late 1990s containing the suggestion that 'If you can't taste the chlorine, don't drink the water' – which can only have brought comfort to the manufacturers of bottled water – today's water engineers have to control the level of chlorine very carefully. It is necessary to be able to manage the levels of chlorine within the distribution system to achieve a balance between the competing demands of public acceptability and public health. To do that, it is necessary to be able to predict the flow regimes within pipe networks, which is the subject of the fourth and fifth parts of this review of the historical development of water distribution practice.

1.3   Evolution of pipeline materials

There has been almost continuous development of pipe materials for at least the last 150 years.

But 1750 years before that, Frontinus makes clear that lead was used to manufacture pipes (some of which can be seen in museums today), and also to make discharge devices to control flows from aqueducts and cisterns. It is also interesting to note in passing that, 100 years earlier, Vitruvius appears to report (his writings are not always clear because he presumes knowledge of then-current practice) problems with what seems to be an inverted siphon, where the available material technology could not withstand the bursting forces. It may therefore be the case that Roman preference for aqueducts rather than pipes across valleys related to limitations in pipe materials rather than to ignorance of the hydraulic phenomena which allow inverted siphons to function.

It is also clear that clay pipes were available to Roman engineers and, indeed, earlier civilisations, but they also would have limitations in terms of bursting resistance and were probably used for sewers (where such existed). It would be incorrect to believe that all Roman houses had running water. A few very rich individuals may have had water piped to their houses to a single point of discharge, but, as Frontinus makes clear, the vast majority depended on the collection of water from cisterns at the end of aqueducts, or the services of water sellers who (supposedly) collected water from cisterns and distributed it around Rome in carts.

With the fall of the Roman Empire, much of this technology was lost, and the provision of public water supply did not reappear until the middle ages.

Early water mains were made from hollowed-out tree trunks, connected with a socket and spigot joint which was sealed by wrapping in lead or some other material. There are numerous examples of this technology in museums. One example found in Oxford Street, London, dates back to the early 17th century. In 2004, a wooden water pipe was found in the ruins of a Roman fort in Northumberland, UK, emphasising just how much technology was lost during the Dark Ages.

1.3.1 Iron pipe

It is not clear when the first cast iron water pipes were introduced, but it is known that cast iron pipe was used to distribute water to the various fountains of the Palace of Versailles in 1672. It is of passing interest that there was, and still is, insufficient water to feed all of the fountains at the same time, and so they had to be switched on or off as the king approached or passed them.

In all probability there were earlier uses of cast iron pipes, but they are less celebrated, and sadly not recorded, although the American Ductile Iron Pipe Research Association reports their first use in 1455 in Siegerland, Germany, and suggests that they were a development of the technology used for casting the barrels of canons. However, it is clear that cast iron pipe began to be used for water supply in the 18th and 19th centuries. The first recorded use of cast iron pipe for water supplies in the USA appears to be in the early 1800s in the New Jersey and Philadelphia areas, where they replaced traditional wooden pipes. Initially, these pipes were cast horizontally, and were of variable quality, but vertical casting was developed around 1850, which produced a far more reliable product. There was no standardisation of sizes for these early pipes, and engineers would specify not only the internal diameter of the pipe they required but also its wall thickness.

The next major advance was the invention of the centrifugal spinning process in 1918. In this process, the mould which defined the internal and external diameters of the pipe was replaced with a hollow circular mould that fixed the external diameter. This was rotated at speed about its longitudinal axis, and molten metal was poured into it. The centrifugal force took the molten metal to the internal surface of the mould, and the quantity of liquid iron applied defined the internal diameter. This provided a far more uniform pipe material with less possibility of air bubbles or casting flaws. This process gradually replaced vertical and horizontal casting, but, of course, could not be used for pipe fittings such as T junctions and bends.

In 1955 a further advance was made with the development of ductile iron pipes. Traditionally, pipes had been made of grey iron, which was strong and durable but brittle. By modifying the metallurgical composition a material was developed that was far less brittle and could accept a small amount of deformation. Although being developed in the USA in the mid-1950s, it did not reach the UK until the 1970s. Initially, it appeared to be a far superior material, and was manufactured in part from scrap steel (although still technically an iron); however, experience showed that it was more prone to corrosion than the grey iron that it had replaced. There then followed a series of developments aimed at corrosion prevention, starting with on-site wrapping in polythene sheets, through to today's standard of factory-applied multicoated protection systems.

A number of other materials were developed that challenged the traditional approach of iron water mains, and we shall look at those shortly. However, before doing so we should look at the jointing of pipes.

The early cast iron pipes were sealed with what was known as a run lead joint. Molten lead was poured into the annulus between the spigot of one pipe and the socket of the other pipe. As the lead cooled, it solidified and shrank; it then had to be compressed by the use of a series of chisel-like tools and a hammer to form a water-tight joint. It was, of course, necessary to insert caulking before the molten lead was introduced, in order to prevent it from running down the length of the spigot. Some pipes also had a small channel cut into the inner circumference of the spigot to assist in making the joint, providing a path and holding channel for the lead. This highly skilled and somewhat dangerous technique was replaced by the invention of the sealing ring. There are several different styles of ring, generally protected by patents, but for the purpose of this introduction they can be viewed as an O ring that is compressed between the spigot and the socket at each joint. Flanged joints and various mechanical restrained joints are also available, but push-fit joints are most commonly used for external infrastructure (buried pipes).

1.3.2   Asbestos cement pipe

An early rival to cast iron pipe was the development of asbestos cement pipe. Essentially, it is a cement pipe that gets its tensile strength from the incorporation of 11% asbestos fibres. The pipes were used from the 1920s to the 1980s, and many kilometres are still in use around the world, including the UK, mainland Europe and the USA. They were lighter and lower cost than conventional cast iron pipes, but suffered a bad press over alleged high burst rates: however, an investigation by the UK Drinking Water Inspectorate (DWI) in 1985 (DWI, 1985) showed that this was a fallacy and that burst rates were very similar to those of cast iron and plastic water pipes.

A further DWI report in 2002 (DWI, 2002) suggests that there is little if any health risk from the pipes in terms of water supply, but the hazards involved in their manufacture and, particularly, in on-site cutting of the pipes meant that most countries in the world ceased installing these pipes.

So, although an early contender to cast iron pipe, asbestos is no longer a competitor.

1.3.3 PVC pipes

The next real contender to the supremacy of cast iron pipe was the introduction of unplasticised polyvinyl chloride (uPVC) pipe. Although PVC as a compound had been discovered in Germany in 1835, it was not used to produce water pipes until approximately 100 years later. Some of these early pipes were installed as water supply pipes in Germany on a trial basis, and performed well, but development was interrupted by the Second World War.

Large-scale production of PVC pipes started in the 1950s, and the product became increasingly popular in the 1960s, and started to emerge as a major competitor to traditional grey iron. Its main advantages were that it was light and easy to handle, it did not corrode and it was competitively priced. Unfortunately, these advantages were overtaken by operational issues, as failures of pipes were reported. Investigations in the UK showed that class B uPVC pipe (a low-pressure pipe) when used as a pumping main suffered failures caused by surge pressures generated on starting and stopping pumps, which, although within the pressure rating of the pipe, led to fatigue failure of the material. As a result, class B pipes were largely shunned by the UK water industry. The second problem came with a number of failures of larger-diameter pipes, which this time were linked with poor handling of the material. Pipes of uPVC deteriorate if left in sunlight, and also have to be laid on a bed which does not contain sharp stones – a sand bed and haunch are often specified.

These problems and the emergence of alternative materials led to a significant decline in the popularity of the product in the UK water market. The product was reformulated and relaunched as PVCu, which overcame the early technical problems, but the early experience has hindered its adoption in the UK. In contrast, the American Water Industry has embraced PVC pipe, and it is one of the most popular water pipe materials in the USA. It is available in a range of sizes up to 600 mm (24 inch).

1.3.4 MDPE water pipe

With the problems associated with both uPVC and asbestos cement becoming apparent, and the desire to find a corrosion-free alternative to the then-dominant grey iron pipe, attention turned to polyethylene. Polyethylene had been discovered in 1933 by ICI, but major development and use was again post-war. The gas industry had been the first major user of polyethylene pipe, but the UK water industry of the 1970s was in need of an alternative material, and began to explore the possibilities. It was believed that medium-density polyethylene (MDPE) was more robust and better able to withstand the less-than-ideal handling it would receive on site than the stronger high-density (HDPE) material. This may have been a correct assessment, but there were still problems with rapid crack propagation caused by minute levels of site damage which spread once the pipe was put into service.

There were also considerable problems with pipe jointing: the promise of a continuous pipe with perfect butt-welded joints, being leak free, soon gave way to a reality of less-than-perfect joints leaking or even breaking. Gradually, the technology improved, becoming more tolerant of site conditions, and the industry itself grew accustomed to the need for high standards of workmanship compared with the very abuse-tolerant cast and ductile iron pipe that had traditionally been used.

MDPE was very cost effective, and as confidence in its performance grew, it largely displaced the traditional cast iron pipe for sizes up to 300 mm diameter. Above this figure the thickness of the MDPE material is such that economics start to move against it, but the actual break-even point varies with time.

In the UK and many parts of mainland Europe, MDPE is probably the market-leading material for water pipe up to at least 300 mm diameter, but this is not the case in the USA, where it has trailed behind both PVCu and ductile iron.

1.3.5 Other pipe materials

Having been squeezed out of the smaller-diameter market for water pipe by MDPE and PVC, it might be thought that ductile iron would still enjoy a virtual monopoly in terms of both the medium- and large-diameter market (>450 mm) and the market for fittings for internal pipe work in pumping stations and treatment works, but even here its position is under threat.

For large-diameter underground pipelines, there is growing use made of steel pipe, usually protected with a factory-applied epoxy coating. For internal pipe work, there is a growing use made of welded stainless steel pipe.

Other alternatives for large-diameter pipes include glass-reinforced plastic (GRP), which is a generic term for a number of different pipe materials, each requiring slightly different laying techniques. GRP materials can also be used for pipe renovation work.

Additionally, prestressed concrete pipes are available for sizes above 600 mm diameter, and come in two forms: those with a steel tube cast into the concrete before prestressing and those that rely solely on the prestressed concrete.

A more detailed exposition of the practical issues surrounding utilisation of today's pipe-line materials is given in Chapter 5, 'Distribution network elements'.

So, it can be seen that pipeline material technology has evolved in a similar way to water treatment. Even the traditional cast iron pipe has undergone tremendous development in the last 100 years, but, additionally, there has been continuous innovation and development of alternative materials, and that development will undoubtedly continue into the future.

1.4 Development of pipe flow calculations

Having traced the development of water treatment science and pipeline technology up to the present day, it is now necessary to do the same for the hydraulic calculations that are required to enable the design of water distribution systems. The reader should remember that these developments have been taking place in parallel with the development of water treatment science and pipeline material science.

The starting point is once again Frontinus and his treatise on the aqueducts of Rome. In it he describes how he came to doubt the figures produced by the water sellers who took water from the tanks at the end of the aqueducts and conveyed it around Rome, selling it to the population in small quantities (very much the system that can be found in many emerging countries today). He therefore arranged to have measuring vessels deployed at the end of each aqueduct (not simultaneously), and in this way measured the flow being delivered by each aqueduct. Not surprisingly, he found that the water sellers were severely underestimating the amount of water available, and had been defrauding the city for many years in respect of the amount of water they actually sold. It is interesting that the reputation of water sellers in the developing countries is about the same as their Roman predecessors in terms of both quantity and quality. However, that is a diversion: what this event demonstrates is that although the Romans were able to construct aqueducts to carry water over very long distances, they did not have the mathematical tools to allow them to predict accurately what volume of water would actually be carried by the aqueduct. And that state of affairs was to continue for a further 1670 years.

The first published formula for the calculation of flow of water in pipes was by Antoine Chezy (1718–1798) *ca.* 1770, although the actual date is uncertain. His initial formula was for flow in open channels, but pipe flow was also covered as a special case. It is worth remembering that the start of the French Revolution was 1789, which may explain why Chezy's work was lost until Prony, one of his students, published a paper referring to it in 1800.

Today, the Chezy formula is usually written as

$$V = C\sqrt{rS} \qquad (1.1)$$

where V = the average velocity in the pipe
r = the hydraulic radius (the area of flow divided by the wetted perimeter)
S = the hydraulic gradient of the pipe
C = a dimensionless constant representing the friction for a particular pipe

However, the work of Fourier (1768–1830) introduced, among other things, the concept of 'dimensional analysis': this came to be a major contribution to the practice of building physical models in fluid mechanics (although the major breakthrough in that sphere was the work of Lord Rayleigh at the end of 19th century and of Edward Buckingham in the early part of 20th century), but as far as pipe hydraulics are concerned, it can be simply stated as requiring the fundamental physical units of mass (M), length (L) and time (T) to be the same on both sides of the equation. A quick inspection of the Chezy formula reveals that on one side of the equation we have a velocity (L/T) while on the other we have the square root of L and no T term. This indicates that the formula as conceived must be incorrect, and, indeed, C must have dimensions. However, this was not appreciated for many years, and the Chezy formula was (and to some degree still is) used for calculation. The value of C is based on specific experiments for varying hydraulic structures, and Chezy himself derived values for it for designs that he was undertaking. The Chezy formula remains as an empirical tool for hydraulic calculations, and this also applies to derivatives from it. Like all empirical relationships, they are only valid for the range of experiments on which they are based, and extrapolation outside that range can produce very misleading results.

In 1845, Weisbach (1806–1871) published a formula which was dimensionally correct. It is usually written in the form

$$H = \frac{4fl}{D}\frac{v^2}{2g}$$ (1.2)

where f = a friction factor
l = the length of the pipe
D = the diameter of the pipe
v = the average velocity of flow

Checking the above equation with dimensional analysis shows that both sides are consistent, and therefore f is indeed a dimensionless number.

In 1857, Henry Darcy independently published work which added to the understanding of the Weisbach equation, and the relationship is now commonly referred to as the Darcy–Weisbach equation.

Unfortunately, practical use of the equation showed that f was not a constant but instead varied with the pipe geometry (diameter) and roughness of the inside of the pipe, and it was in this area that Darcy made his contribution. But even that did not resolve the problem of a dimensionally consistent equation which failed to predict pipe flows adequately.

Readers should note that in Chapter 2, 'Basic hydraulic principles', $4f$ is replaced by λ: this is a more modern notation which was adopted because f is used to denote a function of other parameters.

Around this time the engineering profession started to split into two camps, divided between those who continued to work on the theory of pipe flow, seeking to find a totally rational relationship on which calculations could be based, and more pragmatic engineers who believed that although theoretically inferior, empirical equations based on extensive experimentation could produce reliable practical tools for practising engineers.

Many empirical relationships were published. These generally were of the form

$$V = Kr^x S^y$$ (1.3)

The values of K, x and y were derived from experiments, with extrapolation and interpolation used to provide missing values.

Two of these empirical formulae became the foundation for practical hydraulics for the first 60–70 years of the 20th century.

In 1889 the Irish engineer Robert Manning published his formula for open channel flow, and pipe flow (Manning, 1891). Not only was it found to be accurate over a wide range of conditions, it was also simple in structure and lent itself to the publication of tables, so that engineers could simplify the iterative nature of design calculations by use of the tables.

Manning's formula had several iterations, but is now generally given as

$$V = Mr^{2/3}S^{1/3} \qquad (1.4a)$$

where M = Manning's constant

It was soon realised that Manning's M was in fact the reciprocal of another constant developed earlier by Kutter and given the symbol n. As a result, Manning's formula is often expressed as

$$V = \frac{1}{n}r^{2/3}S^{1/3} \qquad (1.4b)$$

In 1905 the Hazen–Williams formula was proposed in the USA:

$$V = Cr^{0.63}S^{0.54}(1.32) \qquad (1.5)$$

It should be remembered that the fitting of curves to experimental results was carried out without the benefit of computers, and was laborious, and also approximate compared with modern computer-based regression analysis.

In the absence of even simple electronic calculators, use of even these simplified empirical equations was extremely time-consuming, often requiring the use of log tables. As a result, design tables and charts were developed and published, based on specific equations, in order to allow practising engineers to iterate to a solution, having assumed appropriate roughness coefficients.

Because Manning's equation (equation (1.4)) and that of Hazen–Williams (equation (1.5)) are both empirical, in addition to requiring caution when used outside the range of supporting experimental work (which was seldom referred to in the published tables), great caution was also required in terms of which units were used when listing appropriate coefficients (C and M), as these were not dimensionless and would therefore vary significantly between imperial, US and metric systems. This explains the different formulations that appear in Chapter 2.

Despite their drawbacks, these two empirical formulae became the most widely used, and have remained in (now declining) use for around 100 years.

The emergence of these empirical formulae did not in any way stop the continued search for a more dimensionally correct approach, as represented by the Darcy–Weisbach equation. Rather, it was the sheer complexity of the search for f within that equation which resulted in the emergence of the empirical formulae.

14

It had been recognised quickly in the 1830s that there was a difference between slow and rapid flow in pipes. It was therefore recognised that the Weisbach equation (equation (1.2)) required different values of f for these two cases. At that time, the concepts of laminar and turbulent flow were not understood, although Darcy's contribution was in studying flows in what we would now describe as being in the turbulent range. As knowledge increased, it became clear that f, rather than being a simple coefficient, was, in fact, an extremely complex factor. It was a function of the Reynolds number (Reynolds, 1883), but the complexity did not stop there. As research continued, the concepts of boundary layer theory had to be incorporated, and the complexity of the function increased as it became apparent that relationships only held good within certain ranges of the Reynolds number.

Colebrook and White published research into commercial pipes using the theories that had been developed to explain the behaviour of f in 1939, and demonstrated that the relationship set out below had very good correlation with measured physical experiments over wide ranges of flows (Colebrook, 1939):

$$V = -2\sqrt{2gDS}\log(k_s/3.7D + 2.51\lambda\nu/D\sqrt{2gDS}) \tag{1.6}$$

where S = the hydraulic gradient
ν = the viscosity

Unfortunately, it is immediately apparent that this is not a user-friendly relationship for practising engineers. Apart from the sheer complexity of the equation, including square roots and log functions, the parameter which usually requires to be determined, S (the hydraulic gradient), cannot be isolated algebraically to one side of the equation, and therefore a laborious iterative solution technique must be employed.

Further exploration of the pipe flow formulae is to be found in Chapter 2, 'Basic hydraulic principles'; the purpose of this introduction is simply to demonstrate the evolution of hydraulic theory.

Matters improved slightly in 1944, when Moody published his now-famous diagram (see Chapter 2, 'Basic hydraulic principles'), which was an extension of earlier work (Rouse, 1943) produced by Hunter Rouse (1906–1996), a leading US hydraulics engineer. However, despite the availability of both Moody's diagram and Rouse's own work, the use of these complex relationships lay well outside the practical design experience of most water distribution designers and operators, who depended on slide rules and log tables to perform their calculations. Undoubtedly, the engineers could have used the equations with the help of the Moody and Rouse charts to assist in the calculations, but using the Manning or Hazen–Williams equations was far simpler, and there was a wealth of experience in the use of these formulae by the time that the Moody diagram was developed. The Moody diagram remained, therefore, something that was taught at universities to assist in the understanding of pipe flow, but was not generally taken up by practising engineers. An additional hurdle was that the solution of pipe flow equations was by no means the end of the analysis process for water distribution networks, and that was yet another factor that militated against the introduction of a more-complex method for calculating pipeline flows.

Thus, for around 15 years, the Moody diagram, which should have given access to the Colebrook–White equations, remained of largely academic interest. However, the availability of computers was to revolutionise previous practice.

In the late 1950s and early 1960s, the Hydraulics Research Station at Wallingford used a then large computer to produce, initially, design charts based on the Colebrook–White equation, and then produced a series of design tables. In theory, these put the Colebrook–White equation on an even footing with both the Manning and Hazen–Williams equations, but, of course, there was by that time over 50 years of practical experience of using both of these latter formulae, and despite the more rigorous theoretical basis of the Colebrook–White equation, the industry was slow to adopt the use of this 'new entrant'.

As the use of computers became more common in engineering offices through deployment of, at first, time-sharing terminals and even a few mini computers, so software began to be developed to assist in hydraulic calculations. Initially, this was based on the tried and trusted Manning and Hazen–Williams formulae, but gradually the Colebrook–White equation began to be offered by vendors as an alternative. As the power of computers increased, and dedicated mini computers were replaced by desktop PCs, the use of the more complex but also more rigorous Colebrook–White equation grew in popularity: when feeding parameters into a computer, there was no extra work involved for the designer whichever formula was adopted; additionally, technical papers were being published about the continued development of the Colebrook–White equation; and, finally, for possibly the preceding 20 years, graduates had left university having been taught the theoretical benefits of the Colebrook–White equation but had to accept the impracticalities of adopting the equation for use in a design office. Thus, the advent of technical computing on PCs at last allowed the practicalities of design office practice to catch up with the technical advances that had been made over the preceding 60 years in theoretical hydraulics.

However, as indicated above, there remained a further problem in the design and operation of distribution systems, and this relates to the analysis of pipe networks rather than simple pipelines.

1.5   Analysis of pipe networks

So far, hydraulics has been restricted to the analysis of a single pipeline; however, distribution networks are not single pipes in isolation but, rather, many pipes joined together at nodes, as shown in Figure 1.1. In fact, the diagram shows two distinct types of network problem, the first (Figure 1.1a) for a branched system and the second (Figure 1.1b) for a looped system. In both cases the problem is to determine how much water flows down each of the pipes, which, of course, are not necessarily of uniform diameter or roughness factor.

Figure 1.1 Two distinct types of network: (a) a branched system and (b) a looped system

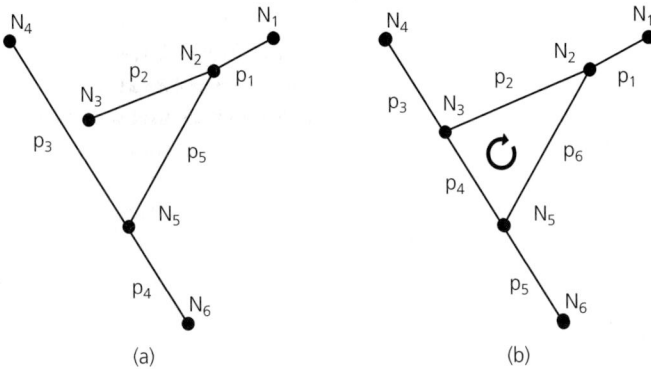

The solution to this problem is well beyond the scope of this introduction, and the reader will find a comprehensive treatment in Chapter 2, 'Basic hydraulic principles'. However, in order to understand the evolution of network analysis it is necessary to distinguish generically between the two idealised problems and to understand that with looped systems there are always more unknowns than available equations, and thus an iterative solution technique has to be adopted.

A number of mathematical techniques were developed over the years to assist in solving these iterative problems, including the use of the Hardy Cross relaxation method, which many engineers have encountered as 'moment distribution' when applied to structural analysis. However, the sketch in Figure 1.1 is a very oversimplified network compared with a normal distribution system. The only way of solving the equations was to simplify the network by restricting the calculations to large pipes only. This resulted in the majority of small pipes never being analysed, and network analysis calculations were only a minor part of the distribution engineer's armoury. They had to be used for new supply calculations and for major reinforcements, but were just too cumbersome to be deployed for day-to-day operational decisions.

The distribution system was therefore managed on the basis of the combined knowledge of operatives and engineers which had been built up from years of practical operational experience. There was a long-established tradition of operators keeping 'little black books' giving details of how their part of the network appeared to operate, and tales of these books being sold to successors at the end of a career to fund retirement parties are not all apocryphal.

By the mid-1970s, some distribution engineers had access to dedicated mini computers such as the PDP-11, which was the size of a kitchen table and had 8 kB of RAM. However, machines such as this could run programs to automate the solving of simple networks. As computers grew more powerful, following Moore's law (the power of computer chips doubles every 2 years), the complexity of networks that could be analysed continued to increase, and by the mid-1980s large network models could be solved, although again restricted to 'large pipes' and well removed from the all-pipe models that would have been the ideal.

Even with these simplified models, it was becoming clear that much of the information contained in the little black books was erroneous. However, the programs were the province of expert network modellers, and the outputs were not in graphical form but sheets of numerical output that required to be interpreted by the modeller. It is hardly surprising, therefore, that network modelling remained a tool to be used by experts when considering large capital expenditure, but continued to play almost no part in the day-to-day operation of distribution networks.

By the mid-1990s, the power of computers was such that it was possible to approach the all-pipe-modelling ideal by omitting small pipes, rather than simplifying the principal pipes in the physical network. However, the really big breakthrough was the development of graphical user interfaces (GUIs). This now meant that the network could be represented on the screen as a network, and pipes could be added or isolated on the screen by the operator. The model user no longer had to be a specialist, and by the turn of the millennium, network modelling had finally become an operational tool: at long last the little black books could be seen by all to be of little further value.

Over the last 10 years, further refinements have emerged: all-pipe models are now a reality, links are routinely made to geographic information systems to update models, and new model build has been greatly simplified by these links. The functionality of the models has increased to allow the decay of chlorine to be predicted, thus assisting with control of chlorine levels in the distribution system. Pollution incidents can be tracked and predictions made about the spread of the polluting substance. Isolation of mains can be explored in terms of the pressure reduction experienced by all properties within the model, rather than simply the properties that will not receive water. Links can be made to customer databases to issue warning notices. Chapter 6 is devoted to a detailed explanation of the modern network modelling techniques now available to the distribution engineer.

The advances over the last 10 years have been incredible and will continue, but all of them are reliant on the advances that have been made over the last 220 years of hydraulic engineering.

1.5.1 Water distribution engineering in the 21st century

Having established that water distribution has an innovative and progressive technical history, it will be helpful to also consider the challenges facing today's engineer and how this book may be of assistance. The history of water supply and distribution is one of over 200 years of innovation and development. Where there have been lulls, it has not been through lack of effort, but rather waiting for science and technology to develop and sometimes waiting for those concepts to be capable of deployment. The advent of the digital age has facilitated huge strides forward, with a result that today's technical sophistication is now beyond the wildest imaginings of those who worked in the industry only 50 years ago.

Today, the engineer's ability to calculate pipe flows and predict the consequences of operational decisions is higher than it has ever been: there can be no justification for failing to use these tools for all aspects of water distribution engineering, be it routine isolation of pipes for cleaning or reinforcement of the existing system.

Our modern world is extremely complex, and few engineers can expect to work in total isolation. Most engineers are expected to work as part of a team, whether they be employed by consulting engineers, or directly by client organisations such as water service providers. These teams will themselves be multidisciplinary, and it is essential that the engineer is capable of presenting his or her knowledge and advice to other team members in a clear and understandable form. The idea of a single engineer as an infallible source of technical wisdom is long past. Within the discipline of water engineering there are numerous specialists, who have sophisticated and evolving techniques at their disposal: Chapter 7, 'Design of water distribution systems', explores some of the most recent ideas, while Chapter 8 covers a modern approaches to operation, maintenance and performance.

The concept of asset management is a relatively recent one, perhaps evolving as a formal discipline over the last 25 years. It is now vital to all water service providers that have to recognise the disciplines of the financial markets, and Chapter 9, 'Asset planning and management', is devoted to exploring these ideas and associated techniques.

If the engineer is to succeed in this element of his or her work, he or she must have at least an understanding of the role of others as well as command of his or her own technology. Chapter 10, covering finance and project appraisal, establishes the basic mechanisms available for financing projects by government, municipalities and private companies; at the very least it is hoped that this chapter will establish that raising finance is a complex business and that no asset owner has access to unlimited funds. It also sets out the various approaches to project appraisal, ranging from those adopted for major international projects funded by institutions such as the World Bank to more modest appraisals carried out routinely by engineers requiring to reinforce an existing distribution system.

There is much debate about sustainability and climate change, and Chapter 11 sets out the views of an eminent engineering practitioner. However, this is a rapidly developing field, and one where it is not possible to provide unequivocal advice on all aspects. Issues such as the conflict between 'sustainability' and conventional economic theory can be identified and discussed, but the actual mechanisms will have to be resolved at governmental level before definitive solutions can be given. For example, much increased energy use is associated with meeting higher treatment standards required by EU Directives; however, this conflict between the mutually exclusive need for higher treatment standards and lower CO_2 emissions can only be resolved by the lawmakers who imposed both objectives, albeit one hopes informed by advice from the scientific community. This chapter differs, therefore, from others in that its purpose is to inform the debate rather than provide well-proven technical and operational solutions.

In concluding this introduction, it is tempting to try to predict which areas of water distribution will be the subject of the next major change, but with such rapid development on so many fronts, it is not possible to provide a comprehensive forecast. However, one issue does appear to be emerging, driven by separate although linked objectives: this is the question of water metering. This is not a reference to household metering but, rather, measurement of flows at points throughout the distribution system. There are two main drivers for this.

The first is the political focus on water leakage, particularly as electricity usage in the water industry is significant, and if 30% of water is lost as leakage, then 30% of the energy is also probably wasted. This could make a very significant contribution to the country's Kyoto targets (United Nations, 1997), and with no increased cost or loss of service to customers. The level of metering in the typical water supply and distribution system is well below that found in the hydrocarbons industries, where their products have higher monetary value. It is likely, therefore, that some techniques established in those industries will find their way into the water industry.

Box 1.2 *Eldorado* by Edgar Allan Poe, 1849

Gaily bedight,
A gallant knight
In sunshine and in shadow,
Had journeyed long,
Singing a song,
In search of Eldorado.

But he grew old –
This knight so bold –
And o'er his heart a shadow
Fell as he found
No spot of ground
That looked like Eldorado.

And, as his strength
Failed him at length,
He met a pilgrim shadow –
'Shadow,' said he,
'Where can it be –
This land of Eldorado?'

'Over the Mountains
Of the Moon,
Down the Valley of the Shadow,
Ride, boldly ride,'
The shade replied –
'If you seek for Eldorado!'

The second driver is the 2009 Cave Review (Cave, 2009), which supports the theoretical concept of a vertical disaggregation of the water industry. Whether or not this will actually happen is unclear, and Professor Cave himself recognises that the practical barriers are significant, but one of the issues that he uncovers is the paucity of information about what actually happens between water treatment works and the customer. Before decisions can be reached, this gap will have to be filled, and that will, in turn, also require more information on flows.

However, there are other major challenges facing the industry that will also require innovative approaches, not the least of which is the ever-increasing challenge of replacing life-expired infrastructure.

In closing this introduction the reader's attention is drawn to the poem published in 1849 by Edgar Alan Poe, in Box 1.2. It tells the story of a knight setting forth on a quest to find the fabled city of gold, Eldorado. He starts off full of enthusiasm, but as the poem progresses he ages without fulfilling his quest. He becomes melancholy, and then in the final verse he receives less than clear advice from a mysterious stranger who tells him that he must 'Ride boldly ride ... If you seek for Eldorado'.

Like Poe's knight, it most unlikely that engineers will ever achieve their Eldorado of a perfect water distribution system, and the road to further improvement is not always clear, but, nonetheless, that is no reason for failing to pursue the objective with all the tools available, both currently existing and yet to be developed; it is hoped that this book will provide some assistance in that quest.

REFERENCES

Barty-King, H. (1992) *Water: The Book – An Illustrated History of Water Supply and Wastewater in the United Kingdom*. London: Quiller Press.

Cave, M. (2009) *Independent Review of Competition and Innovation in Water Markets: Final Report (Cave Review)*. London: Defra.

Chadwick, E. (1842) *Report on the Sanitary Condition of the Labouring Population of Great Britain*. London: HMSO.

Colebrook, C. F. (1939) Turbulent flow in pipes with particular reference to the transition region between the smooth and rough pipe laws. *Proceedings of the Institution of Civil Engineers* 12: 393–422.

Council for Science and Technology (2009) *Improving Innovation in the Water Industry: 21st Century Challenges and Opportunities*. London: The Stationery Office.

Darcy, H. (1857) *Recherches Experiméntales Relatives au Mouvement de l'Eau dans les Tuyaux*. Paris: Mallet-Bachelier.

DWI (1985) *Usage and Performance of Asbestos Cement Pressure Pipe*, DWI0033. London: Drinking Water Inspectorate.

DWI (2002) *Asbestos Cement Drinking Water Pipes and Possible Health Risks*. London: Drinking Water Inspectorate.

Frontinus, Sextus Julius (1980) *Stratagems and Aqueducts*. London: Heinemann.

Manning, R. (1891) On the flow of water in open channels and pipes. *Transactions of the Institution of Civil Engineers of Ireland* 20: 161–207.

Moody, L. F. (1944) Friction factors for pipe flow. *Transactions of the ASME* 66(8): 671–684.

Reynolds, O. (1883) An experimental investigation of the circumstances which determine whether the motion of water shall be direct or sinuous, and of the law of resistance in parallel channels. *Philosophical Transactions of the Royal Society* 174: 935–982.

Rouse, H. (1943) Evaluation of boundary roughness. *Proceedings of the 2nd Hydraulics Conference, The University of Iowa Studies in Engineering. Bulletin 27*, pp. 105–116. New York: Wiley.

Shelley, P. B. (1839) *Peter Bell the Third* [publisher unknown].

Snow, J. (1849) *On the Mode of Communication of Cholera* [publisher unknown].

Snow, J. (1855) *On the Mode of Communication of Cholera*, 2nd edn [publisher unknown].

United Nations (1997) *Kyoto Protocol*. United Nations Framework Convention on Climate Change website: http://unfccc.int/kyoto_protocol/items/2830.php [accessed 01.09.10].

Vitruvius (1970) *De Architectura*. London: Heinemann.

REFERENCED LEGISLATION

Metropolis Water Act 1852 (15 & 16 Victoria c. 84). London: HMSO.

Water Distribution Systems
ISBN: 978-0-7277-4112-7

ICE Publishing: All rights reserved
doi: 10.1680/wds.41127.023

ice

Institution of Civil Engineers

publishing

Chapter 2
Basic hydraulic principles

Dragan A. Savić University of Exeter, UK
Rob Casey Thames Water, Bourne End, UK
Zoran Kapelan University of Exeter, UK

2.1   Introduction

This chapter provides a brief review of basic hydraulic principles, including fluid properties and fluid dynamics, necessary for understanding the hydraulics of water distribution systems. The review encompasses fluid properties, the governing equations of conservation of mass and energy and the framework in which they are solved. Then, network analysis equations are introduced, and methods for solving them are briefly described.

This chapter is not intended as a substitute for a course in fluid mechanics or hydraulics. It is, rather, assumed that the reader already has a fundamental understanding of fluid statics and fluid dynamics. The primary aim of this chapter is to remind and update the reader of the concepts used in the modelling of water distribution systems.

The chapter starts with an introduction to basic fluid principles, before moving on to describe the basic flow equations and losses in pipes. The second part of the chapter covers network analysis methods, including steady and unsteady flow modelling and water quality analysis.

2.2   Basic fluid properties
2.2.1   Density

The density of a fluid ρ $(\mathrm{M/L^3})$ is defined as its mass m (M) per unit volume V $(\mathrm{L^3})$. The SI unit for density is $\mathrm{kg/m^3}$, and the density of water at 5°C (reference temperature) is $1000\,\mathrm{kg/m^3}$, and at 20°C it is $998.2\,\mathrm{kg/m^3}$.

2.2.2   Viscosity

Viscosity is an internal measure of resistance to flow in a liquid, or, more precisely, it is a measure of the susceptibility of a fluid to shear deformation. Informally, we can think of viscosity as 'thickness', but for fluids only. For example, water can be considered 'thin', having a low viscosity, while honey and treacle are 'thick', having a high viscosity. Generally, fluids resist the relative motion of immersed objects through them, and act like different sets of layers with relative motion between them (Figure 2.1a shows how friction between the fluid and the moving boundary causes the fluid to shear). In the case of fluid flowing between stationary plates, the velocity of flow varies from zero

23

Figure 2.1 Viscosity

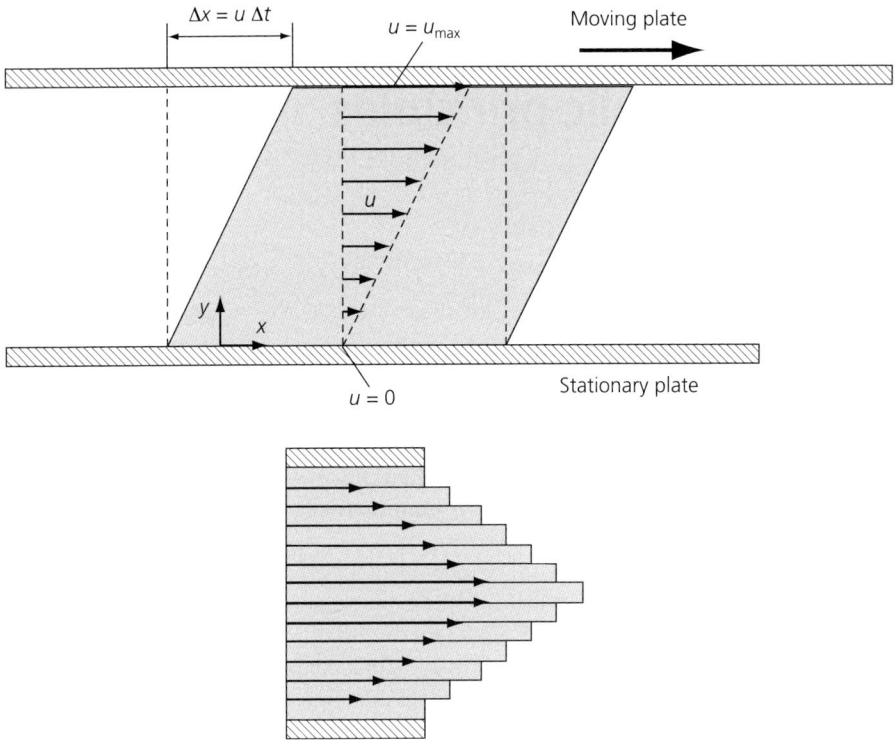

for the layer next to the plate to a maximum along the centreline (Figure 2.1b). During that process, viscosity opposes the flow of fluid, which is attributed to viscous force.

Formally, viscosity (sometimes called dynamic viscosity) is the ratio of the shear stress to the velocity gradient in a fluid. More often, this relationship, also called Newton's equation because of its similarity to Newton's second law of motion, is given as follows:

$$\tau = \mu \frac{du}{dy} \tag{2.1}$$

where $\tau =$ the shear stress $(M/L/T^2)$
$\quad \mu =$ the absolute (dynamic) viscosity $(M/L/T)$
$\quad \dfrac{du}{dy} =$ the velocity gradient perpendicular to the direction of shear $(1/T)$

Water is considered a Newtonian fluid, as the shear stress versus velocity gradient relationship is linear. In the SI system the dynamic viscosity unit is $N\,s/m^2$, Pa s or $kg/m\,s$.

Kinematic viscosity ν is the ratio of the dynamic viscosity to the density of the fluid:

$$\nu = \frac{\mu}{\rho} \tag{2.2}$$

Note that no force is involved with this quantity. In the SI system the theoretical unit is m^2/s or, commonly used, the stoke (St), where $1\,St = 10^{-4}\,m^2/s$. The viscosity of a fluid is highly temperature dependent, and for either dynamic or kinematic viscosity to be meaningful, the reference temperature must be quoted.

2.3 Basic flow equations
2.3.1 Flow and velocity

Flow through a pipe of uniform bore running completely full with the same velocity at each consecutive cross-section is called *uniform flow*. This is the most basic type of flow in pipes. If, however, the cross-sectional area and velocity of the fluid vary from cross-section to cross-section, but the flow rate does not change over time, the flow is called *steady*. An example is flow through a tapering pipe. The mean velocity V (m/s) at any cross-section of area A (m^2) when the volume passing per unit of time is Q (m^3/s) is given as

$$V = \frac{Q}{A} \tag{2.3}$$

where Q is also called the discharge. The cross-sectional area of a circular pipe can be expressed directly by using the diameter D (m), and thus the velocity equation becomes

$$V = \frac{4Q}{\pi D^2} \tag{2.4}$$

Finally, if both the cross-sectional area and velocity of the flow vary with time, the flow is said to be *unsteady*. A pressure wave travelling along a pipe after a valve has suddenly closed is an example of unsteady flow.

2.3.2 Flow regime

When calculating losses in pipes, it is important to know if the fluid flow is *laminar* or *turbulent* (Box 2.1). Osborne Reynolds (1842–1912), an English scientist, found that the type of flow is determined by the pipe diameter D, the density of the liquid ρ, its

---

Box 2.1 Laminar and turbulent flow

Laminar flow is usually associated with slow-moving, viscous fluids in small-diameter pipes. *Turbulent* flow is much faster and chaotic because vortices, eddies and wakes make the flow unpredictable. Laminar flow is relatively rare in real situations, while turbulent and *transitional* flow commonly occur. Transitional flow is a mixture of laminar and turbulent flow, with turbulence typically occurring in the central region of the pipe, and laminar flow near the edges.

Figure 2.2 A control volume

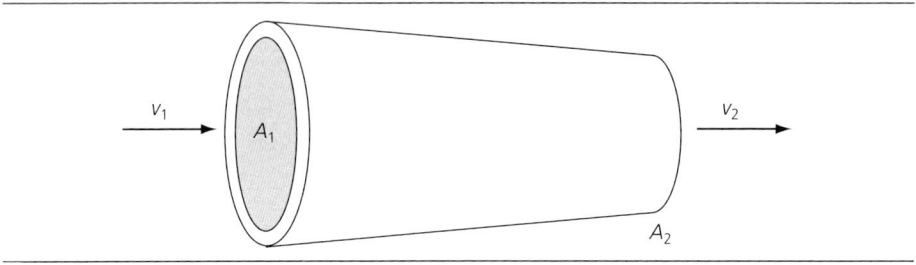

dynamic viscosity μ and the mean velocity V (Reynolds, 1883). The dimensionless parameter Re, called the *Reynolds number* after him, gives a measure of the ratio of inertial forces to viscous forces, and consequently can be used to distinguish between laminar, transitional and turbulent flow:

$$\text{Re} = \frac{\rho V D}{\mu} \qquad (2.5)$$

As a very general guide, laminar flow of water in pipes occurs for $\text{Re} < 2000$, turbulent flow occurs for $\text{Re} > 4000$ and transitional flow occurs for values of Re between 2000 and 4000.

2.3.3 Mass conservation (continuity) law

The *mass conservation law* states that mass can be neither created nor destroyed, i.e. any mass entering a system must either accumulate in it or leave it. Applying this law to a steady flow through a control volume (Figure 2.2), where the stored mass m in it does not change, gives

$$m_1 = m_2 \qquad (2.6)$$

where m_1 = the mass coming in (M)
$\qquad m_2$ = the mass going out (M)

Assuming that water is an incompressible fluid, the mass conservation law can be applied to volumes or discharge, hence

$$Q_1 = Q_2 \qquad \text{or} \qquad v_1 A_1 = v_2 A_2 \qquad (2.7)$$

This is then called the *continuity equation*.

2.3.4 Energy conservation law

Water in a pipe system may possess three forms of internal energy: *potential*, *pressure* and *kinetic*. If we consider a system, such as the one given in Figure 2.3, the *potential energy* arises due to the position of the fluid above some reference (datum) level. Thus, if a weight W of liquid is at a height z above datum, then:

$$\text{potential energy} = Wz \qquad (2.8)$$

Figure 2.3 A static pipeline water system

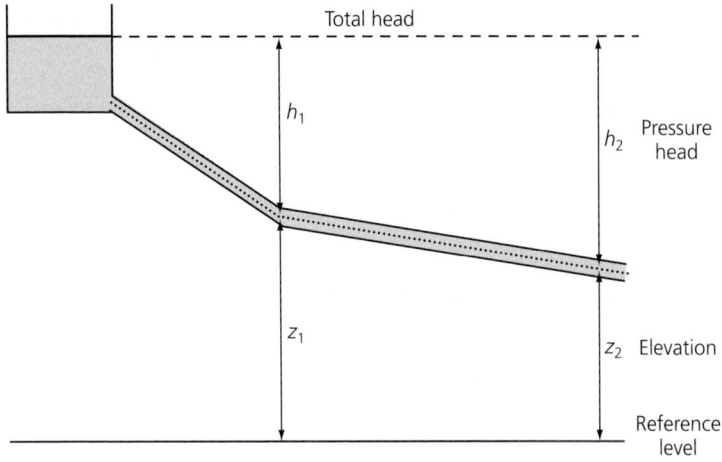

Note that at two points 1 and 2 along the pipe in Figure 2.3 the potential energy changes as the elevation changes (i.e. $z_1 > z_2$). If we now consider the *pressure energy* that arises due to hydrostatic pressure (the internal energy of a fluid due to the pressure exerted on the pipe), the liquid can thus do work when travelling through a distance L. Knowing that the force due to the pressure p on the cross-sectional area A is pA, then the work done is

$$\text{pressure energy} = pAL \tag{2.9}$$

If we express the volume of fluid AL in terms of weight W as $W/\rho g$ (where g is the gravitational acceleration), we get

$$\text{pressure energy} = p\frac{W}{\rho g} \tag{2.10}$$

The *kinetic energy* is the energy possessed by a liquid of weight W moving at a velocity v. Then,

$$\text{kinetic energy} = \frac{1}{2}\frac{W}{g}v^2 \tag{2.11}$$

It is normal for pipe systems to express energy per unit weight, thus resulting in

potential energy per unit weight (or elevation head) $= z$

pressure energy per unit weight (or pressure head) $= \dfrac{p}{\rho g}$ (2.12)

kinetic energy per unit weight (or velocity head) $= \dfrac{v^2}{2g}$

Note that all three types of energy in equation (2.12) are given as the head in terms of metres of fluid column. If we add the three equations (2.12) together to get the total energy (per unit weight), the following expression for the total energy head is obtained:

$$z + \frac{p}{\rho g} + \frac{v^2}{2g} = \text{total energy per unit weight} \qquad (2.13)$$

The three components of the total energy head shown in equation (2.13) are the main building blocks of the *energy* (*Bernoulli*) *equation*, which is a statement of the conservation of energy principle. It states that the total energy of each particle of a body of fluid is the same provided that no energy enters or leaves the system at any point. The division of this energy between potential, pressure and kinetic may vary, but the total remains constant, as shown in the following Bernoulli equation:

$$z + \frac{p}{\rho g} + \frac{v^2}{2g} = \text{constant} \qquad (2.14)$$

For a static open water system, such as the one in Figure 2.3, the total head (comprised of only two types of energy, potential and pressure, also called the piezometric head) is the same at any point along the pipeline. However, the pressure will change as the elevation of the pipeline changes. The piezometric head represents the height to which the water will rise in a piezometer (or a stand-pipe in a pipeline).

Equation (2.14) can be written for two cross-sections of a pipe as follows:

$$z_1 + \frac{p_1}{\rho g} + \frac{v_1^2}{2g} = z_2 + \frac{p_2}{\rho g} + \frac{v_2^2}{2g} \qquad (2.15)$$

The equation assumes that the flow is one-dimensional (i.e. assumes constant velocity and pressure across a cross-section), steady (i.e. velocity does not change over time), frictionless (i.e. no losses) and ideal (i.e. no energy losses) and that the fluid is incompressible (i.e. ρ is constant). If we now write the energy equation for a real fluid (with energy losses, e.g. water) between any two sections of a pipe, the equation becomes

$$z_1 + \frac{p_1}{\rho g} + \frac{v_1^2}{2g} = z_2 + \frac{p_2}{\rho g} + \frac{v_2^2}{2g} + h_f \qquad (2.16)$$

where h_f = the energy head losses (L)

The equation can be used to evaluate energy head losses, which can be caused either by *friction* (resistance) along the pipe walls or by *geometry or shape configuration changes* (minor or local losses). There will be more discussion about these two types of energy losses later.

Figure 2.4 illustrates the relationship between velocity, pressure and losses in a simple system consisting of a reservoir and a pipeline comprising two segments.

Figure 2.4 A static pipeline water system

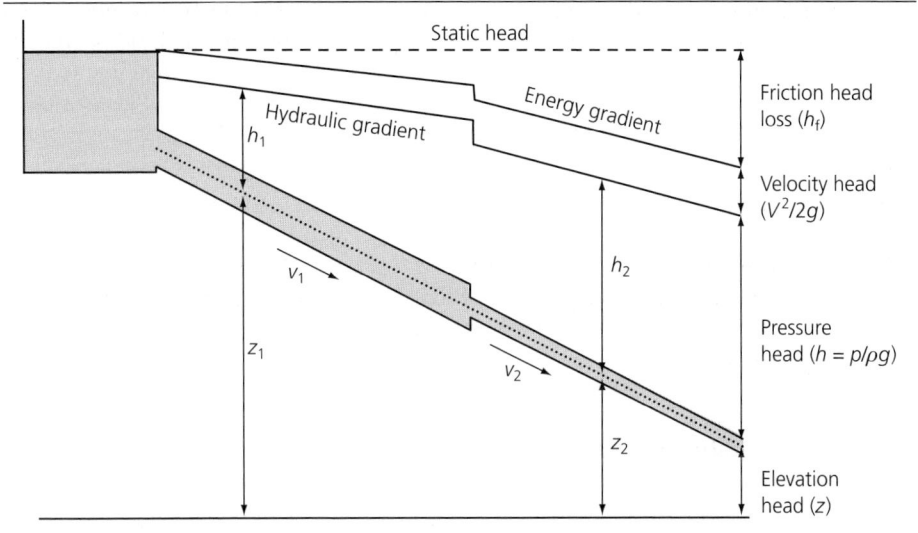

It is worth pointing out that in real-life water distribution systems, flow velocities rarely go above 1–2 m/s, hence velocity heads are typically very small (0.05–0.20 m), i.e. negligible when compared with pressure heads (can be anything from 10 m to over 100 m, usually 20–50 m). This is the reason why many practitioners often equalise the EGL with HGL by referring to HGL as the 'total head' (see Box 2.2). Another interesting point to note is that in real-life water distribution systems, the elevation of the pipe centreline z is often not too far from the ground elevation, as the top of the pipe is typically buried less than 1 m (perhaps 2 m in very cold climates) below ground level. This is the reason why practitioners often approximate the pressure head as the distance between the 'total head' and the ground elevation (which is often more easy to obtain than the pipe centreline elevation).

Box 2.2 Static head, energy grade line and hydraulic grade line

The *static head*, which also represents the total energy line for an ideal liquid with no energy losses, is always horizontal. The *energy grade line* (EGL) or *energy gradient* represents the total energy if head losses are taken into consideration. The energy grade line always falls in the direction of flow (unless there is a pump along the pipeline), because there must always be a head loss due to friction and minor losses. The *hydraulic gradient* or *hydraulic grade line* (HGL) is obtained when the piezometric levels along the pipe sections are joined together by a straight line. The hydraulic grade line generally falls in the direction of flow, but it can also go up (e.g. when at a sudden expansion in pipe cross-section). The sudden vertical step downward in the hydraulic grade line and the energy grade line (see Figure 2.4) at the sudden contraction of the pipeline or where water enters the pipe connected to the reservoir represents a local loss in energy.

2.4   Losses in pipes

The energy head loss term in equation (2.16) results from two different phenomena. The head loss due to friction between the moving water and the pipe wall is called *friction loss*, and is always present throughout the length of a pipe. Additional head loss due to local disruption of the fluid flow (e.g. due to valves and junctions) is called *local loss*. Friction losses are often dominant in long pipes, whereas minor losses may be significant in short pipes. It is commonly considered that, in pipe networks, local losses do not contribute significantly to overall losses and can be neglected.

2.4.1   Friction losses

The two most commonly used equations for calculating friction losses in pressurised pipes are the Darcy–Weisbach and Hazen–Williams equations.

2.4.1.1   Darcy–Weisbach (Colebrook–White) equation

The Darcy–Weisbach equation is named after the French engineer Henry Darcy (1803–1858) and the German engineer Julius Weisbach (1806–1871), who independently discovered it around 1850, and relates the energy head loss due to friction along a pipe to the average velocity of the fluid flow:

$$h_f = \lambda \frac{L}{D} \frac{V^2}{2g} \tag{2.17}$$

where λ = the non-dimensional friction factor

The equation can be used for all pipe flow categories (i.e. laminar, turbulent and transitional) and, as such, it is a function of the Reynolds number, Re, and relative pipe roughness, k/D, where k is the absolute pipe roughness height, also called Nikuradse's equivalent sand-grain roughness, expressed in mm (Nikuradse, 1933). The functional behaviour of λ is shown in the Moody diagram (Moody, 1944) in Figure 2.5.

On the left of the diagram the value of λ for laminar flow conditions is

$$\lambda = \frac{64}{\text{Re}} \tag{2.18}$$

This shows that λ is a function not only of the pipe material but also of the Reynolds number, i.e. it depends on viscosity, density and flow velocity. The laminar flow equation holds for Re < 2000. For Reynolds numbers above 2000 the flow changes from laminar to weakly turbulent, and beyond approximately 4000 it becomes turbulent, but is characterised by three characteristic flow zones:

■ Smooth turbulent flow (the lower envelope line in Figure 2.5), which represents flow in a hydraulically *smooth pipe*. This happens at low velocities in the turbulent flow region where there is a layer of laminar flow adjacent to the pipe wall. If this layer is thicker than the roughness of the pipe internal surface, then the actual roughness has no effect on the nature of the flow.
■ Rough turbulent flow (where the curves become horizontal), for which the friction factor λ is a function only of the relative roughness k/D and not of Re.

Figure 2.5 Example Moody diagram for the Darcy–Weisbach friction factor λ
Reproduced with permission from Purcell (2003)

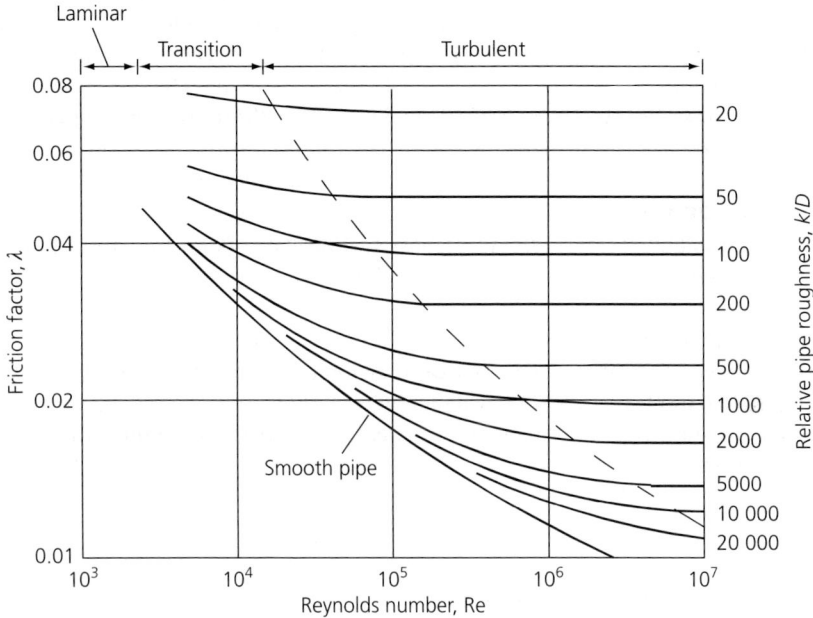

- Between the smooth and rough turbulent flow zones there is a transitional turbulent region for which λ decreases as Re increases.

One equation that is applicable for Re > 4000 (i.e. over the whole turbulent flow region) is the Colebrook–White formula (Colebrook and White, 1937; Hager, 2010), which is used to solve for the Darcy–Weisbach friction factor λ:

$$\frac{1}{\sqrt{\lambda}} = -2\log\left(\frac{k}{3.7D} + \frac{2.51}{\text{Re}\sqrt{\lambda}}\right) \qquad (2.19)$$

It relates the friction factor, λ, implicitly to the roughness, k, and the Reynolds number, Re. Because of its implicit nature (i.e. λ appears on both sides of the equation), iterative numerical methods have to be used to determine λ from equation (2.19). This is not always practicable, especially with today's modelling tools when λ needs to be determined for a large number of pipes. The other important feature of this equation is that it uses the base 10 logarithm, whereas in many computer languages the computation is based on the natural logarithm, thus requiring a base change. Therefore, the equation is often written in the natural base as

$$\frac{1}{\sqrt{\lambda}} = -0.8686\ln\left(\frac{k}{3.7D} + \frac{2.51}{\text{Re}\sqrt{\lambda}}\right) \qquad (2.20)$$

Swamee and Jain (1976) came up with an explicit approximation of the implicit Colebrook–White equation:

$$\lambda = \frac{0.25}{\left[\log\left(\dfrac{k}{3.71D} + \dfrac{5.74}{Re^{0.9}}\right)\right]^2} \tag{2.21}$$

The relative simplicity and accuracy (typically, it provides a friction factor estimate only 1–2% different from the Colebrook–White equation) of the explicit Swamee–Jain equation has influenced software developers to implement this equation in a number of modelling tools for water distribution system analysis. For example, the widely available hydraulic analysis software EPANET (Rossman, 2000) uses this equation to solve for the Darcy–Weisbach friction factor.

2.4.1.2 Hazen–Williams equation

The Hazen–Williams equation, originally introduced in 1902 (Liou, 1998), is still widely used by water supply engineers to calculate the head loss along a pipe based on the flow in the pipe and the physical properties of the pipe. The equation is empirical, and in SI units could be expressed as

$$h_f = 10.67 \frac{Q^{1.85}}{C^{1.85} D^{4.87}} L \tag{2.22}$$

where C = the Hazen–Williams friction coefficient, which indicates the roughness of the interior surface of a pipe

Note that equation (2.22) implies that the coefficient is dimensional and would change for different units of h_f, Q and D. However, the C value is considered to be a pipe constant, and a numerical conversion coefficient needs to be introduced if different units for head loss, flow and diameter are used in the equation. However, being empirical, the equation is not dimensionally homogeneous, and its range of application is limited. Note also that lower C values mean higher head losses in the pipe (opposite to relative roughness used in the Darcy–Weisbach equation).

Both the Darcy–Weisbach and Hazen–Williams equations can be expressed using the following single head loss equation:

$$h_f = RQ^n \tag{2.23}$$

where R = the pipe/flow resistance coefficient
$\quad\quad n$ = the head loss equation exponent

Note that for the Darcy–Weisbach equation (and for turbulent flow), $R = 0.8106\lambda L/(gD^5)$ and $n = 2.0$, and for the Hazen–Williams equation, $R = 10.67L/(C^{1.85}D^{4.87})$ and $n = 1.85$.

2.4.2　Local and minor losses

Local losses are generally caused by increased turbulence as the flow passes through valves or other pipe fittings such as bends, tees or tapers. Figure 2.6 illustrates the

Figure 2.6 The head loss generated by a partially shut valve

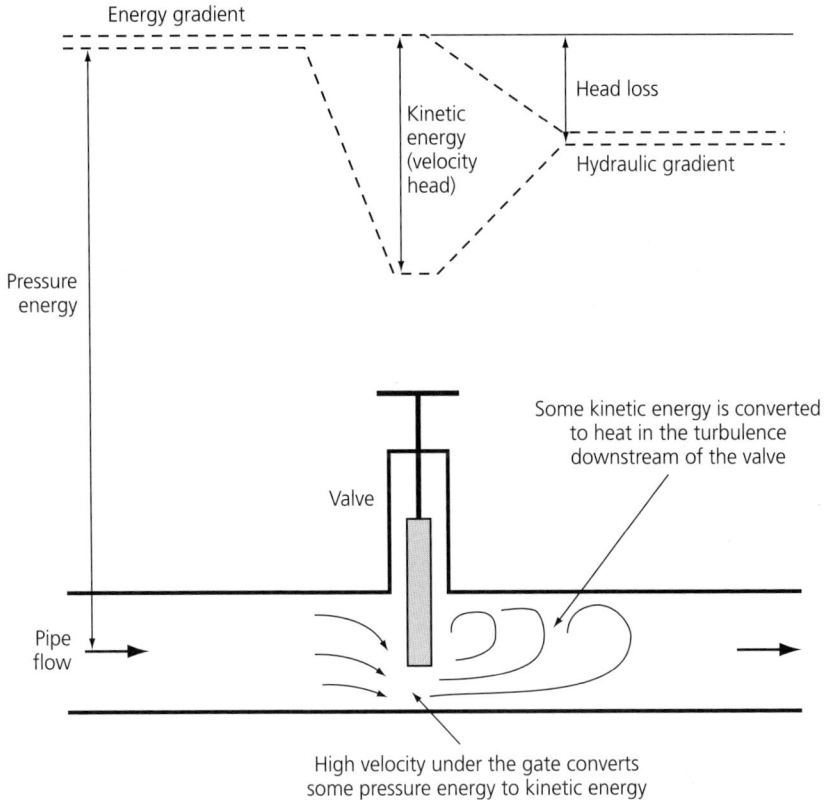

head loss generated by a partially shut valve. In this case, the velocity under the gate of the valve is increased significantly by the reduced cross-sectional area, as the flow at all pipe sections must be same (conservation of mass). This velocity increase results in some of the pressure energy being converted into kinetic energy, as the overall energy must remain constant (conservation of energy).

If all the pressure energy is converted to kinetic energy, then the pressure at the reduced cross-section under the valve gate may become sub-atmospheric, and vapour cavities may form due to water evaporating at the ambient temperature. The violent collapse of these cavities downstream as the pressure rises causes cavitation, which can damage the valve and the pipeline.

Downstream of the valve, massive eddies are created as the velocity slows. This results in some of the energy being transferred to eddies and eventually converted to heat, as not all the kinetic energy is reconverted to potential energy downstream of the valve, and an overall head loss occurs across the valve.

Figure 2.7 Head loss due to turbulence through pipe fittings

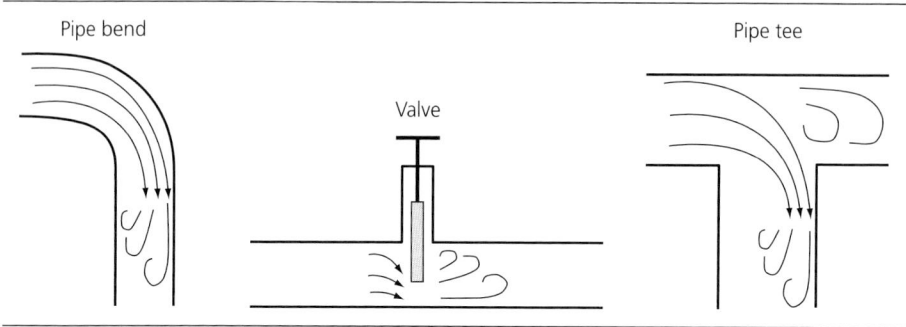

As the head loss is related to the kinetic energy lost, this head loss is often expressed as part (fraction) of the initial kinetic energy of the pipe flow:

$$h_l = k\frac{v^2}{2g} = KQ^2 \qquad (2.24)$$

where h_l = the local head loss (L)
k = the local head loss coefficient (–)
K = the local resistance coefficient (T^2/L^5)
v = the characteristic velocity (typically downstream of a local head loss) (L/T)
Q = the flow rate (L^3/T)
g = the acceleration due to gravity (L/T^2)

Figure 2.7 illustrates that the mechanism of head loss through increased turbulence is the same for other pipeline fittings such as bends and tees. The value of K for various types of pipeline fittings can be obtained from standard hydraulic textbooks such as the *Manual of British Water Engineering Practice*, Table XXXIV (Skeat and Dangerfield, 1969).

For most water transmission pipelines the fitting losses are normally small compared with the overall friction losses along the pipe, but they may be significant at points where there are a number of fittings and the diameter of the pipe is constrained, e.g. at pumping stations, bridge crossings, etc.

2.5   Steady flow analysis in networks

A typical water distribution network is a complex system normally consisting of thousands of simpler individual elements (e.g. pipes, valves, pumps and reservoirs). As with many other complex systems, computer simulation is normally the only viable way of analysing the complex hydraulic behaviour in a water distribution network, i.e. to determine the flow in every pipe and the pressure at every node in the system. With many network elements linked together to form loops, computer models are used to calculate how much water takes which of the numerous alternative routes and what are the head losses along each route. The two equations used to compute the flows and energy heads are the continuity equation and the energy equation.

When applied to a network node connecting n_p pipes the continuity equation (2.7) gives

$$\sum_{j=1}^{n_p} Q_{ij} - q_i = 0 \qquad (2.25)$$

where Q_{ij} = the flow in pipe ij – positive for inflow and negative for outflow (L^3/T)
q_i = the nodal discharge at node i (L^3/T)

Note that although demands are actually distributed along a pipe in reality (via property connections), these demands are lumped at nodes for modelling purposes and represented as the nodal discharge. The above equation can then be written for every node in the system.

Similarly, the energy balance equation can be written between any two nodes (e.g. nodes 1 and 2) in the water distribution network:

$$H_1 - H_2 = \sum_{\text{path}_{1-2}} h_f = \sum_{\text{path}_{1-2}} R_{ij} Q_{ij} |Q_{ij}|^{n-1} \qquad (2.26)$$

where path_{1-2} includes the set of pipes on the path from node 1 to node 2
H_1, H_2 = the heads at the upstream (1) and downstream nodes (2) of the path
R_{ij} = the resistance coefficient for each pipe ij of path_{1-2}
n = the exponent from the head loss equation (e.g. 2 for the Darcy–Weisbach equation and 1.85 for the Hazen–Williams equation)

Network nodes can be classified into the following three groups:

- fixed-head nodes
- variable-head nodes
- ordinary nodes or junctions.

The main feature of a fixed-head node is that its water level does not change during the analysis, e.g. a fixed-head node represents external sources or sinks of water to the network, such as lakes, rivers or groundwater aquifers. Variable-head nodes have associated storage capacity, where the volume of stored water can vary with time during the analysis, e.g. service or elevated reservoirs. Junctions represent all other points in the network where pipes branch out (with or without demand associated with them) or where there is a net export (or import) of water from (to) the system, i.e. water enters or leaves the network (see Figure 2.8).

Figure 2.8 shows a branched system and a looped system, where a loop is defined as a closed circuit that has no interior crossing pipes – also called a simple, independent or 'natural' loop (Epp and Fowler, 1970). It is important to note that in a branched network there are no loops. Furthermore, for a closed loop (e.g. pipes p_2–p_4–p_6), i.e. one beginning and ending at the same node (N_2), the energy loss is zero.

Steady flow analyses in branched networks are performed directly, without iterations, and fall into two main problem categories: (1) for known demand at nodes and

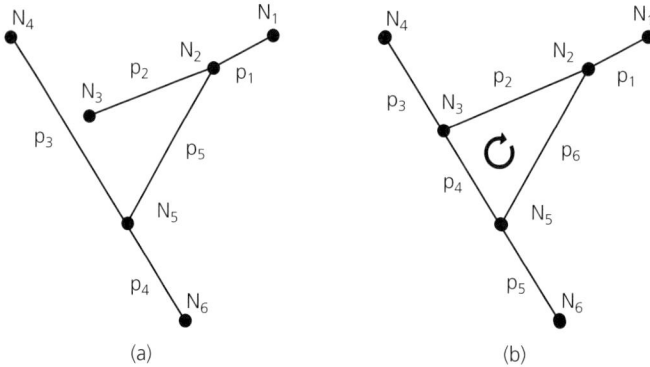

Figure 2.8 An examples of (a) a branched system and (b) a looped system

known pipe characteristics, find pressure at nodes or (2) for known head loss along pipes and known pipe characteristics, find flows in the network. For example, a steady-state analysis for a branched system (see Figure 2.8) consisting of three pipes, p_1 ($D = 250$ mm), p_2 ($D = 100$ mm) and p_5 ($D = 200$ mm), each 1000 m long and having Hazen–Williams coefficient $C = 100$, can be performed using the continuity equation (2.25) and the energy balance equation (2.26). If the demand at nodes is known to be $q_2 = 2$ l/s, $q_3 = 3$ l/s and $q_5 = 12$ l/s and the known fixed head at N_1 is 20 m, the computations are made in the following manner. From equation (2.25):

$$Q_{12} - Q_{23} - Q_{25} - q_2 = Q_{12} - 3 - 12 - 2 = 0 \Rightarrow Q_{12} = 17 \text{ l/s}$$

By using equation (2.26):

$$H_1 - H_2 = h_f = R_{12}Q_{12}|Q_{12}|^{0.85} = 10.67 \frac{1000}{100^{1.85}0.25^{4.87}}0.017|0.017|^{0.85} = 0.97 \text{ m}$$

$$H_1 - H_2 = h_f = 0.97 \Rightarrow H_2 = 20 - 0.97 = 19.03 \text{ m}$$

Similarly,

$$H_3 = H_2 - 10.67 \frac{1000}{100^{1.85}0.10^{4.87}}0.003|0.003|^{0.85} = 19.03 - 3.39 = 15.64 \text{ m}$$

$$H_5 = H_2 - 10.67 \frac{1000}{100^{1.85}0.20^{4.87}}0.012|0.012|^{0.85} = 19.03 - 1.51 = 17.52 \text{ m}$$

Once all pipes, nodes and independent loops are identified prior to performing a steady-state flow analysis for a looped network, a system of non-redundant continuity and energy equations is developed (for the entire network). The number of these equations is related directly to fundamental relations between the number of pipes, the number of nodes (including all types of nodes, i.e. fixed-head, variable-head and ordinary nodes/junctions) and the number of closed simple loops in the network system (Larock *et al.*, 2000).

Various formulations of the governing equations exist for the solution of network analysis problems (Larock *et al.*, 2000). These are named:

- *Q equations* or flow equations, in which pipe discharges are treated as unknowns
- *H equations* or node equations, in which nodal heads are treated as unknowns
- *Q-H equations* or combined flow and node equations (e.g. the Todini and Pilati algorithm)
- the ΔQ equation, in which a loop flow correction of ΔQ is computed for each of the independent loops and each of the paths between fixed-head nodes in the network.

Irrespective of the formulation of the governing equations chosen for implementation, an approach based on the Newton–Raphson method can be applied to solve the system of equations (Press *et al.*, 2007).

2.5.1 Hardy Cross method

Prior to the introduction of computers, engineers had to simplify the network into a few strategic pipe loops and use an iterative method developed by Hardy Cross (1936). The so-called Hardy Cross method is the first systematic approach for solving a system of equations resulting from writing a set of continuity equations for each node and energy equations for each independent loop in the network.

The method starts with an initial guess of the direction and magnitude of pipe flows, such as the principle of continuity being satisfied at each node. It then calculates the head losses around each independent loop, which normally results in the need to rebalance the network (i.e. change assumed flows) until the sum of the head losses around each loop is close to zero. A correction of the assumed flows, ΔQ, is computed for each independent loop in the network. Iterations are performed such that

$$\Delta Q^{(i)} = \frac{-\sum h_f^{(i)}}{n \sum \dfrac{h_f^{(i)}}{Q^{(i)}}} \tag{2.27}$$

where $Q^{(i)}$ = the flow in a pipe at iteration i (M^3/T)
$\qquad \Delta Q$ = the corrective flow passing through a pipe in a loop (M^3/T)
$\qquad \sum$ = the summation operator applied to all pipes in a loop
$\qquad n$ = the exponent from the head loss equation

This gives the new flow estimate in a pipe for the next iteration:

$$Q^{(i+1)} = Q^{(i)} + \Delta Q^{(i)} \tag{2.28}$$

It is important to note that the corrective flow is equal for each pipe in a loop and that, for pipes belonging to multiple loops, the corrective flow is a sum of individual loop corrective flows. The value of n in the Hardy Cross method is assumed to be constant.

After the flows in each pipe are found, i.e. the corrective flow becomes lower than a predefined error threshold, and the head loss in each pipe is also known, the nodal heads are simply calculated by starting at known (i.e. fixed) heads.

The Hardy Cross method depends on the initial estimate of the flows, and could suffer from slow convergence. Instead of adjusting flows around each individual loop in the network, the Newton–Raphson method adjusts the flows (Q equations), heads (H equations), or flows and heads (Q-H equations) along all loops simultaneously. These simultaneous corrections have improved convergence over the Hardy Cross method, and the current modelling software relies almost exclusively on the Newton–Raphson approach. Therefore, the Hardy Cross method is of historical interest rather than an approach that can be used for the large water distribution models preferred by today's practitioners.

Flow analysis in large water distribution networks has been made possible by the advances in computer technology, particularly in the 1960s and 1970s when the first research-based simulation tools were developed and then in the 1980s when personal computers were first introduced. A number of methods implemented in software packages, such as GINAS (Coulbeck *et al.*, 1991), PICOLLO (Jarrige, 1993), WATNET (developed by the WSA/WCA Engineering and Operations Committee (WRc)), STONER (now SinerGEE), etc., have advanced beyond the Hardy Cross method and allowed sophisticated steady-state and dynamic analyses to be performed. The development of the gradient method (Todini and Pilati, 1988) and its implementation within the widely available freeware EPANET (Rossman, 2000) have led to the widespread use of the algorithm by software developers and vendors.

2.5.2 Todini–Pilati method

At a conference in 1987, Todini and Pilati (1988) presented an elegant formulation (Q-H equations) and an efficient gradient solution methodology for the system of non-linear equations. The gradient solution method has excellent convergence characteristics and is incorporated into a number of software packages currently available as commercial or public-domain software (e.g. EPANET).

Following from equations (2.24) to (2.26), the flow–head loss and continuity relations can be derived for any water distribution system as

$$H_i - H_j = RQ_{ij}^n + KQ_{ij}^2 \quad \text{for all pipes, } n_\text{p}$$

$$\sum_{j=1}^{n_\text{p}} Q_{ij} = q_i \quad \text{for all junction nodes, } n_\text{n} \tag{2.29}$$

where q_i = the demand at node i

The continuity equation is developed for each junction node connecting n_p pipes, while the head loss equation is developed for each pipe between nodes i and j. Note that for pumps, the non-linear head loss equation can be represented by a power law expression (Todini, 2003).

For the gradient solution method, the system of equations (2.29) can be written in a matrix form. Thus, the simulation of a network of n_p pipes with unknown discharges/

flows, n_n nodes with unknown heads (internal nodes) and n_0 nodes with known heads (e.g. tank levels) can be formulated as

$$\begin{bmatrix} \mathbf{A}_{pp} & \mathbf{A}_{pn} \\ \mathbf{A}_{np} & \mathbf{0} \end{bmatrix} \begin{bmatrix} \mathbf{Q}_p \\ \mathbf{H}_n \end{bmatrix} = \begin{bmatrix} -\mathbf{A}_{p0}\mathbf{H}_0 \\ \mathbf{q}_n \end{bmatrix} \tag{2.30}$$

where $\mathbf{Q}_p =$ the $[n_p, 1]$ column vector of unknown pipe flows (M^3/T)
$\mathbf{H}_n =$ the $[n_n, 1]$ column vector of unknown nodal heads (M)
$\mathbf{H}_0 =$ the $[n_0, 1]$ column vector of known nodal heads (M)
$\mathbf{q}_n =$ the $[n_n, 1]$ column vector of demands lumped at nodes
$\mathbf{A}_{pn}$ $(=\mathbf{A}_{np}^T)$ and $\mathbf{A}_{p0} =$ topological incidence sub-matrices of size $[n_p, n_n]$ and $[n_p, n_0]$, respectively, derived from the general topological matrix $\bar{\mathbf{A}}_{pn} = [\mathbf{A}_{pn}|\mathbf{A}_{p0}]$ of size $[n_p, n_n + n_0]$ as defined by Todini and Pilati (1988)
$\mathbf{A}_{pp} =$ a diagonal matrix whose elements are given by the entry-wise product $\mathbf{R}_p|\mathbf{Q}_p|$, with $\mathbf{R}_p$ being the vector of the pipe hydraulic resistances

The above matrix equation can then be written as the following non-linear and linear system of equations, respectively:

$$\begin{aligned} \mathbf{A}_{pp}\mathbf{Q}_p + \mathbf{A}_{pn}\mathbf{H}_n &= -\mathbf{A}_{p0}\mathbf{H}_0 \\ \mathbf{A}_{np}\mathbf{Q}_p &= \mathbf{q}_n \end{aligned} \tag{2.31}$$

Thus, the non-linear mathematical problem has $n_p + n_n$ unknowns $(\mathbf{Q}_p; \mathbf{H}_n)$, and its boundary conditions are $(\mathbf{R}_p; \mathbf{q}_n; \mathbf{H}_0)$, i.e. the total number is equal to $n_p + n_n + n_0$.

The topological incidence matrix $\bar{\mathbf{A}}_{pn}$ is defined as

$$\bar{\mathbf{A}}_{pn}(i,j) = \begin{cases} -1 & \text{if the flow in pipe } j \text{ leaves node } i \\ 0 & \text{if pipe } j \text{ is not connected to node } i \\ +1 & \text{if the flow in pipe } j \text{ enters node } i \end{cases} \tag{2.32}$$

The global gradient algorithm (GGA) for solving the system of equations (2.31) was given by Todini and Pilati (1988) as

$$\begin{aligned} \mathbf{B}_{pp}^{iter} &= (\mathbf{D}_{pp}^{iter})^{-1}\mathbf{A}_{pp}^{iter} \\ \mathbf{F}_n^{iter} &= -\mathbf{A}_{np}(\mathbf{Q}_p^{iter} - \mathbf{B}_{pp}^{iter}\mathbf{Q}_p^{iter}) + \mathbf{A}_{np}(\mathbf{D}_{pp}^{iter})^{-1}(\mathbf{A}_{p0}\mathbf{H}_0) + \mathbf{q}_n \\ \mathbf{H}_n^{iter+1} &= -(\mathbf{A}_{np}(\mathbf{D}_{pp}^{iter})^{-1}\mathbf{A}_{pn})^{-1}\mathbf{F}_n^{iter} \\ \mathbf{Q}_p^{iter+1} &= (\mathbf{Q}_p^{iter} - \mathbf{B}_{pp}^{iter}\mathbf{Q}_p^{iter}) - (\mathbf{D}_{pp}^{iter})^{-1}(\mathbf{A}_{p0}\mathbf{H}_0 + \mathbf{A}_{pn}\mathbf{H}_n^{iter+1}) \end{aligned} \tag{2.33}$$

where iter $=$ a counter of the iterative solving algorithm
$\mathbf{D}_{pp} =$ a diagonal matrix whose elements are the derivatives of the head loss function with respect to $\mathbf{Q}_p$

The gradient solution method begins with an initial estimate of flows in each pipe that normally does not satisfy flow continuity. In the next step, the GGA methodology is applied, thus tackling simultaneously unknown flows and heads. The iterative process continues until a convergence criterion is reached.

Further details on the numerical methodology employed by the gradient solution method can be found in Todini and Pilati (1988) and Todini (2003).

2.5.3 Demand-driven versus head-driven analysis

Most of the commercial water distribution analysis software packages, including EPANET (Rossman, 2000), are built on the assumption that demand at a node is fixed (predetermined) and fully satisfied regardless of the pressure at the node, i.e. outflow at the node is equal to its demand. This is the so-called *demand-driven analysis* (DDA), which determines the nodal pressures and pipe flow rates that correspond to the specified nodal demands (regardless of whether or not they can be satisfied). This assumption is reasonable in well-designed systems and when they operate under normal conditions. In these cases, residual heads at demand nodes are sufficient to supply the demand. However, DDA gives unrealistic results in cases where networks operate under abnormal conditions, e.g. due to pipe burst or excessive demands.

Under abnormal conditions the distribution of flow in pipes and the deliverable discharge at the nodes cannot be calculated correctly using DDA. For example, nodal heads or pressure may be insufficient to deliver the required water demand such that the real outflow at the nodes is only a fraction of the original demand. *Head-driven* (or pressure-driven) *analysis* (HDA) is needed to correctly account for the actual outflow at nodes where the pressure head falls below the minimum service level required to supply the full demand.

Bhave (1981) was the first to investigate the pressure dependency of flow in water distribution systems. Since then, a number of head–outflow relationships have been suggested and incorporated into network analysis methodologies. For example, Wagner *et al.* (1988) suggested the following parabolic relationship:

$$q = \begin{cases} q^{(d)} & \text{for } H \geqslant H_{\mathrm{ser}} \\ q^{(d)} \left(\dfrac{H - H_{\min}}{H_{\mathrm{ser}} - H_{\min}} \right)^{0.5} & \text{for } H_{\min} \leqslant H \leqslant H_{\mathrm{ser}} \\ 0 & \text{for } H \leqslant H_{\min} \end{cases} \quad (2.34)$$

where q = the actual outflow at a node (M^3/T)
$q^{(d)}$ = the demand at a node (M^3/T)
H = the actual head at node (M)
H_{ser} = the minimum service head required to supply the full demand $q^{(d)}$
$H_{\min}$ = the minimum head below which no outflow is possible

In the UK, normally, the minimum service pressure head $H_{\mathrm{ser}} = 15\,\mathrm{m}$ is assumed to be sufficient to supply demand in full (WRc, 1994). At any pressure head below this threshold the node would supply only a fraction of the demand based on equation (2.34).

Todini (2003) showed how the gradient solution method can be extended to perform HDA by either adopting a head–outflow relationship or performing an iterative analysis to determine the possible outflow at demand nodes as a function of the prevailing system pressure. Giustolisi *et al.* (2008) demonstrated how leakage from pipes can be incorporated in an HDA approach by using the model proposed by Germanopoulos (1985). Recently, Giustolisi and Todini (2009) and Giustolisi (2010) showed how to deal with an actual demand pattern along mains which resulted in preserved model parsimony and energy balance.

2.6   Unsteady flow analysis in networks
2.6.1   Extended period simulation
When applied to a service reservoir for dynamic (unsteady) conditions the continuity equation states that the difference between the flow in and out of the reservoir is equal to the change in storage, or

$$\sum Q_{in} - Q_{out} = \frac{dS}{dt} = A\frac{dH}{dt} \tag{2.35}$$

where Q_{in} = the sum of the inflows into the reservoir (L^3/T)
$\quad Q_{out}$ = the flow out of the reservoir (L^3/T)
$\quad S$ = storage (L^3)
$\quad H$ = the water level in the reservoir (L)
$\quad A$ = the reservoir cross-sectional area (L^2)
$\quad t$ = time (T)

This equation is used when it is necessary to take into consideration the slow variation of flow conditions over time. This may result in changes of water level in a reservoir, changes in nodal demand, or changes in the state of valves and pumps. This type of network analysis is generally known as extended period simulation (EPS) or dynamic analysis.

Extended period simulation is normally computed as a sequence of steady-state (snapshot) analyses performed at fixed times, each followed by a mass balance computation based on equation (2.35). In solving the set of hydraulic equations for the unknown heads and flows at each snapshot, i.e. equations (2.24)–(2.26), the changes that occurred at the end of the previous time step are taken into consideration. These changes include the dynamic state of reservoirs and the predefined changes at the current simulation time, such as nodal demands, and the status of the pipes, pumps and valves. Further details of the methodology can be found in Walski *et al.* (2003).

2.6.2   Transient flow analysis
Pressure transients occur when there is a rapid change in the fluid velocity within a pipeline. These pressure changes are often generated by the rapid opening/closure of valves or

Figure 2.9 Propagation of a transient pressure wave following valve closure

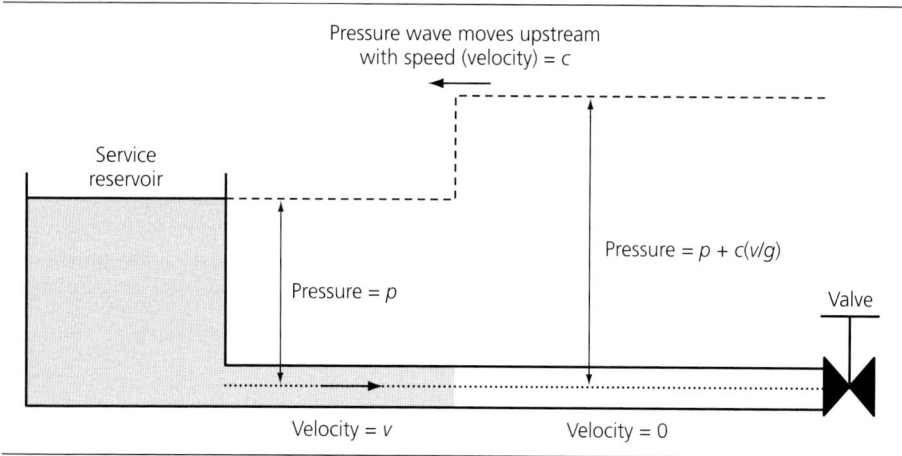

switching/failure of pumps within the pipe system. An initial estimate of the pressure change generated by a sudden or instantaneous change in velocity can be calculated using the Joukowsky equation (Thorley, 2004):

$$\Delta h = c\frac{\Delta v}{g} \tag{2.36}$$

where Δh = the change in pressure head (L)
   Δv = the sudden change in velocity (L/T)
   c = the wave speed at which the pressure transient travels along the pipeline (L/T)

The speed of the transient wave is partly determined by the fluid carried by the pipeline and partly by the pipeline material, diameter and thickness. However, the wave speed is significantly reduced by the presence of free gas in the fluid.

Figure 2.9 illustrates the propagation of a typical transient wave in a gravity pipeline from a reservoir following the instantaneous closure of a downstream valve. As the velocity of the water in the pipeline is reduced to zero by the valve, the pressure increases in accordance with the Joukowsky equation, and a positive-pressure wave is sent upstream to the reservoir. On reaching the free surface of the reservoir, the increased pressure energy is then released, creating a return wave at the original pressure but with a velocity in the reverse direction towards the valve. When this negative velocity hits the valve, a negative-pressure wave is generated which travels upstream to the reservoir as the velocity of flow is again reduced to zero. On reaching the free surface of the reservoir, the pressure again returns to the initial steady-state value, and a positive velocity is created which travels back to the valve again.

To help visualise this cycle of operation, imagine a goods locomotive pulling a train of wagons. When the locomotive stops rapidly, all the couplings between the wagons

Figure 2.10 Typical transient pressure wave recorded by a data logger

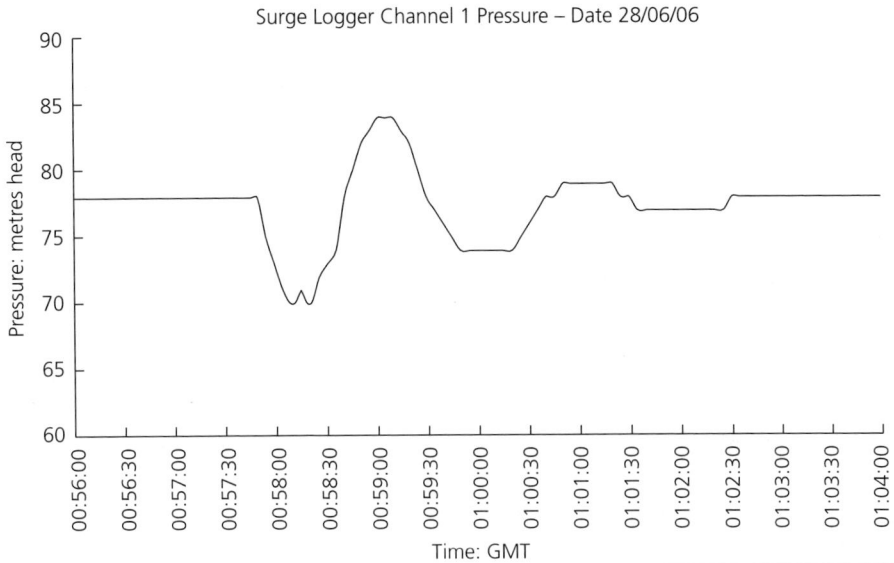

Surge Logger Channel 1 Pressure – Date 28/06/06

compress, as each wagon in turn is stopped, until the last wagon is reached. The last wagon then bounces back, pulling the couplings between each of the wagons taut with a velocity in the opposite direction.

In a frictionless pipeline system of ideally elastic material the pressure wave would continue back and forth, but in a real system the energy lost through friction gradually reduces the amplitude of the wave, creating the classic pressure surge wave often recorded by transient pressure loggers (Figure 2.10).

The Joukowsky equation is often used to initially calculate the maximum or worst-case pressure change due to an instantaneous velocity change to assess the potential magnitude of the surge problem. However, where the velocity change is not instantaneous the pressure change will be less severe. In this case the behaviour of a fluid in a slightly deformable pipeline can be described by momentum and continuity equations. Details of these equations can be found in standard textbooks on pressure surge (e.g. Wylie and Streeter, 1978).

In the past, graphical methods of analysis, such as those developed by Schnyder and Bergeron (Bergeron, 1961), have been used to provide a solution to the momentum and continuity equations as an estimate of surge pressures in most practical situations. More recently, solution by the method of characteristics has become the most popular method of analysis, as it can be applied to most practical systems and can easily be adapted for analysis by computer (Thorley and Enever, 1979). The method of characteristics can also include an allowance for non-linear terms such as friction, column separation and air bubbles.

2.7   Water quality analysis in networks

Drinking water quality is of great importance due to its impact on public health. It is not surprising, then, that considerable attention has been given to the development of algorithms for water quality analysis in water distribution systems. Water quality modelling applications in water distribution systems include the simulation of chlorine decay, the simulation of blending of multiple sources, the estimation of water age, the tracing of contaminant propagation, the prediction of the formation of disinfection by-products and total trihalomethanes in water, etc. Because of the improvement in public health due to the use of chlorine as a water disinfectant, the simulation of chlorine decay in water distribution systems is one of the most often encountered applications of water quality analysis.

Similarly to hydraulic network models, water quality analysis can be formulated as either steady-state or dynamic models. Steady-state models are based on the principles of mass conservation, which is used to determine the ultimate distribution of dissolved substances with constant hydraulic conditions. Thus, conservation of constituent mass for node j is given as

$$\sum_j \overbrace{(QC)}^{\text{incoming}} - \sum_j \overbrace{(QC)}^{\text{outgoing}} = 0 \tag{2.37}$$

where Q = the volumetric flow rate (L^3/T)
  C = the volumetric constituent concentration (M/L^3)

The nodal concentrations can be obtained for conservative and zero-order reacting constituents by an explicit analytic approach (Boulos and Altman, 1993). These models assume that either the constituent concentrations do not change over time (conservative) or the time rate of change of a constituent is constant (zero order). For general nth-order reaction kinetics the rate of decay of a constituent is given as

$$\frac{dC}{dt} = -kC^n \tag{2.38}$$

where t = time (T)
  k = the reaction rate constant ($1/T$)

The analytic approach by Boulos and Altman (1993) for solving the steady-state water quality modelling problem was later extended by Chung et al. (2007) to include first- and second-order reactions. The rate of the first-order reaction is proportional to the concentration of the constituent, and depletes a constant percentage of a constituent with each period of time that elapses (exponential decay), while the rate of the second-order reaction is proportional to the concentration of the constituent squared ($n = 2$ in equation (2.38)). The first-order model for chlorine decay is most commonly employed to simulate the disappearance of residual chlorine due to reactions with materials in the aqueous phase at different residence times, t, in the network. Thus, integrating equation (2.38) for $n = 1$ gives an exponential decay model:

$$C(t) = C_0 e^{-kt} \tag{2.39}$$

where $C_0 =$ the initial chlorine concentration (M/L^3)

With particular reference to chlorine, several studies concluded that chlorine undergoes reactions with particulates in the bulk of the water as well as at and near the pipe wall at separate reaction rates (Rossman *et al.*, 1994). Therefore, much better agreement between observed and simulated chlorine concentration are obtained when the simulated decay of chlorine is separated into a bulk and wall decay model:

$$\frac{dC}{dt} = \overbrace{-kC}^{\text{bulk decay}} - \overbrace{R}^{\text{wall decay}} \tag{2.40}$$

where $R =$ the pipe wall demand, e.g. due to the corrosion rate $(M/L^3/T)$

The most commonly encountered dynamic water quality analysis is based on a one-dimensional modelling approach, single- or extended-period simulation hydraulics, instantaneous and complete mixing at nodes, negligible longitudinal dispersion and a single constituent (e.g. chlorine) with one or more feed sources (Vasconcelos *et al.*, 1996). These models simulate three processes: (1) advection in a pipe; (2) the kinetic reaction mechanism; and (3) mixing at nodes. The advection process of a conservative constituent is given (in one-dimensional form) as the following mass conservation differential equation:

$$\frac{\partial C}{\partial t} = \overbrace{-v\frac{\partial C}{\partial x}}^{\text{spatial dispersion}} \quad \overbrace{-r}^{\text{temporal dispersion}} \tag{2.41}$$

where $v =$ the velocity of flow (L/T)
$r =$ the rate of decay of chlorine in the pipe $(M/L^3/T)$

For a first-order kinetic reaction equation, the rate of decay r becomes k, as seen in equation (2.39). Finally, mixing at nodes is given by the nodal mass balance principle in equation (2.37).

The analytical solution to the problem described in the above equations becomes intractable, and numerical solution techniques are employed to simulate the advection process. Within the EPANET toolkit (Rossman, 2000), a modified version of the Lagrangian time-based approach developed by Liou and Kroon (1987) is employed to simulate the variation of chlorine throughout a water distribution system. The method starts with the known direction of flow and velocities in the network provided by the hydraulic model simulation. It then stores the chlorine concentration information in discrete (water) volume elements that are progressed through the network at short time intervals, which are typically 5 minutes or less. For each time step, the water

45

segments travel through the pipe with the flow velocity. This approach avoids the numerical diffusion commonly associated with numerical solutions (e.g. finite-difference or finite-element schemes) for the advection equation (2.41).

Shang *et al.* (2002) developed a general framework for modelling the reaction and transport of multiple, interacting chemical species in drinking water distribution systems. The framework accommodates reactions between constituents in both the bulk flow (through pipes and storage reservoirs) and those attached to pipe walls. The software implementation is provided as an extension to the EPANET programmer's toolkit (a library of functions that simulates hydraulic behaviour and water quality transport in pipe networks).

The US Environmental Protection Agency (2008) also released EPANET-MSX (a Multi-Species eXtension of EPANET), which can be used for the simultaneous modelling of multiple species via interacting chemical and biological reactions in water distribution systems.

REFERENCES

Bergeron, L. (1961) *Waterhammer in Hydraulics and Wave Surges in Electricity*. New York: Wiley.

Bhave, P. R. (1981) Node flow analysis of water distribution systems. *Journal of Transportation Engineering, ASCE* 107(4): 457–467.

Boulos, P. F. and Altman, T. (1993) Explicit calculation of water quality parameters in pipe distribution systems. *Civil Engineering Systems* 10: 187–206.

Chung, G., Lansey, K. E. and Boulos, P. F. (2007) Steady-state water quality analysis for pipe network systems. *Journal of Environmental Engineering, ASCE* 133(7): 777–782.

Colebrook, C. F. and White, C. M. (1937) Experiments with fluid friction in roughened pipes. *Proceedings of the Royal Society of London, Series A* 161(906): 367–381.

Coulbeck, B., Orr, C. H. and Cunningham, A. E. (1991) *GINAS 5 Reference Manual. Research Report No. 56, Water Software Systems*. Leicester: De Montfort University.

Cross, H. (1936) Analysis of flow in networks of conduits or conductors. *University of Illinois Bulletin*, No. 286.

Environmental Protection Agency (2008) *Epanet Extensions*. www.epa.gov/nhsrc/water/teva.html#_epanet [accessed 01/09/10].

Epp, R. and Fowler, A. G. (1970) Efficient code for steady-state flows in networks. *Journal of the Hydraulics Division, ASCE* 96: 43–56.

Germanopoulos, G. (1985) A technical note on the inclusion of pressure dependent demand and leakage terms in water supply network models. *Civil Engineering Systems* 2: 171–179.

Giustolisi, O. (2010) Considering actual pipe connections in WDN analysis. *Journal of Hydraulics Engineering, ASCE* 136(11): 889–900.

Giustolisi, O. and Todini, E. (2009) Pipe hydraulic resistance correction in WDN analysis. *Urban Water Journal* 6(1): 39–52.

Giustolisi, O., Savić, D. A. and Kapelan, Z. (2008) Pressure-driven demand and leakage simulation for water distribution networks. *Journal of Hydraulics Engineering, ASCE* 134(5): 626–635.

Hager, W. H. (2010) Cedric Masey White and his solution to the pipe flow problem. *Proceedings of the ICE – Water Management* 163(10): 529–537.

Jarrige, P. A. (1993) *PICCOLO – Users Manual. Potable Water Distribution Network Modelling Software*. Nanterre: SAFEGE Consulting Engineers.

Larock, B. E., Jeppson, R. W. and Watters, G. Z. (2000) *Hydraulics of Pipeline Systems*. Boca Raton, FL: CRC Press.

Liou, C. P. (1998) Limitations and proper use of the Hazen–Williams equation. *Journal of Hydraulics Engineering, ASCE* 124(9): 951–954.

Liou, C. P. and Kroon, J. R. (1987) Modeling the propagation of waterborne substances in distribution networks. *Journal of the American Water Works Association*, 54–65.

Moody, L. F. (1944) Friction factors for pipe flow. *Transactions of the ASME* 66: 671–684.

Nikuradse, J. (1933) Laws of flow in rough pipes. *Forschungsheft Verein Deutsche Ingenieure*, No. 361 [in German].

Press, W. H., Flannery, B. P., Teukolsky, S. A. and Vetterling, W. T. (2007) *Numerical Recipes: The Art of Scientific Computing*. Cambridge: Cambridge University Press.

Purcell, P. (2003) *Design of Water Resources Systems*. London: Thomas Telford.

Reynolds, O. (1883) An experimental investigation of the circumstances which determine whether the motion of water shall be direct or sinuous, and of the law of resistance in parallel channels. *Philosophical Transactions of the Royal Society* 174: 935–982.

Rossman, L. A. (2000) *EPANET 2, Users Manual*. Cincinnati, OH: Environmental Protection Agency.

Rossman, L. A., Clark, R. M. and Grayman, W. M. (1994) Modeling chlorine residuals in drinking water distribution systems. *Journal of Environmental Engineering, ASCE* 120(4): 803–820.

Shang, F., Uber, J. G. and Polycarpou, M. M. (2002) Particle backtracking algorithm for water distribution system analysis. *Journal of Environmental Engineering, ASCE* 128(5): 441–450.

Skeat, W. O. and Dangerfield, B. J. (eds) (1969) *Manual of British Water Engineering Practice*. Cambridge: Heffer.

Swamee, P. K. and Jain, A. K. (1976) Explicit equations for pipe-flow problems. *Journal of the Hydraulics Division, ASCE* 102(5): 657–664.

Thorley, A. D. R. (2004) *Fluid Transients in Pipelines*, 2nd edn. London: Professional Engineering Publishing.

Thorley, A. R. D. and Enever, K. J. (1979) Control and suppression of pressure surges in pipelines and tunnels, England. *CIRIA Report 84*.

Todini, E. (2003) A more realistic approach to the 'extended period simulation' of water distribution networks. In: C. Maksimovic, D. Butler and F. A. Memon (eds), *Advances in Water Supply Management*, pp. 173–184. Leiden: Balkema.

Todini, E. and Pilati, S. (1988) A gradient algorithm for the analysis of pipe networks. In: *Computer Applications in Water Supply*. Letchworth: Research Studies Press.

Vasconcelos, J. J., Boulos, P. F., Grayman, W. M., Kiéné, L., Wable, O., Biswas, P., Bhari, A., Rossman, L. A., Clark, R. M. and Goodrich, J. A. (1996) *Characterisation and Modeling of Chlorine Decay in Distribution Systems*. Denver, CO: AWWA Research Foundation.

Wagner, J. M., Shamir, U. and Marks, D. H. (1988) Water distribution reliability: Simulation methods. *Journal of Water Resources Planning and Management, ASCE* 114(3): 276–294.

Walski, T., Chase, D. V., Savić, D. A., Grayman, W. M., Beckwith, S. and Koelle, E. (2003) *Advanced Water Distribution Modeling and Management*. Waterbury, CT: Haestad Methods Press.

WSA/WCA Engineering and Operations Committee (1994) *Managing Leakage: UK Water Industry Managing Leakage Reports A–J*. London: WRc/WSA/WCA.

Wylie, E. B. and Streeter, V. L. (1978) *Fluid Transients*. New York: McGraw-Hill.

Water Distribution Systems
ISBN: 978-0-7277-4112-7

ICE Publishing: All rights reserved
doi: 10.1680/wds.41127.049

ice
Institution of Civil Engineers

publishing

Chapter 3
Water demand: estimation, forecasting and management

Adrian McDonald Environment Faculty, University of Leeds, UK
David Butler Centre for Water Systems, University of Exeter, UK
Clare Ridgewell Essex & Suffolk Water, Chelmsford, UK

3.1   Introduction and context

Water is fundamental to life on earth. We cannot address the food security, health needs, education provision and governance of people unless they first have water services. The world is rapidly urbanising and so part of the water issue is to provide large, dense communities who do not (perhaps cannot) have a sufficient local supply. Thus, an estimate of demand is required so that supply can be sourced and provision made to transfer and distribute the water resource.

We need to balance supply and demand. This is a reasonably straightforward task for short periods and modest areas. However, for large conurbations the scale of works needed, plus the needs for inquiry and local governance, results in a very long lead time. So, demand is often estimated for 20–40 years ahead (certainly for major resource schemes). The demand planner is therefore catering for the water needs of age cohorts not yet born in a climate not yet experienced and within government policies not yet thought of.

It is possible that in arriving at future water balances we will be required to recognise further water demand components. For example, every community and every activity has a water footprint – the direct use of water plus an amount of water required to provide other goods and services (see Chapter 11, 'Sustainability and climate change'). In the UK we each use about 150 litres of water per day, but it is claimed that the goods and services we require have a hidden water requirement of 3400 litres per person per day, making the UK the sixth largest water 'importer' in the world (Orr and Chapagain, 2009). There may be pressure to recognise and account for these virtual flows in future (Chapagain and Hoekstra, 2008). In addition, the natural environment has a right to water to maintain landscapes, habitats and species. The rights of the channel are effectively enshrined in the EU Water Framework Directive, which requires the achievement and maintenance of good ecological status for most river reaches. Attaining this, especially in the timeframes currently envisaged, is ambitious, and many EU countries will fall back on the derogations and cost exemptions in the Directive.

So, the water industry competes for raw water resources with agriculture, the channel and industry. Of course, the UK abstracts only 1% of rainfall and 2% of run-off, so nationally it appears that there is an ample water balance, but for some seasons and years and in some high-demand regions such as the south east of England this is a very misleading picture! In the future, with climate change probably reducing the available supply (UK Climate Projections – Defra, 2009) and encouraging an upward shift in demand, it may not be possible to balance all the demands with the supply, and the exercise will become one of prioritisation of the allocation of scarce resources. Under current demand structures, compliance, at some sites, under some conditions, may not be achievable. This is an issue only recently being recognised in the UK and in Europe, as evidenced by the growing requirement in the UK to consider company mutual support in asset plans, and in Europe to consider, for example, water transfers between Languedoc Roussillon and Barcelona.

Water is fundamental to life, as we have seen, but it is also a fundamental characteristic of the business system for a water service provider (WSP). The majority of WSPs source their supplies from within the company boundary or from historically defined external sources. The demand for water in a region and in adjacent regions will partially govern the export/import potential, the investment needed to meet demand, and the actions and initiatives needed to control demand and meet efficiency targets, to ensure compliance with standards of service and to meet customer expectations even in the highest-demand periods of a prolonged drought. Assessing and forecasting how supply and demand will be maintained in a safe balance, given future uncertainties, is a core element in the UK of the economic regulator's (Ofwat) price reviews, particularly the asset management plan (AMP) process. These are plans drawn up every 5 years by every WSP. This process has as its foundations a strategic direction statement and a water resource management plan – in essence, the vision and the practical proposals. These are subject to consultation, revision and possibly inquiry before a final accepted plan for the next 5 years is agreed. Since the plan controls the water price for each company and contains firm commitments, it is in effect the financial foundations for the WSP for the ensuing period.

Demand is not a simple measure nor is it static. Figure 3.1 indicates the scale of the components that require to be measured or estimated if a thorough and robust understanding of the water components is to be attained. This is the analysis process map as required by the regulator for the AMP reporting. Starting at the catchments, be they ground or surface waters, the company determines the collected raw waters as the balance of abstracted raw water plus or minus imports and exports. Only a proportion of the raw water collected is transferred to the water treatment works, as some is lost or used as raw water. Some of the water is used or lost at the treatment plant, but the majority (with any imported potable water) provides the available potable water stock. This amount, less any exported potable water, is the distribution input. Some is lost in distribution (the leakage usually cited in discussions, but note that there are several other losses in the overall process), and the remainder is delivered to 'customers' and to distribution system operational use. Not all customers pay for water, so the final distribution is charac-terised as unbilled or billed to different customers (all of whom also leak some of the water that they have received). Even this detailed water accounting is, however,

Figure 3.1 Water balances and components from raw sources to delivered water (DSOU, distribution system operational use; USPL, underground supply pipe losses)

From Environment Agency (2007). © Crown Copyright, reproduced with permission of the Controller of Her Majesty's Stationery Office

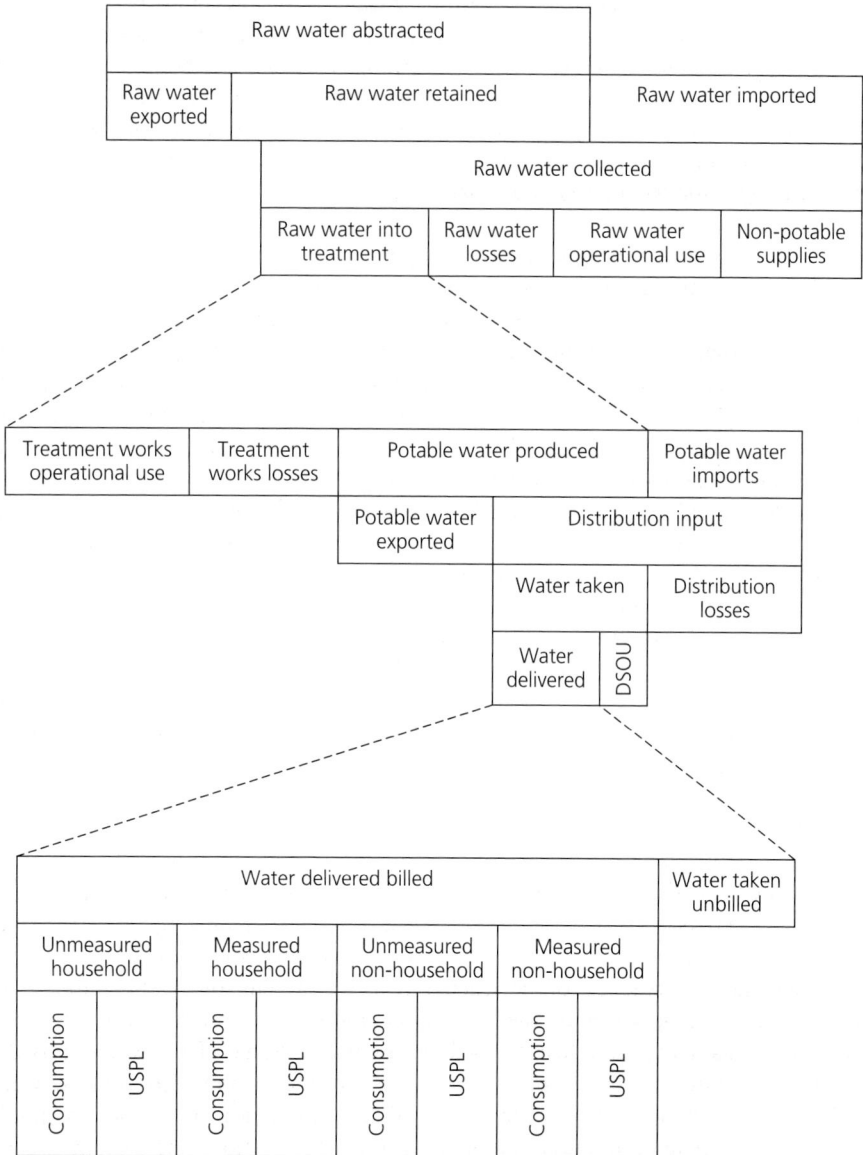

Raw water abstracted		
Raw water exported	Raw water retained	Raw water imported

Raw water collected			
Raw water into treatment	Raw water losses	Raw water operational use	Non-potable supplies

Treatment works operational use	Treatment works losses	Potable water produced	Potable water imports

Potable water exported	Distribution input
	Water taken	Distribution losses
	Water delivered	DSOU

Water delivered billed				Water taken unbilled
Unmeasured household	Measured household	Unmeasured non-household	Measured non-household	
Consumption / USPL	Consumption / USPL	Consumption / USPL	Consumption / USPL	

a summary, as we shall see later in this chapter. The International Water Association (IWA) Water Loss Task Force (Lambert, 2003) examines the water balance post-distribution-input only, but does so at a slightly finer scale by accounting for meter inaccuracies.

Demand can be measured as the aggregate total volume of demand over the long term, as discussed, or it can be measured as peak demand. This peak demand is not so instantaneously expressed as is the case in, say, the electricity industry, as there is a 'storage' in the distribution network that does not exist to quite the same extent in some other utilities. The long-term demand for water, whether measured per capita or per household, is a significant component of the overall supply required. So, this 'aggregate' demand defines the adequacy of the river flow resource, the storage available, the abstraction licensing agreements and so on. Peak demand, on the other hand, is more strongly related to the capacity of the infrastructure to treat, pump and transmit the peak flows demanded. Thus, systems (and so companies or resource zones) can either be resource or asset constrained.

3.2   Variations in water demand

Customer demand for water varies over time. There is a marked diurnal variation, with demand falling off significantly overnight. There is both a morning (sharp) and an evening (broader) peak in domestic demand, which is more obvious during the working week and less so at weekends, public holidays and vacation periods. Over the year, domestic demand increases in the growing season and in warmer weather (although only very modestly), but demand on resources can peak in winter, as the reduced domestic demand is masked by the increases in leaks (and in the difficulty of addressing leaks when the ground is frozen).

Almost half of the water abstracted from England and Wales is used for public water supply: 52% is for household use, 23% is for non-household use and 23% is due to leakage. The remaining 2% is for other uses such as fire fighting, operational and illegal water use (Defra, 2008).

Demand for potable water varies between countries. OECD (Organisation for Economic Co-operation and Development) countries, for example, consume three times more water per person than persons in Latin American, East Asia, Africa or India. Households in the most developed countries use only a small proportion of the water abstracted (around 8%), and in continental Europe the dominant raw water users are industry and agriculture. There are limited UK data on raw consumption, as there are big differences between licensed and actual abstractions. Households are the largest consumers of public potable water supplies, but, as we noted above, the water consumed by households is a small component of overall river flow. Thus, *generalised* arguments that domestic water saving is required to address lowered river flows arising from climate change are misleading. Instead, we need to concentrate on the water balance interventions needed in specific areas at times of water stress. Saving water is much less appropriate in (wetter) Northumberland than in (drier) Kent, for example.

Per capita consumption varies even in the developed world from 100 to 300 litres per person per day. The USA and Canada lead this table, while the Netherlands and Germany are lower users. Of this water use, only 5% is for cooking and drinking (European Environment Agency, 2001).

3.3   Components of demand

The water balance attempts to balance supply to a number of demand components, the majority of which are unquantified. Each is estimated separately, and there are few independent measures of error and uncertainty. The main demand components are:

- household
- commercial
- industrial
- operational
- unbilled
- illegal
- leakage.

Domestic water use is that used by households. It can be either measured (metered) or unmeasured. One might assume high levels of accuracy for measured consumption. However, the WSP does not know exactly how much this group uses because data are typically collected initially for billing purposes, and many providers still have to take a *sample* from their billing computers to gain an estimate of measured demand. The measurement of demand for unmetered users is more complex yet, and is discussed in the next section. Unmetered users are still in the majority in the UK, being about 70% of customers, but this varies significantly between companies, being driven by business philosophy and resources stress. Some companies have permission to compulsorily meter (in water-stressed areas such as Kent) all their customers, but in the main the growth in metering is through customers opting to have a meter or moving into a new house with an existing meter. New houses are now always metered.

Domestic demand, measured over the longer term, has been rising, but in recent years there is a suggestion of a demand decline. This small signal currently towards a decline in demand may well be a managed decline driven by the WSPs, who are themselves being encouraged by regulators to encourage demand reductions by advice and subsidised devices such as shower timers and water butts. Clearly, there is a growing number of lower water use appliances, particularly in new homes, and, as a result, even without behavioural change, there will be a trend to a lowered demand.

Commercial demand tends to be stable. Over the longer term it is declining marginally, but the real reduction in water demand has been in raw water not potable water commercial demand. In practical terms, all commercial demand is measured. Commercial demand itself can be subdivided into service and non-service components, and for the needs of specific companies into Standard Industrial Classification (SIC) classes. If refined to this level, the discussion starts to focus on industrial demand. In some industries, water demand is for raw waters, but even in these cases there remains a residual of potable water needed for the workforce. In effect there is a domestic demand component within the commercial and industrial water use. In some food-processing industries, say a bottled drinks manufacturer, the demand will be entirely for potable waters. So, despite the significant decline in manufacturing output, the demand for potable water in this sector has not reduced to the same extent because the heavier manufacturing industry used raw waters (sometimes partially treated at their own plants). For some WSPs the industrial water demand is dominated by a small number of very large users (e.g. the Royal Navy dockyards with a small number of large bases, food-processing and -freezing plants, major car plants). Here, the WSPs tend to account for these as separate lines in the water accounting and to determine the impact of their closure on the water and financial balances.

Some water is used by the WSPs themselves as part of the operation of the water treatment and distribution system. Typically, the water is used (1) to flush mains when they are being replaced or repaired, (2) as a pump lubricant in water treatment works and (3) as a normal potable supply in the company offices. There appears to be differences in the practices between companies as to where that third component should appear in the water balance accounting. Principally, this relates to whether it should be billed internally or remain unbilled, and this choice is governed in part by whether the water is taken before the works output meter or is taken after the output meter and has entered the distribution system. If not accounted for, this latter water use might well be interpreted as leakage, a value that every company would not wish to see increase.

There are two main components of the unbilled water: that which is legally unbilled and that which is illegally taken unbilled. Legally unbilled is, for example, the use of water for fire-fighting, while illegally taken water is stolen by standpipe in the case of some building sites or by tapping into a neighbour's supply pipe (Lambert, 2003). There is a grey area between these two extremes that relates to houses that use water but for which the WSP has no occupant details. Known in the industry as void properties, a significant minority are not actually void of people but void of the details of the occupants. These voids are growing (Figure 3.2), are concentrated in particular house types, ages and postcodes, and are typically new builds, rented flats and student accommodation. Vacant properties are widely distributed through the regions of England and Wales (Table 3.1). It is possible to take the view that it is up to the company to find its customers, and that until they do so the water is taken unbilled but legally. However, it is clear that many customers deliberately withhold details and act to avoid payment by moving without forwarding details. If the owner rather than the tenant could be billed, this may ease this situation. Water companies tend to have much less information on their customers than do the other utilities simply because they have to supply water as a right and have no contract, whereas an electricity company has a specific contract and will seek considerable customer information before agreeing a tariff and contract.

Finally, there is the question of that demand component called leakage, and this is discussed in detail in Chapter 8, 'Operation, maintenance and performance'. Methods are currently widely used to determine the so-called 'economic level of leakage' (Ofwat, 2002), which is the point at which it costs more to repair leaks than to source new water. Further improvements are surely needed, as current methods are static whereas they need to be dynamic, reflecting changes in resource availability or competition or prioritisation of abstraction or change in abstraction licensing. The

Figure 3.2 Growth in void households relative to household numbers (data anonymous and indexed to 100 in 2001)

McDonald *et al*. (2008a); data aggregated from ONS

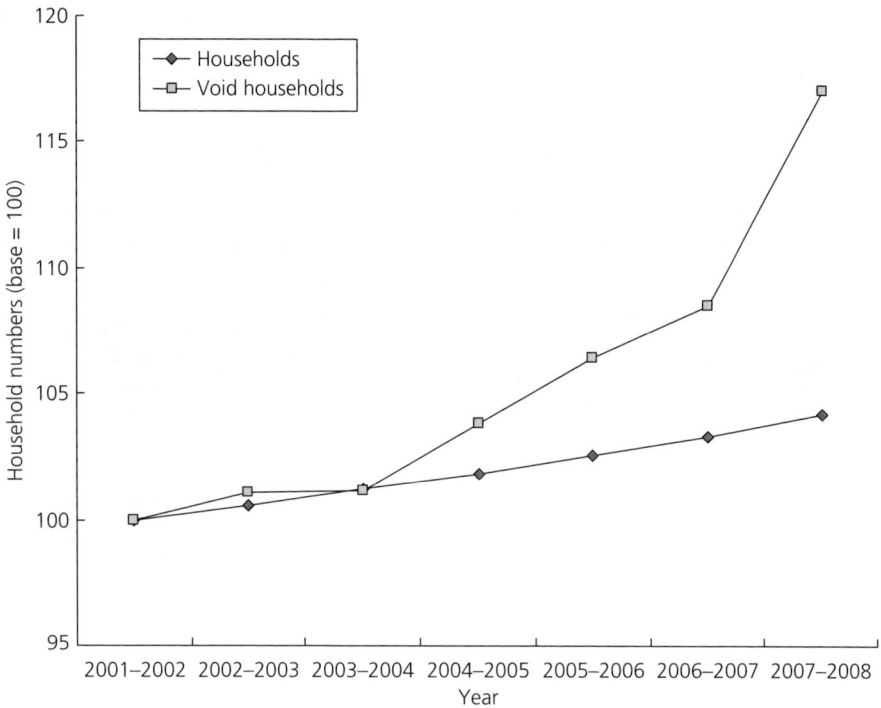

Table 3.1 Vacant property numbers, excluding second homes, holiday lets and student accommodation (data from 2005/2006)

Region	Properties	Vacant	%
North West	3 063 183	127 449	4.2
North East	1 137 976	41 570	3.7
Yorkshire and the Humber	2 237 381	76 154	3.4
West Midlands	2 299 011	74 440	3.2
East Midlands	1 887 023	60 624	3.2
London	3 219 048	86 730	2.7
East of England	2 423 286	64 227	2.7
South West	2 277 546	58 774	2.6
South East	3 541 679	85 557	2.4
England	**22 086 133**	**675 525**	**3.1**

McDonald *et al*. (2008a)

methods need to address the commonsense perception of customers that (1) visible leaks must be repaired whether economic or not and (2) water should not leak!

3.4   Drivers of demand

What drives water demand? The key influencing factor affecting domestic demand is population change (Sim *et al.*, 2007; Defra, 2008). Westcott (2004) has identified several further drivers, namely water policy (e.g. metering and water regulations), technology (white goods and other water-saving devices), behaviour (attitudes to water consumption) and economics (e.g. personal affluence). Studies from around the world have identified similar factors (Jeffrey and Geary, 2006).

Domestic demand breaks down into two key components – that associated with the house and that associated with the occupants; the former is near constant while the latter changes in proportion to the number of occupants. Thus, a five-person household uses much more water than, say, a two-person household, but, measured as per capita consumption, the five-person household will have a much lower individual demand (McDonald *et al.*, 2003; Memon and Butler, 2006; Sim *et al.*, 2007). These studies note, in addition, the role of the ages of the household residents and the time of year. Therefore, given the demographers' forecast of significantly smaller household size in the UK by 2025, even if the total population does not change, water demand will increase. Given that there is expected to be a significant increase in population as well (DCLG, 2007), there are then two separate population influences likely to drive up demand.

Herrington (1996), Downing *et al.* (2003), Roaf (2006) and others have also noted and attempted to quantify the potential effect of climate change, with the climate forecast to become more variable with drier summers and wetter winters – the extent of this disparity related to regions. Thus, we can expect more extremes in water demand with lower winter demand, less exacerbated by frost-related leakage increases, and summer demand increases. To achieve a continued resource balance in the face of this volatility may well require additional storage. A key question remains, namely the reliability of the increased winter rainfalls. We have in the last 30 years seen a greater frequency of two sequential winter droughts. This places considerable stress on the water resource system. We have not yet determined whether a three-dry-winter drought is a realistic and quantifiable prospect, and, if so, whether the system can cope with it.

3.5   Estimating current demand

There are two approaches to the estimation of water demand in unmeasured house-holds – the use of either domestic consumption monitors (DCMs) or area meters (AMs). For example, in the East of England, where population growth will be marked over the next 20 years, the incumbent WSP, Anglian Water, uses AMs, whereas Thames Water uses DCMs. So, the choice of approach appears unlinked to location and customer attributes, and appears to relate more to the historic approach developed by the company.

A DCM is a sample of 2000 households (although, in practice, it may be smaller than this) selected at random to be offered a place in the DCM, which, should they opt to

Table 3.2 Example output from a 10-household DCM

Household consumption: l/days	Number of residents	Per capita consumption: l/person/day
400	4	100
150	1	150
280	2	140
360	3	110
1190	10	119/127.5

join the survey, are then metered but continue to pay on a fixed tariff related to the rateable value of the property. DCM populations are therefore self-selected and subject to a range of biases (financial advantage, staff inclusion, self-selection – see McDonald *et al.*, 2003). Most of these biases result in underestimation of demand. In addition, DCMs decay through births, deaths, migrations and opt-out, and so DCM maintenance is a complex and seldom successful operation, particularly if the companies are required (by the regulator) to report outcomes in per person consumptions rather than per household consumption values. The population of households (billing addresses) is known with more certainty than the populations within households, and resurveying repeatedly to determine household numbers in the DCM adds to some aspects of bias. Detailed information about occupants is also increasingly difficult to collect, as people have become more wary of disclosing personal information. Calculation of demand from DCM data can offer misleading alternatives, as illustrated in the very simplified example in Table 3.2, in which data for 10 households are presented. The two demand estimates shown (in bold) in the final entry of the third column of the table are significantly different. The first estimate is determined from the column totals of water use and population, and the second is derived from the average demand of each household size. The former is, of course, the correct approach.

This absolute value of 2000 for the DCM sample was the outcome of an industry-wide study. Normally, to arrive at a sample size, you need to know the variability in the population and the accuracy needed of the estimate. Since the former will certainly vary between regions, the size requirement must vary by region and thus between companies. So, all DCMs are inherently biased. A good DCM will recognise the possibility of bias and address it by:

- Removing a source of bias if that is an option – thus, a DCM that has had water company staff encouraged into the membership to get the process started will remove members (but not all members – there should be equality of representation of the water employees).
- Correcting a bias by adapting recruitment – for example, if a local authority has refused to allow its tenants to take part, it is necessary to recruit only from social housing to replace these potential members. If the managers simply return to the list of customers and draw further potential members at random, then council-rented groups will be under-represented.

■ Analytical correction, which may be required if a specific group is under-represented and particularly if they appear to be higher-water users, for example Asian households. Here, you would need to correct for the under-representation.

In practice, many corrections will be required to deal with, for example, under-representation of specific house types, such as flats, or social groups, such as ethnic minority communities. It is both a complex and sensitive task.

AM-based determination of household demand is more complex and less direct. An AM gives the water use in a relatively large area. It is therefore the sum of many demands, few of which are known. Initially, the nightline is determined. The nightline is the demand in the early hours of the morning, and assumes that household use is at a minimum. From the nightline an allowance for legitimate night use is removed, as is night-metered usage and night commercial and industrial use (also metered); the residual is leakage. From the average demand on the AM, commercial and metered demand is subtracted, as is the now 'known' leakage. Further allowances for operational use, theft and the lowered leakage under the lower pressure, daytime regime may also be entered, but these are all estimates. The result is average demand exercised by the unmetered population. There is obviously considerable error in this process. Meters typically under-register, but the under-registration varies by meter type, size and age, none of which tend to be corrected in an area-specific manner. What constitutes legitimate night-time use is a generalised figure and one that changed following a water industry investigation which included the beneficial effect of reducing leakage. Household numbers in the area will be derived from the census, but these numbers will be several years out of date. As a result of such issues, there is an understandable tendency to seek simple uncomplicated sites, but these tend to be 'cul de sacs' and urban edge sites which host a different and probably unrepresentative population, and so, as with DCMs, the challenge is to understand how the results relate to the overall customer base.

The use of the term 'unmetered population' perhaps implies a precision that should now be questioned. Populations are derived either (1) from census material released through the UK Office for National Statistics (ONS) or through commercial firms that pass on this information, somewhat modified, or (2) from company billing records for the households in the area under scrutiny. In applying both DCM- and AM-derived water demands there are issues related to the population figures for the supply area as a whole – indeed, the ONS recognises that there are inaccuracies in its summary figures, figures with which WSPs often try to reconcile with their own population components. More fundamentally, however, is the lack of recognition and accounting for hidden and transient populations which exist in a region and use water, but which are unlikely to be recognised in the water balance and population accounting. Hidden populations are those that do not appear in the census records. They comprise a range of categories – from visa over-stayers, through asylum seekers who are rejected but not returned, to true clandestines. They are a category of the population that is typically poorly quantified by governments. Transient populations are those that live and are recorded in one region but who work in another, but are not in a metered establishment and who are not captured by normal accounting (McDonald *et al.*,

Table 3.3 Examples of the tabulations of hidden and transient populations in a water company area

Category	ONS capture	Legal	Population × 1000		
			Low	Medium	High
Transients					
Students	Yes	Yes	0.8	5.1	9.3
Second home householders	Partly	Yes	5.1	9.0	13.0
Accession country migrants	No	Yes	1.1	2.8	4.5
Hidden					
Clandestine	No	No	8.5	11.8	15.7
Human traffic	No	No	0.1	0.8	1.5
Totals			50.5	64.9	80.5

The totals given here are real but the components are incomplete

2008b). Table 3.3 provides an anonymised example of part of the categorisations used. In a full analysis there are 14 categories of hidden and transient populations to be considered. Here, we simply give some examples limited by commercial confidentiality and by the sensitivity of the material in some regions. Each category requires analysis. Take, for example, the student population. The total number in higher education in a region is a matter of record. Assume all students with a home post code within 50 km of the university will commute and reduce the student number accordingly. Assume all halls of residence and university flats are metered, and so their water use is captured. A figure can then be determined for the number of students who do not live at home or in student-identifiable halls/flats. To identify the water use of these groups – who will be part-year occupants but in higher household numbers than expected for the size of the property – requires further assumptions: that they will not opt to be DCM members and that they will have a lower water use. There is a danger of double counting in this process, as students are also a major component of the void property problem discussed earlier.

3.6   Forecasting demand
3.6.1   Commercial
There are two approaches to forecasting commercial demand: (1) the empirical statistical and (2) the process deterministic. The statistical approach uses the empirical data of the commercial water demand history, and plots a trend through the data. Although a simple and easily explained approach, it suffers from the assumption that the drivers over the period of the data will be continued into the future. The process models look either at the inputs (employment) or outputs (gross product volume or value) and establish a relationship between numbers employed, for example, and water demand. Such models tend to work better if there is some degree of disaggregation. Mitchell (1999) developed an alternative approach to forecasting: a more rigorously tested and refined

process model which explored a range of possible drivers of commercial demand and which then used employment, price and soil moisture deficit (as an index of drought). The model incorporated future water-saving activity designed to minimise waste. This component of the model differentiated between large and small firms, differences in rate of uptake and differences in potential saving between industrial sectors.

All of the industrial forecast models have to estimate the significance of business structural changes. This incorporates the changes in the predominance of some industrial types and the changes in the processes involved that use water. Businesses more than households are sensitive to the costs of the water used. A particular difficulty arises when demand changes arise from factory closures driven by international and sometimes national policies. Thus, a WSP with a large water demand associated with a Navy establishment in its area will need to pay close attention to defence reviews and possible rationalisations, as these could affect its planning. Recession-driven short-term closures at car factories in Swindon, Derby or Teeside would similarly have created a sudden and largely unforecastable change in water demand.

3.6.2   Domestic

There are several alternative approaches to forecasting domestic water demand based on demographics, behaviour, micro-components of demand and amalgamations of these approaches (e.g. see Williamson *et al.*, 1996, 2002; Clarke *et al.*, 1997; Memon and Butler, 2006; Bellfield, 2001; Jin, 2009). All these approaches may or may not choose to adhere to policy predictions or incorporate the effects of tariffs, smart meters and information.

The key determinant of water demand in a region is the population. Population forecasts are based on the fecundity of the age cohorts that comprise the population, modified by the balance of inward and outward migration. More sophisticated population models will incorporate fertility differences between different social groups. Thereafter, for water forecasts, it is necessary to estimate the household size class distribution that the population will form. However, because WSPs are required to report per capita consumption for both measured and unmeasured populations, the companies need a measure of occupancy in both measured and unmeasured properties. This is information that is not available through the census and is highly dynamic, being influenced by household break-ups, deaths, births, migration, etc., at a household level. Arriving at robust figures for these two occupancies is one of the most challenging measures facing demand planners. Added to this complexity is the effective discounting of household water supplied to blocks of flats. Where a block of flats is supplied as a single supply it is deemed 'non-household'. Quite how this influences the final population reconciliation is unclear, but it must lead to imprecision at least. Total water demand is accounted for most commonly from either household size or house type stratification. In the example in Table 3.4, for one resource zone of a UK water company, we use household size for stratification. The advantage is that the DCM no longer needs to mimic the population as a whole. Instead, the DCM is simply a sample of one-, two-, three-, etc., person households, and the household demand is calculated for each separately.

Table 3.4 Demand forecasting using a stratified household size approach

Household size	Number in WSP area	Domestic consumption monitor demand: l/day/household	Total demand: m³/day	Total population	Demand: l/person/day
1	545 196	221	120 615	545 196	221
2	636 634	322	205 109	1 273 268	161
3	290 886	392	114 163	872 658	131
4	247 324	465	115 121	989 296	116
5+	126 257	538	67 886	631 285	108
Total	1 846 297		622 895	4 311 703	144

Micro-component-based demand forecasts, although mechanistic, are widely used. A micro-component water demand is simply the aggregation of individual water-using components such as clothes washing, dish washing, toilet flushing, personal washing and garden watering (Gardiner and Herrington, 1986; Butler, 1991; Herrington, 1998) that are themselves the product of three estimates, namely ownership, volume used per 'usage' and frequency of use. It is based on averages, and so does not fully capture the range of behaviour found in the real world. It also requires detailed data on behaviour – e.g. for how long does a person shower or bathe on average? Micro-component analysis has no explicit spatial element, although it could have, provided data were available. However, it is useful in building up a picture of the underlying cause of demand, and hence is helpful when establishing the best way to manage demand (discussed later in the chapter).

At its simplest, the micro-component approach is the product of ownership, frequency of use and volume per use, as shown in Table 3.5 for a single component – an older WC in a domestic setting. However, there are at least six main toilet types, each with a different volume characteristic. The ownership percentages are not uniform (there are fewer pre-1935 single-flush toilets than there are first-generation dual flush, for example), and these proportions differ between metered and unmetered households.

Usefully, Twort et al. (2000) have divided domestic water use into in-house and out-of-house use. Typical in-house household uses are personal hygiene, drinking and food preparation and toilet use, while examples of out-of-house use are garden and car washing. It is useful because demand management interventions typically focus 'in' or 'out' of house, and these two macro-components are greatly influenced by house type,

Table 3.5 Example micro-component breakdown for domestic WCs

Ownership	Frequency of flushes: l/day	Volume per use: litres	Daily volume: litres
1.0	4.65	9.25	43.01

a driver that can be identified externally and which features in planning scenarios and permissions, and can therefore contribute to forecasting.

At the heart of effective forecasting is population attributes. Let us assume that the gross population forecasts are accurate (but see earlier caveats). The most important characteristics remaining are the household size and the affluence and lifestyle of the household. While the census examines many of these attributes, confidentiality requires that the data be aggregated into large spatial bundles. Microsimulation, more properly micro-analytic simulation, 'unpicks' these bundles and creates small-area units of hypothesised household characteristics – a synthetic micro-population. This is done through sets of conditional probabilities that reflect the likelihood of family size and class x living in region and property type z. While some individual simulated households may be incorrect, the entire attributes, when summed, match with those of the aggregate information in the census. This allows us to use known water demand relationships to forecast and assess demand for smaller areas than would otherwise be possible. While this is a complex and computationally intensive procedure, it is well documented and of considerable provenance, being used in both the UK and USA as a basis for economic estimation by the Treasury (Roe and Rendell, 2009). Micro-analytic simulation can be both static (i.e. a cross-section in time – and thus appropriate for demand estimation) or dynamic (in which the population is grown over time – and thus more appropriate for demand forecasting).

3.7   Managing demand

Current UK government policy is to halt or reverse the trends towards higher water demand that have been apparent for several decades – or even, it might be argued, since records began (Defra, 2008). However, this aspirational target also needs to be backed by an implementation plan if aspiration is to be turned into reality. The basis of this policy is being driven by information on aggregate abstraction licence totals granted rather than actual abstractions, and, perhaps more importantly, is based on all abstractions, when it appears that agricultural abstractions will be the most significant component of the withdrawal balance in the future. Figure 3.3 shows UK abstractions by major industry based on data from 1997 to 1998 available from ONS (2010).

Methods for demand control fall into two broad categories of technical measures and behavioural adjustment. We will address, firstly, the technical measures.

3.7.1   Technical methods

We can displace demand by encouraging the use of alternatives to potable waters or by reusing potable water following its first use. Such reuse can be on site or integrated into the overall supply system of a region. Clearly, the former is simpler, as it has the implicit consent of the householder.

Rainwater is both a resource and a potential problem. It is a problem when excess rainfall occurs in impervious urban areas and results in flooding or in a problematic combined sewer overflow operation. Harvesting this 'green' water for use in garden watering or toilet flushing displaces the demand for potable water for these functions.

Figure 3.3 Components of abstracted water
Based on data from 1997 to 1998 available from ONS (2010)

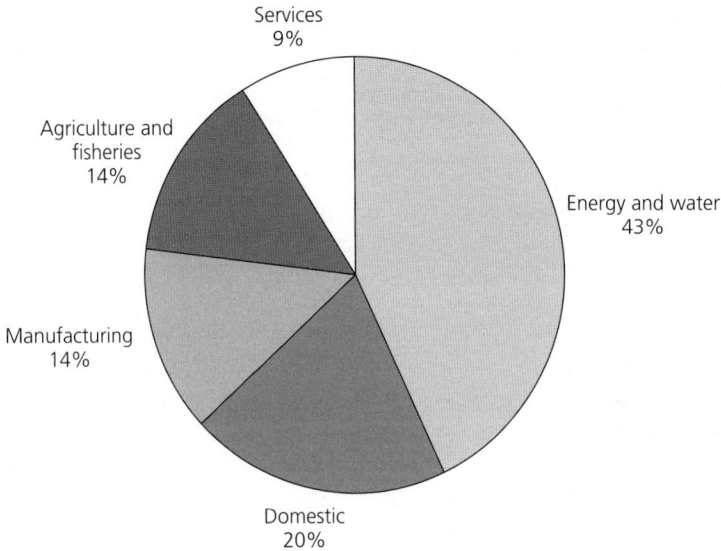

Services
9%

Agriculture and
fisheries
14%

Energy and water
43%

Manufacturing
14%

Domestic
20%

It is not a complete solution, however. Green water must be stored in relatively large volume to be effective either in terms of urban drainage amelioration or for demand displacement (approximately $2\,m^3$ per household if used for water-saving purposes only). If stored in the roof space, it may require stronger joists to bear the additional weight, and so there is additional embedded carbon, as every building material has a carbon cost for extraction, refining and transportation. If stored externally below ground, it uses electricity for pumping, which leads to carbon emissions. Although called 'green', with connotations of purity, etc., rainwater will have an avian-derived faecal component, and so care must be taken to ensure that the green water cannot be inadvertently used, for example, to fill a paddling pool (Ward et al., 2009). Care might also be required if aerosols result from flushing (see Fewtrell et al., 2008).

Light grey water (water from a bath or shower) or somewhat more contaminated dark grey water (from kitchen sinks, etc.) might also be used for toilet and garden water displacement. Grey water recycling systems require much smaller collection tanks (approximate $0.5\,m^3$ per household), but need more intensive water treatment. Such systems can also be used in combination with rainwater harvesting (Dixon et al., 1999). As the scale of potential contamination increases, so does the available resource but so also must the diligence in ensuring that no improper use occurs. Black water (toilet flush water) has no role in direct reuse, but has the potential to displace new abstraction if the black water is sufficiently treated and if the product is publicly accepted. Schemes such as that at Langford (UK) which re-use sewage effluent have, for public acceptability, a process whereby the water effectively is treated twice. Clearly, in severely water-stressed areas this approach has a significant role to play, although it is clearly not without a carbon cost. Whether that would count as demand reduction or as resource development is open to question and interpretation.

There is some doubt whether rainwater harvesting and grey water recycling are viable options for single households based on financial or carbon accounting (Butler and Dixon, 2002). Installation and, especially, retrofit are far from perfected, and we do not as yet have a 'fit and forget' technology available.

Water-using appliances grow more efficient each year as lowered water requirement is perceived as a marketable attribute (but, possibly, this is simply another side effect of energy efficiency). Of course, the simplest appliance efficiency increase is through the installation of a cistern displacement device, which reduces the capacity, and thus water usage, of a toilet cistern by 10–15%. The future role of cistern displacement devices is limited, however, as a 6 litre flush is now the mandatory standard. There is a growing number of alternative water efficiency solutions, and Butler and Memon (2006) provide a more comprehensive review of available approaches. A good source of up-to-date information is available on Waterwise's website (www.waterwise.org.uk).

3.7.2   Metering and tariffs

Moving next to issues of behavioural adjustment, we consider the impact of metering and tariffs. In the UK, the majority of water supplied is unmeasured, and so there is no 'cost' to the customer beyond the annual, flat-rate charge. Globally, water is most commonly metered and is charged in relation to the unit price and the volume used. The key assumption in advocating metering as a demand management tool is the assumption that, by adjusting prices, consumption will be moderated. In fact, water demand has generally been found to be inelastic (i.e. does not change in direct proportion) to changes in price (Renwick and Archibald, 1998; Nauges and Thomas, 2000; Arbues *et al.*, 2003; Young, 2005), and widely varying (Mays, 2001). Peak demand appears to be more elastic than off-peak demand (Lyman, 1992).

In the UK, it is widely asserted that metering of water supplies will reduce demand by approximately 10%, based on a mix of demand suppression and leakage identification. This is based on limited evidence derived, for the most part, from the National Water Meter Trials and small, unconnected studies conducted by several WSPs. The trials themselves took place on the Isle of Wight between 1989 and 1993. Around 50 000 households were compulsorily metered, and the ensuing demand profiles were compared against water demand in communities in Hampshire as a control. That important study is now ageing, the study site is certainly atypical of the UK and, of course, as with any study of behavioural response, the study is partly a product of awareness and perception at the time. The response to metering as seen in the Isle of Wight study is not predominantly a price-related response but an observational and behavioural response.

In the minor studies conducted subsequently, the subjects are meter optants (i.e. they have chosen to have a meter fitted) and so are a self-selected group. They tend to be smaller households which were already lower water users. Using this group will inherently overestimate the potential savings from fitting meters.

There is a difficult conundrum to be faced if a 10% reduction in demand driven by a range of initiatives but underpinned by metering was universal and sustained. The reduction would, in turn, cause an approximately 10% reduction in WSP income while expenditure would not fall by a similar amount, as asset investment is driven more by demand peaks than by average consumptions. As a result, a price increase across the board would be required or a commensurate decline in investment activity accepted. Neither of these options appears likely to be acceptable, and so some more complex, adaptive tariff structures seem likely.

3.7.3 Achievement

To attain the water savings or water efficiencies discussed above is a difficult task. For example, in England, although there are many existing water bylaws and water-related building regulations (CLG/Defra, 2007), and a growing body of new regulations such as the *Code for Sustainable Homes* (CLG, 2009), there is also a body of evidence that the regulations achieve only modest compliance. Some reports suggest that up to 30% of newly built houses are not regulation-compliant, a figure set to increase as public sector employment declines (UKWIR, 2008). Even were they compliant at the outset, the question of the sustainability of that compliance arises. 'Regression' is the term often used to indicate a slow return to the pre-existing conditions. A cistern displacement device removed by a plumber or the householder during maintenance is unlikely to be retained and replaced. A shower delivering a low flow rate of water in a new bathroom may be replaced or re-drilled. Aerating taps may be replaced, as has been observed by the authors in dwellings compliant with the *Code for Sustainable Homes*. The energy efficiency drives of the last 40 years have resulted in numerous well-documented and quantified failures in energy efficiency initiatives, and water efficiency will go the same way unless we learn from energy and, for example, make the water equivalent of an incandescent bulb unavailable on the market. In the energy industry, improved efficiency did not lead to a decline in demand but to a general increase in appliance capacity or house mean temperatures. Efficiency was exchanged for comfort rather than for cost reduction. It will be interesting to see if similar paradoxes occur in the search for more-efficient water. The extent of regression to less-water-efficient but more-customer-satisfying fitments and appliances is difficult to determine. While the energy initiatives of the late 20th century were rather fragmented, the more unified approach of Waterwise, a UK non-governmental organisation, focused on decreasing water consumption and building the evidence base for large-scale water efficiency, may yet result in effective conservation interventions (see, in particular, Waterwise, 2008).

3.8 Water neutrality

A water-neutral development is defined as one in which water consumption post-development is no greater than that before development. This is achieved by minimising water consumption in the development itself and in other surrounding areas. Clearly, this concept makes a number of assumptions about the scale and sustainability of water efficiency, assumptions that need to be considered carefully. Further, savings made in another water region will have no impact on the water balance of the region issuing the new abstraction licence unless and until there are potable water transfer agreements between regions covering the volume of the new abstraction, at least in terms of extreme drought situations. Such an approach has been shown to be theoretically possible in the Thames Gateway development to the east of London (Environment Agency, 2007), but did illustrate how difficult it would be to achieve in practice (Sim *et al.*, 2007; Butler and Herrington, 2007).

3.9 Modifying lifestyles

So, how do we influence the user so that we modify the demand for water? Fundamentally, we must modify behaviour both in the style of their water use and the nature of the water-using facilities installed and retained. At the crudest level, this can, in theory, be driven by price through tariffs, but this can only have an impact if water is metered and is allied to an escalating tariff (see the earlier discussion in this chapter on price elasticity). The tariff structure would need to very significantly raise water bills by having greatly increased unit prices for demand levels above the average. However, a strongly escalating tariff raises immediate questions about water justice and affordability. To attain the twin goals of influencing behaviour and delivering social justice requires that the tariff is both dissuasively progressive and socially just. In other words, a tariff that rises rapidly from a low base to deliver protection for the poor while punishing the profligate (we refer to this later as the 4P tariff). This may work in theory, but, for a resource that is strongly related to numbers in the household, you are likely to markedly penalise larger families.

In the longer term, information both on the cost of water used that might have been saved and on the current state of the water resource will encourage savings. In Victoria, Australia, such information now forms part of the bill. In the UK, Sutton and East Surrey employ similar information on current and past water usage. Information will also justify regulatory interventions prohibiting specific water uses. The public acceptance of this requires that there is a degree of 'ownership' of the state of the resource. Certainly, in the USA, Canada and Australia water reserves are published weekly so that the logic of progressive water-saving requirements is obvious and unsurprising. Smart meters, already being tested for energy, provide immediate cost information for the consumer, and may influence immediate reactions – turning off a light or running a kettle for a shorter period. Water has a problem in this respect in that bills are about a third or less than those of energy, so the cost per usage is far less dramatic and so less likely to promote a beneficial reaction, although trials are currently taking place.

In the long term, education is the key to reducing demand, but there has to be a case that will withstand scrutiny. If you can influence children, then you influence many future generations.

3.10 Visions for the future
3.10.1 Ultra low use systems

The humble toilet 'consumes' a large fraction of the household potable water demand. However, new approaches are becoming available that use much less water for the same or better performance. One such is the toilet that works by locally pressurising air, which is then expelled through the appliance under pressure during the flushing cycle instead of water. Just 1.5 litres per flushing cycle is used to cleanse the bowl and refill the water seal trap, and it can be connected to a standard gravity sewer system (Littlewood *et al.*, 2007).

3.10.2 Dual systems

We use 150 litres of potable water per person per day in the UK. Of this, perhaps 30 litres needs to be of this potable quality, perhaps less. Looking 50 years ahead, should we be planning for a dual system, which, in its extreme manifestation, might be bottled water for drinking with somewhat lower-quality washing water and recycled water distributed for all other functions? Dual systems are not new and have existed in North America and Europe (Leamington Spa and the Rive Gauche (Left Bank) in Paris). The cost of maintaining separate systems, however, appears to be prohibitive, and examples of this approach have largely been discontinued.

3.10.3 Advanced tariffs

Some people believe that a 4P tariff that '*p*rotects the *p*oor and *p*enalises the *p*rofligate' is needed urgently. However, development of such a tariff is not as simple as tariffs for energy companies, for water use is strongly driven by individual needs whereas the energy tariff is much less occupancy dependent. Therefore, the tariff needs to be set either on the occupancy numbers or on the house type as a surrogate for occupancy. For practical efficiency reasons, the tariff needs to be set at an upper expectation for occupancy, and an appeal procedure that allows evidence to be presented to initiate a lower tariff needs to be in place.

3.10.4 Quotas and advanced payment cards

Establishing a quota right to an amount of water could be achieved by having prepayment cards that allocate a free 'social' quota of water with every recharge. However, advanced payment cards are not currently approved for water use, but they may again find favour if a trickle (i.e. limited flow rate) supply becomes the permitted default.

3.10.5 Behavioural change

We have seen major changes in behaviour over the last 30 years – reliance on the web, use of mobile phones, etc. We have also seen a growing environmental awareness and a willingness to engage with the environmental agenda. With information that will stand scrutiny and with a schools-based education on responsible water citizenship, there is every prospect that behavioural changes in our water use will permeate society. However, we must be aware that the prospect of a backlash exists if government policy on water saving is found to be based on dubious arguments – then, public reaction may well ignore the many sound arguments for water saving.

3.10.6 Living with environmental and social change

Our climate is changing. The debate on the balance between natural and anthropogenic drivers remains live. But, for water managers, increased drought prevalence, more variations between 'wet' and 'dry' seasons, and greater variations in climate change across the UK must mean that our management of demand becomes more sophisticated and responsive. Add to this the clear need to use less energy in water provision, and the result is probably a movement to more local sourcing and more engagement with water saving and an acceptance of more risk. Social change is occurring both nationally and internationally. We must expect China, for example, to increase its demand for air conditioning, with consequent energy demand rises distorting the existing energy market with an outcome that more biofuel crops are planted that may require more water. We may indeed anticipate a move to allocation of a scarce resource and expect the days of total water demand satisfaction at a very low price might pass.

3.10.7 Lowered standards of service

The water industry has defined standards of service, for example that use of stand pipes will 'never' arise, that hosepipes will be banned only once in 20 years. Thus, the industry is investing in an infrastructure that will supply the water demanded on effectively all occasions. There is a case for questioning how and why such standards of service arose and whether they continue to be appropriate. If we were willing to accept a higher risk – let us say of standpipes being introduced for a period once in 50 years – then levels of investment could be reduced or retargeted.

3.10.8 Mutual support versus local provision

The industry has evolved from single-source, single-sink local provision to regional WSPs that, with the exception of some edge rationalisation, still source all their resources internally or from 'traditional' sources external to the region. Within a WSP, the benefits of water transfer have become apparent, and in most cases companies have continued to develop resource flexibility and transfer. Each company has a 'headroom' allowance, a reserve capacity, that exists not to meet an unexpected eventuality but to address the situation when all 'uncertainties' work against demand satisfaction – so, headroom addresses the worst-case scenario. If companies agreed, in advance, water transfer agreements (potable or raw as appropriate to local circumstances) with adjacent companies, then they would have a larger headroom to draw upon. This would result in an industry more resilient to external variations or the retention of the same resilience without expanded resource requirements. Of course, at the heart of this argument is the impact of climate change, and currently we have no forecasts of the return periods of droughts in relation to their spatial extent, a measure that would be key to developing rational water transfer agreements.

3.11   Conclusions

The estimation and forecasting of water demand is key to effective resource planning. Potable water demand cannot be separated from other demands for raw water. Climate and demographic change complicates forecasting and resource planning. We do not start with a blank canvas – and our choices are strongly constrained by existing urban structures and infrastructure.

REFERENCES

Arbues, F., Garcia-Valinas, M. A. and Martinez-Espineira, R. (2003) Estimation of residential water demand: a state-of-the-art review. *Journal of Socio-Economics* 32(1): 81–102.

Bellfield, S. L. (2001) Short-term domestic water demand: estimation, forecasting and management. *PhD thesis*, University of Leeds.

Butler, D. (1991) A small scale study of wastewater discharges from domestic appliances. *Journal of Institution of Water and Environmental Management* 5(2): 178–185.

Butler, D. and Dixon, A. (2002) Financial viability of in-building grey water recycling. In: *International Conference on Wastewater Management and Technologies for Highly Urbanized Cities*. Hong Kong.

Butler, D. and Herrington, P. (2007) *Towards Neutrality in the Thames Gateway. Peer Reviews*. London: Environment Agency.

Butler, D. and Memon, F. A. (eds) (2006) *Water Demand Management*. London: IWA.

Chapagain, A. K. and Hoekstra, A. Y. (2008) The global component of freshwater demand and supply: an assessment of virtual water flows between nations as a result of trade in agricultural and industrial products. *Water International* 33(1): 19–32.

Clarke, G. P., Kashti, A., McDonald, A. and Williamson, P. (1997) Estimating small area demand for water: a new methodology. *Journal of the Chartered Institution of Water and Environmental Management* 11(3): 186–192.

CLG (2009) *Code for Sustainable Homes: Technical Guide – May 2009 Version 2*. London: Community and Local Government.

CLG/Defra (2007) *Water Efficiency in New Buildings: A Joint Defra and Communities and Local Government Policy Statement*. London: CLG/Defra.

DCLG (2007) *New Projections of Households for England and the Regions to 2029*. London: Department of Communities and Local Government. www.communities.gov.uk/documents/statistics/pdf/1089402.pdf [accessed 01.09.10].

Defra (2008) *Future Water: The Government's Water Strategy for England*. London: Stationery Office.

Defra (2009) *UK Climate Projections*. http://ukclimateprojections.defra.gov.uk.

Dixon, A., Butler, D. and Fewkes, A. (1999) Water saving potential of domestic water re-use systems using grey water and rainwater in combination. *Water Science and Technology* 39(5): 25–32.

Downing, T. E., Butterfield, R. E., Edmonds, B., Knox, J. W., Moss, S., Piper, B. S., Weatherhead, E. K. and the CCDEW Project Team (2003) *Climate Change and the Demand for Water*. Oxford: Stockholm Environment Institute Oxford Office.

Environment Agency (2007) *Towards Neutrality in the Thames Gateway*. Project No. SC060100/SR3. London: Environment Agency.

European Environment Agency (2001) *Environmental Signals 2001: Environmental Assessment Report No. 8*. European Environment Agency Regular Indicator Report. Copenhagen: European Environment Agency. www.eea.europa.eu/publications/signals-2001/signals2001 [accessed 01.09.10].

Fewtrell, L., Kay, D. and McDonald, A. T. (2008) Rainwater harvesting – an HIA of rainwater harvesting in the UK. In: Fewtrell, L. and Kay, D. (eds), *A Guide to the Health Impact Assessment of Sustainable Water Management*, Ch. 3. Amsterdam: International Water Association.

Gardiner, V. and Herrington, P. (1986) The basis and practise of water demand forecasting. In: Gardiner, V. and Herrington, P. (eds), *Water Demand Forecasting*. Norwich: Geo.

Herrington, P. (1996) *Climate Change and Demand for Water*. London: HMSO.

Herrington, P. R. (1998) Analysing and forecasting peak demands on the public water supply. *Journal of the Chartered Institution of Water and Environmental Management* 12(2): 139–143.

Jeffrey, P. and Gearey, M. (2006) Consumer reactions to water conservation policy instruments. In: Butler, D. and Memon, F. A. (eds), *Water Demand Management*, pp. 305–330. London: IWA.

Jin, J. (2009) A small area microsimulation model for water demand. *PhD thesis*, University of Leeds.

Lambert, A. (2003) Assessing non-revenue water and its components: a practical approach. *Water21*: 50–51. www.iwapublishing.com/pdf/WaterLoss-Aug.pdf [accessed 01.09.10].

Littlewood, K., Memon, F. A. and Butler, D. (2007) Downstream implications of ultra-low flush WCs. *Water Practice and Technology* 2: 2.

Lyman, R. A. (1992) Peak and off-peak residential water demand. *Water Resources Research* 28(9): 2159–2167.

Mays, L. W. (2001) *Water Resources Engineering*, 1st edn. Chichester: Wiley.

McDonald, A., Bellfield, S. and Fletcher, M. (2003) Water demand: a UK perspective. In: Maksimovic, C., Butler, D. and Memon, F. A. (eds), *Advances in Water Supply Management*, pp. 683–692. London: Balkema.

McDonald, A., Boden, P. and Clarke, M. (2008a) Void property management. Unpublished report for Yorkshire Water.

McDonald, A., Boden, P., See, L. and Rees, P. (2008b) Hidden and transient populations in the Severn Trent Region. Unpublished report for Severn Trent Water. Leeds: CREH.

Memon, F. A. and Butler, D. (2006) Water consumption trends and demand forecasting techniques. In: Butler, D. and Memon, F. A. (eds), *Water Demand Management*, pp. 1–26. London: IWA.

Mitchell, G. (1999) Demand forecasting as a tool for sustainable water resource management. *International Journal of Sustainable Development and World Ecology* 6(4): 231–241.

Nauges, C. and Thomas, A. (2000) Privately operated water utilities, municipal price negotiation, and estimation of residential water demand: the case of France. *Land Economics* 76: 68–85.

Ofwat (2002) *Best Practice Principles in the Economic Level of Leakage Calculation*. Birmingham: Ofwat. www.ofwat.gov.uk/publications/commissioned/rpt_com_tripartite studybstpractprinc.pdf [accessed 01.09.10].

ONS (2010) *Water and Industry*. www.statistics.gov.uk/cci/nugget.asp?id = 159 [accessed 01.09.10].

Orr, S. and Chapagain, A. (2009) *UK Water Footprint*. Godalming: WWF-UK. http://assets.wwf.org.uk/downloads/water_footprint_uk.pdf.

Renwick, M. E. and Archibald, S. O. (1998) Demand side management policies for residential water use: who bears the conservation burden? *Land Economics* 74(3): 343–359.

Roaf, S. (2006) Drivers and barriers for water conservation and reuse in the UK. In: Butler, D. and Memon, F. A. (eds), *Water Demand Management*, pp. 215–235. London: IWA.

Roe, D. and Rendell, D. (2009) *Microsimulation at H M Treasury: Methods and Challenges. ESRC Microsimulation Seminar Series*. Brighton: University of Sussex.

Sim, P., McDonald, A., Parson, J. and Rees, P. (2007) *Revised Options for UK Domestic Water Reduction: A Review. Working Paper 07/04*. Leeds: School of Geography, Faculty of

Environment, University of Leeds. www.geog.leeds.ac.uk/wpapers/Water_Conservation_Lit_Review2.pdf [accessed 28/04/08].

Twort, A. C., Ratnayaka, D. D. and Brandt, M. J. (2000) *Water Supply*, 5th edn. London: Arnold/IWA.

UKWIR (2008) *The Cost Effectiveness of Demand Measures*. UKWIR Project Report 25/C. London: UK Water Industry Research.

Ward, S., Butler, D., Barr, S. and Memon, F. A. (2009) A framework for supporting rainwater harvesting in the UK. *Water Science and Technology* 60(19): 2629–2636.

Waterwise (2008) *Evidence Base for Large-scale Water Efficiency in Homes*. http://www.waterwise.org.uk [accessed 01.09.10].

Westcott, R. (2004) A scenario approach to demand forecasting. *Water Science and Technology: Water Supply* 4(3): 45–55.

Williamson, P., Clarke, G. P. and McDonald, A. T. (1996) Estimating small-area demands for water with the use of microsimulation. In: Clarke, G. P. (ed.), *Microsimulation for Urban and Regional Policy Analysis*, pp. 117–148. London: Pion.

Williamson, P., Mitchell, G. and McDonald, A. T. (2002) Domestic water demand forecasting: a static microsimulation approach. *Journal of The Chartered Institution of Water and Environmental Management* 16(4): 243–248.

Young, R. A. (2005) *Determining the Economic Value of Water: Concepts and Methods*. Washington, DC: Resources for the Future.

REFERENCED LEGISLATION

The Water Framework Directive. Directive 2000/60/EC of the European Parliament and of the Council of 23 October 2000 establishing a framework for Community action in the field of water policy. *Official Journal of the European Communities* L327: 1–73.

Water Distribution Systems
ISBN: 978-0-7277-4112-7

ICE Publishing: All rights reserved
doi: 10.1680/wds.41127.073

ice
Institution of Civil Engineers

publishing

Chapter 4
Water supply systems

Howard S. Wheater Imperial College London, UK
Nigel J. D. Graham Imperial College London, UK
Michael Pocock Veolia Water Central, Hatfield, UK
Barrie Holden Anglian Water Services Ltd, Peterborough, UK

4.1   Introduction

To understand the supply of water for human use, for example for domestic supply, industry or agriculture, for energy or for transport, it is necessary to understand firstly the behaviour of water in the natural environment, i.e. the available stores of water and their natural variability, and secondly the potential for management to increase the availability and/or reliability of a water source. For a complete assessment of water resources, we will also need to understand the potential impacts of abstractions on the environment, to balance the competing needs for water of aquatic and riparian ecosystems. And we should be aware of the wider impacts of human activity on the water environment. At the scale of a river basin, this includes the effects of land use and land management on the quality and quantity of water resources, and the role of rivers in diluting waste discharges: for example, sewage effluent. At the global scale, we see the effects of anthropogenic emissions on the climate itself, and the future impacts of a changing climate on water resources remain a source of great uncertainty in strategic planning.

The first part of this chapter on water supply systems therefore begins with an introduction to the natural water environment and its characteristics, before moving on to consider resource development and design, water resource performance, and the regulatory environment for water. The second part provides a general introduction to potable water quality (e.g. parameters and standards) and an overview of water treatment methods and distribution systems; the latter is described in greater detail in subsequent chapters.

4.2   Water resources and their management
4.2.1   The hydrological functioning of the natural environment

Figure 4.1 illustrates the main global stores of water. Most of the world's water is saline, held in the seas and oceans. Only about 2.5% is fresh water (UNESCO, 2009), which circulates in a highly dynamic hydrological cycle. Precipitation of atmospheric water vapour provides the water that maintains river flows and recharges groundwater. Evaporation from the land surface and water bodies changes the liquid water to vapour to rejoin the store of atmospheric water vapour and complete the cycle. The main sources of our freshwater resources are abstractions of surface water, from rivers and lakes, or of groundwater from major or minor aquifers. The key to water resources

Figure 4.1 The hydrological cycle

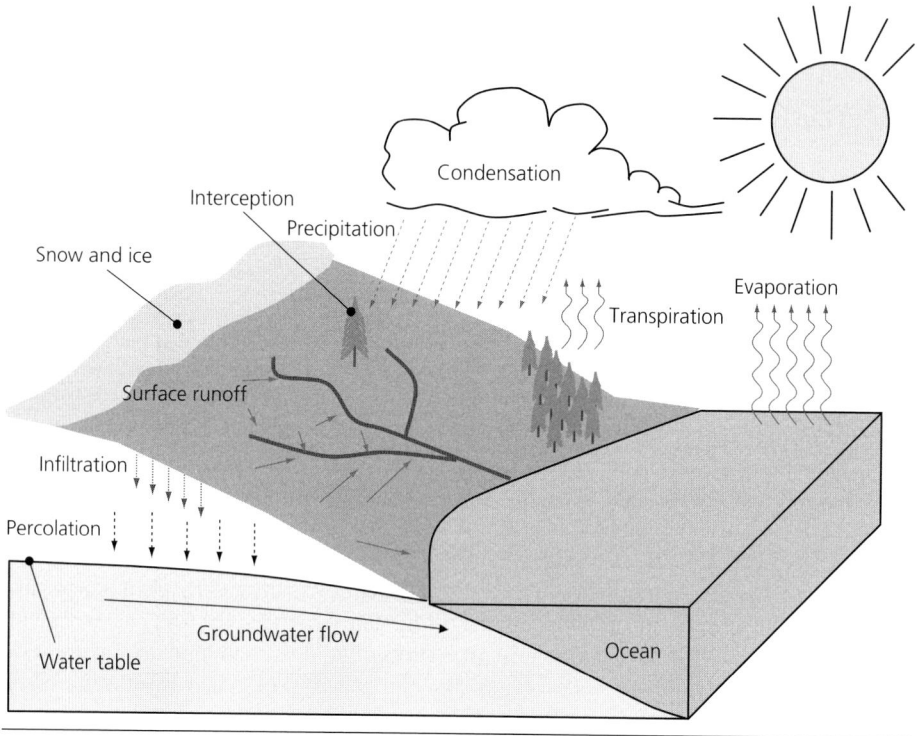

management is understanding the dynamic characteristics of these sources, and in particular their variability in space, and in time.

The origin of freshwater resources is precipitation. In some climates, snow may be significant, accumulating over the winter, and melting in the spring to provide an important resource, for example from the eastern slopes of the Rocky Mountains in North America, the Andes in South America and the Himalayas in the Indian sub-continent. However, in the UK, rainfall is the dominant source of precipitation for water resources. A map of annual precipitation (Figure 4.2) shows a gradient in rainfall from high elevation areas in the west, with annual totals in excess of 3000 mm, to the low-lying topography of eastern England, with totals of less than 800 mm. Clearly, this distribution has important implications for the development and management of the UK's water resources.

To understand hydrological processes, and hence the functioning of potential water resources, it is instructive to follow the potential paths of a (virtual) raindrop as it reaches the ground and travels through the terrestrial environment.

We begin with vegetation, which has a critical role in various aspects of the hydrological cycle, and the interception store indicated in Figure 4.1. Our raindrop may fall onto

Figure 4.2 Annual precipitation (mm) over the UK
Data from Perry and Hollis (2005)

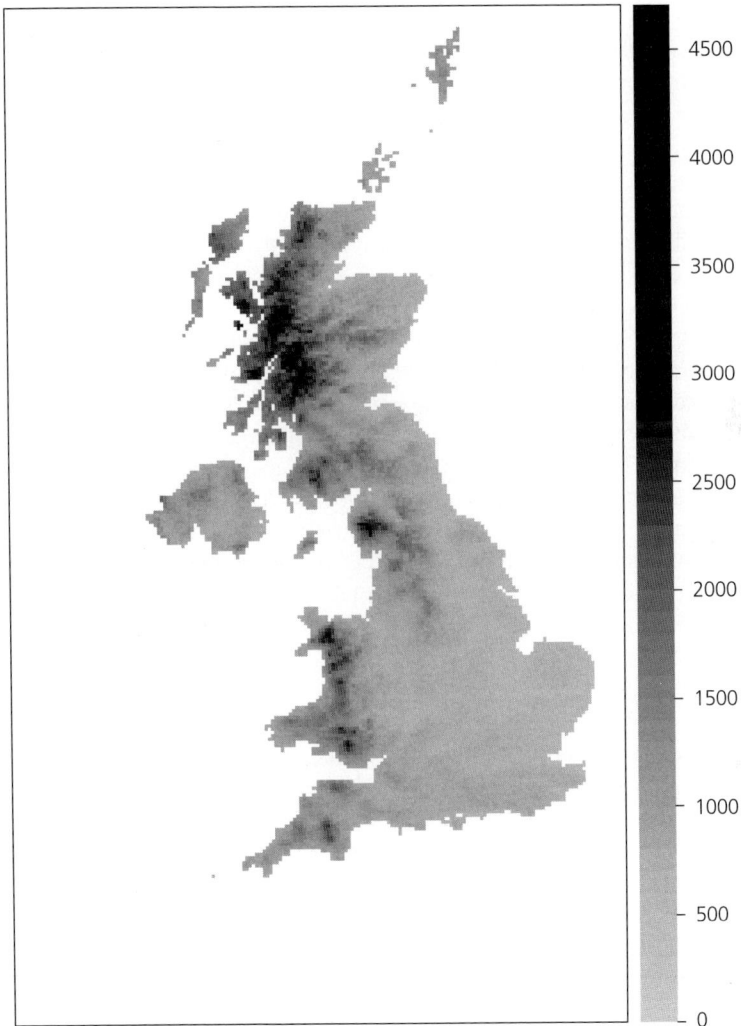

vegetation, wetting the surface of leaves, as *interception*. If the surfaces become fully wet, rainfall will either fall to the ground or run down the stem as *stemflow*. Rain that falls through a vegetation canopy is known as *throughfall*. Values of interception storage depend on the type of vegetation. Typical values for a forest canopy are 1–2 mm. This small store is not enough to play a major role in affecting run-off from a large storm, but in the wetter parts of the UK, it can be an important influence on the availability of water resources. Evaporation rates of intercepted water are relatively high for a forest due to the aerodynamic roughness of the canopy and the associated atmospheric turbulence. In a wet climate (e.g. in Wales, for example, where rain may

75

occur three days out of four), this small interception store can empty and fill repeatedly. An important research question in the 1960s concerned the effects of afforestation on the water balance of the UK uplands. If there were extensive tree planting, what would happen to the streamflow and hence the available water resource? Detailed research (e.g. Gash *et al.*, 1980; Marc and Robinson, 2007) showed that interception was a key process, and that in wet and windy parts of the UK, 370 mm of water could be lost by evaporation of intercepted water from woodland over the course of a year – a significant amount.

If our raindrop does not evaporate from interception storage, but falls to the ground, what then will be its fate? It can either remain on the land surface as *depression storage*, run off over the land surface as *overland flow*, or *infiltrate* into the soil (see Figure 4.1). The pathway taken will depend on the rate at which the soil can accept infiltration (its *infiltration capacity*) in comparison with the rainfall rate (typically measured in units of mm/day). Infiltration capacity will depend on the type of surface. Urban surfaces, such as roofs and pavements, are largely impermeable, and hence mainly generate overland flow, which is capable of transmitting water very rapidly to a nearby stream. The infiltration capacity of soils depends on the soil type (low values for soils such as clays, high values for soils such as chalk) and soil wetness. In general, soil infiltration capacity decreases with increasing soil wetness, so the wetter the soil, the more likely it is that overland flow will occur. Infiltration rates are also affected by the condition of the soil. When soils are compacted – for example by animals or heavy machinery – infiltration capacity may be dramatically reduced (Holman *et al.*, 2003; Carroll *et al.*, 2004).

If we wish to follow our raindrop along the infiltration pathway, some basic principles governing water in the subsurface must be borne in mind. Soil and rock contain pore spaces and fissures. Pores may account for a large part of their volume (as high as 50% for a clay soil). The pores and fissures may contain water or air. If both are present, the soil or rock is said to be *unsaturated*. If the pores are full of water only, the soil or rock is *saturated*. Soils are commonly unsaturated – most vegetation needs air as well as water around its roots. An unsaturated soil is like a sponge – it will absorb and retain water against the forces of gravity due to surface tension effects. This means that water pressures in an unsaturated soil will be less than atmospheric pressure – in effect under suction. Conversely, where the soil or rock is saturated, water pressures will be atmospheric, or greater. The term 'groundwater' refers to *saturated* conditions, in soils or rock. The boundary between unsaturated and saturated conditions is where pore water is at atmospheric pressure, and is known as the *water table*. If we were to dig a hole to below the water table, the hole would fill with water to the level of the water table (demonstrating the atmospheric pressure boundary condition).

Let us return to our raindrop. If it has infiltrated into the soil, the force of gravity will support downward movement, or *percolation*, within the soil profile (see Figure 4.1). In relatively permeable soils and rock, for example the Chalk of southern England, water will move downwards primarily through the fine pores (although under exceptional conditions, flow in the fissures may occur – see Ireson *et al.*, 2006, 2009) until it reaches the water table, and hence the saturated zone that defines groundwater. In the Chalk, the unsaturated zone may be as deep as 100 m at a catchment boundary (see Figure 4.1). Groundwater normally flows slowly under gravity, and may discharge to a river (providing a source to maintain low flows in dry periods), a lake or the sea. Where the water table intersects the land surface, it will emerge as a spring. In fractured rock, water movement is dominated by fissure flow, and under certain circumstances (e.g. karst systems) this can be rapid.

In many upland catchments, which form the headwaters of our major rivers, soils are relatively shallow, overlying bedrock, and often have a structure in which the permeability reduces, often dramatically, with depth. In this case, water will tend to move downslope, as *throughflow* (e.g. see Kirkby, 1978). The near-surface structure of soils can be complex. Large, interconnected pores may exist due to biological activity, due to earth worms or burrowing animals, or to decayed plant roots (Beven and Germann, 1982). Clay soils may shrink and crack, creating a further type of macropore. These macropores provide routes for water to be transported rapidly downslope to a stream, generating an intermediate response – slower than rapid overland flow, but much faster than slow groundwater discharge.

One final process that we need to introduce to complete the set of important hydrological processes is evaporation. The soil provides a store of water for vegetation to draw on, so our virtual raindrop may enter the soil as infiltration and then be lost back to the atmosphere as evaporation. Plants extract water from the soil through root action (soil water is at less than atmospheric pressure, so the plants have to exert a suction to take up the water). The water moves through the plant's vascular system, and into the leaves. Evaporation of liquid water takes place within the leaves, and water vapour then moves out into the atmosphere through small apertures on the leaf surface (*stomata*). This process is known as *transpiration* (sometimes the combined sources of evaporation are termed *evapotranspiration*). Stomata provide an important control on evaporation. If there is plenty of available soil moisture, the plants will evaporate at a rate determined by atmospheric conditions (commonly known as *potential evaporation*). If soil water is limited, the plants will have to apply high suctions to extract water from the soils, and will begin to wilt under the water stress. This effect is associated with closure of the stomata – the plant has a convenient 'tap' with which to control the evaporation process. In this case, the *actual evaporation* will drop below the potential evaporation. This adversely affects plant productivity, so in the case of agriculture, irrigation could be used to replenish soil water to avoid loss of agricultural production or crop failure.

Having introduced the key hydrological processes experienced by our virtual raindrop, we can see how these are integrated at the scale of a river *catchment* (the catchment area is the area of land draining to a river system). Streamflow is characterised by the streamflow *hydrograph* – a time series of the discharge in a stream or river, commonly defined by units of cubic metres per second (m^3/s), or cumecs. Figure 4.3a shows a typical hydrograph from a surface-water-dominated catchment, the South Tyne. This has a very rapid (flashy) response to rainfall, as the streamflow increases in response to rainfall, and subsequently falls rapidly when rainfall stops. Rapid-response pathways are dominant here. Between rain periods, flows are small, supported by the slow drainage of soils and an obviously limited groundwater system. This flashy response is also typical of urbanised catchments – where roofs and paved surfaces generate rapid overland flow, commonly transmitted rapidly to a nearby stream via a piped storm water drainage network.

In contrast, we turn now to a Chalk catchment, the Lambourn, a tributary of the Thames near Reading. The hydrograph (Figure 4.3b) is dominated by a seasonal pattern. Response to individual rainfall events is small (perhaps just 2% of the incident rainfall appears as rapid response). So what gives rise to this seasonal pattern? The answer lies in the balance between rainfall and evaporation. As the weather warms in spring, and plants actively grow and transpire, evaporation exceeds rainfall, so that plants draw on the store of available soil water. During summer, the soils progressively dry, giving rise to a *soil moisture deficit*. In autumn, evaporation rates reduce, rainfall exceeds evaporation, and the soil begins to wet. Once the soils have rewetted to a point known as *field capacity*, they are capable of transmitting infiltrated water as percolation through the unsaturated zone, recharging the groundwater. Water tables rise, and the groundwater feeds the streamflow. In spring, the cycle repeats: evaporation exceeds rainfall, the soils dry and groundwater recharge virtually stops. Water tables fall and streamflows decline. This characteristic is seen in the many Chalk streams of south-east England such as the River Misbourne, and is commonly associated with extension of the flowing length of the river into upstream areas during the winter recharge season, and contraction during the summer; such streams are known as winterbourne rivers.

Figures 4.3a and 4.3b represent extremes of response typical of UK conditions. However, catchments often have a complex mix of conditions, with differing soils and geology in different areas, and possibly urbanisation in some parts. Figure 4.3c shows the hydrograph from the River Pang, which is adjacent to the Lambourn, and also located on the Chalk. However, the downstream areas of the catchment are covered by Palaeogene deposits, which generate a rapid response to rainfall. As can be seen from comparison of Figures 4.3b and 4.3c, this generates a rapid response component of the hydrograph, overlying the seasonal pattern of the Chalk.

Having considered the variability of surface water catchment characteristics, we also need to introduce some further aspects of groundwater systems. Our discussion of groundwater was based on a simple system where the upper boundary of the groundwater body was the water table, an *unconfined* groundwater system. Many groundwater systems are located between upper and lower layers of impermeable rock, in which case the upper boundary of the groundwater system is the overlying rock boundary. In this case, the groundwater is said to be *confined*. Two aspects of this situation are relevant. One is that when conditions in a confined aquifer vary, there is no change in the upper boundary – whereas for a water table aquifer the water location changes. Hence, any changes in stored water relate simply to the compressibility of the water and the elasticity of the rock, as pressures change. The second is that pressures in a confined groundwater system may be high, depending on the elevation of the recharge area, so that if a borehole is drilled, water may emerge under its own pressure, in

Figure 4.3 Typical streamflow hydrographs: (a) surface water dominated (South Tyne);
(b) groundwater dominated (Lambourn); (c) mixed (Pang)

Data courtesy of the National Environment Research Council

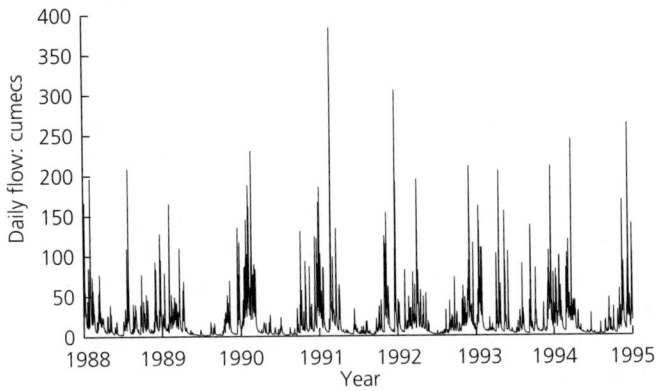

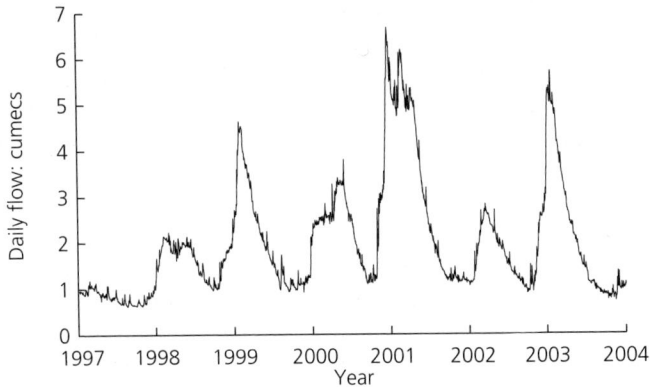

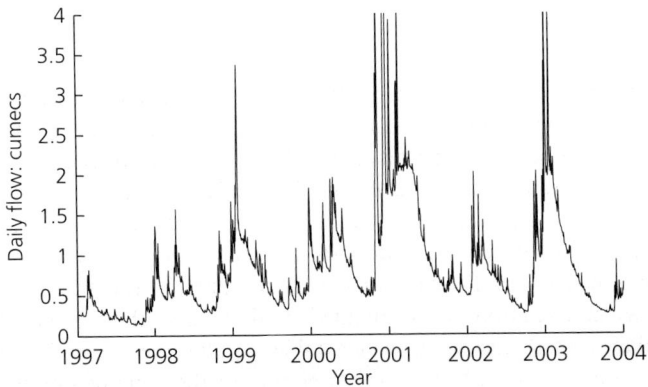

which case conditions are said to be *artesian*. A classic example is the Chalk basin under London (Figure 4.4), which also usefully illustrates the potential effects of human intervention. Historically, groundwater conditions below London were artesian, but during the industrial development of the 19th and early 20th centuries, large-scale abstractions lowered the water table, by some 65 m. London became accustomed to living with lowered groundwater, and there was extensive subsurface construction of services, such as the Underground railway, and deep basements and foundations for tall buildings. However, as industrial abstractions of groundwater declined in the mid-20th century, groundwater levels started to recover, thus threatening a wide range of services and facilities. Currently, groundwater pumping is required, particularly to maintain dry conditions in many of London's subterranean tunnels, including the London Underground transport system (Jones, 2010). Around the world, declining groundwater levels (due to over-abstraction of groundwater) are a major and widespread problem; however, rising groundwater is also an issue elsewhere, for example in Saudi Arabia, where the capital city, Riyadh, now has perennial flows in the previously ephemeral Wadi Hanifah, due to leakage from urban water services.

Having discussed groundwater systems in general, we need to introduce the term *aquifer*. Clay soils have high porosity and can hold large amounts of water, but due to the fine pore structure, the ability of clay to transmit water is extremely limited. While many subsurface materials are capable of holding significant volumes of water, an aquifer is a body of rock that is able not only to store but also to transmit water in useful amounts.

A final point in this discussion of groundwater concerns pollution. Water generally moves slowly in groundwater systems, and hence protection from pollution is particularly important, and the subject of UK legislation and regulations concerning controls on land use and industrial activities required for *aquifer protection* and also *source protection*, i.e. the protection of individual abstraction wells. Methodologies are reviewed in Butler (2010), and current EU legislation in EC (2009).

If an aquifer is contaminated, from surface spillage, drainage systems or diffuse pollution, remediation may require years or decades of effort, and recovery to pristine conditions may not be possible. A particular set of issues concern coastal groundwater. Over-abstraction of groundwater in coastal aquifers is a widespread global problem (Simmons *et al.*, 2010). This can lead to salt water being drawn into the aquifer, with potential long-term contamination of the resource. Water quality matters are considered further in Section 4.3 of this chapter.

4.2.2   The nature and variability of water resources

Water resources can be developed from a range of sources, available at a wide range of scales. We begin with precipitation.

4.2.2.1   Precipitation

Precipitation can be used directly, as a basic form of *rainwater harvesting*. This is not widely used in the UK at present, except for some garden watering, but could be in

Figure 4.4 Schematic cross-section of the London Basin

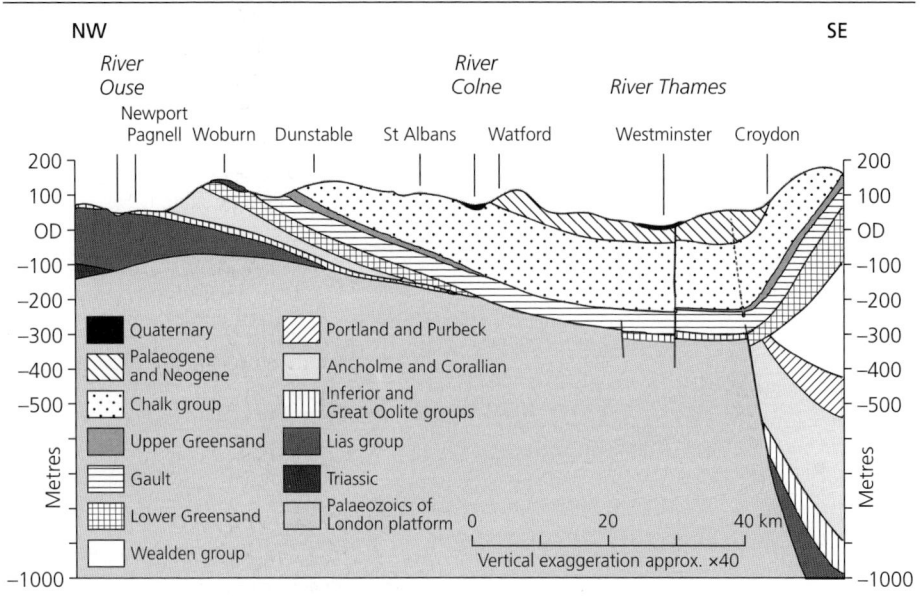

future. In water-stressed areas it can be an important resource. For example, the Greek island of Cephalonia has no surface watercourses and no exploitable groundwater resources; traditionally, households have collected water from roofs and paved areas to provide their sole source of supply. In Australia, where rainfall on the coastal cities is commonly higher than that over the interior catchments that provides the surface water resources, one consequence of recent droughts has been an increased use of rain-water harvesting. Consideration of the design of a rainwater-harvesting system requires us to introduce some basic principles. We need to know the characteristics of the *supply*, i.e. not only how much rain is available but also information about the periods between rainfall, so that appropriate capture systems and storage can be provided. The storage required in turn depends on the *demand* for water. And both rainfall and demand will vary seasonally, so we need information on the *seasonal variability* of both. And, finally, we can expect years of above-average rainfall, and below-average rainfall, so we need to understand the *inter-annual variability* of the resource, and the flexibility to reduce demand if necessary.

4.2.2.2 Surface water resources

The simplest form of supply is direct abstraction from a surface water body, such as a river or lake. As for our simple case of rainwater harvesting, above, we need to consider the seasonal and inter-annual variability of our source. In the UK, river flows are typically high in winter, low in summer. We may therefore wish to store some of the high winter flows for use during summer. This can be done by the construction of a

Figure 4.5 River Severn water resources
Environment Agency (personal communication)

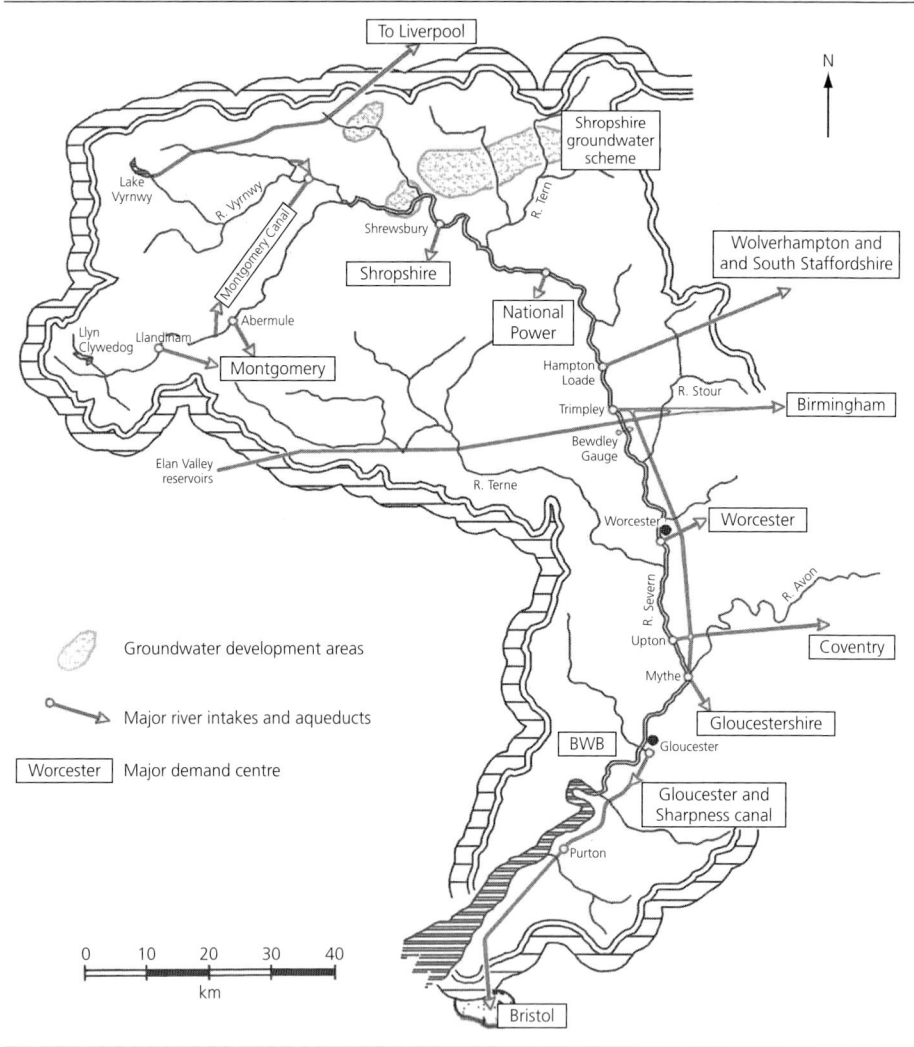

dam, to create a reservoir. Water from a reservoir can be used directly, via pipeline. Alternatively, the reservoir can be used to release additional water into the river to augment summer flows. Figure 4.5 shows a schematic diagram of the River Severn water resources system. Two major reservoirs have been constructed. Craig Goch is a direct-supply reservoir, with water piped to Liverpool. Llyn Clywedog is a regulating reservoir, from which water is released as required to maintain flows in the Severn for direct abstraction, as shown, to provide water resources for the West Midlands. The Severn has two important attributes. The first is high rainfall, in the upper catchment (see Figure 4.2). The second is topographic relief, providing suitable sites for reservoir

Figure 4.6 Typical reservoir control curve showing uncertainty bands (SE, standard error)

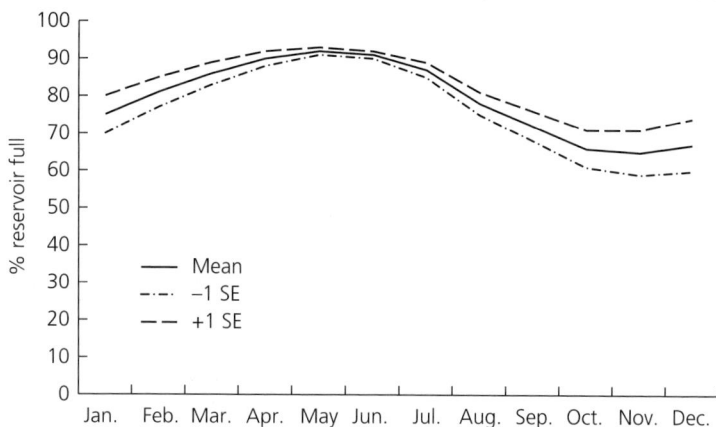

construction. As was the case for rainfall harvesting, to design our water resources we need to consider the seasonal and inter-annual variability of flows. This requires long series of observed or simulated flow data, so that we can simulate the filling and emptying of the reservoir to determine the optimal size of storage, and the risks associated with inter-annual variability. These factors are used to establish operational management rules for the reservoir, which are reflected in control curves. Figure 4.6 shows a typical control curve and the expected range of resource capacity linked to variability in hydrological conditions. Operationally, the current system state is compared with the historic mean and variability for a given time of year, and hence a specified set of management decisions will be invoked. For example, in the case of drought this may mean actions to reduce demand. The range of variability is often represented as including a storage safety margin known as *headroom*.

4.2.2.3   Groundwater resources

If we turn from the high rainfall of Wales to the densely-populated south-east of England, we find a situation with low rainfall and low topography – so, not ideal for surface water resources. However, if we consider the distribution of the major UK aquifers, the Chalk and the Sandstone (Figure 4.7), we see that groundwater systems occur extensively in this area. The London Basin is a classic example, with Chalk outcrops of the Chilterns and North Downs escarpments being the main receptors for infiltration and hence the areas of aquifer recharge. Groundwater systems provide natural storage that can be accessed by wells or boreholes. However, in most aquifers, complex structure and variation in properties exists, so that careful evaluation is needed of the potential effects of resource abstraction.

Two central issues arise – what is the *yield* that can be delivered by an individual well, which depends on the local aquifer properties and the well construction and development, and what is the natural *recharge* of the aquifer, which determines the

Figure 4.7 Major UK aquifers (water company boundaries are not up-to-date; Groundwater Forum, 1998)

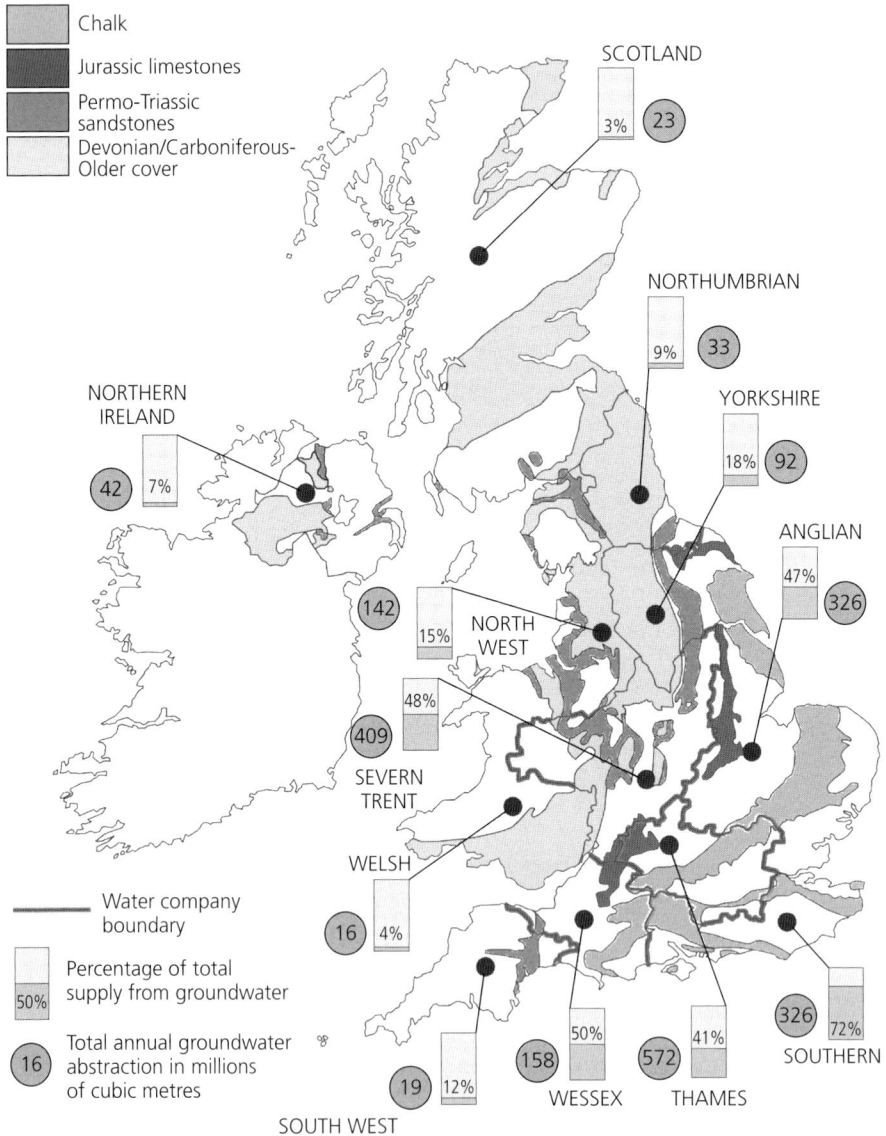

overall resource availability. A critical difference between surface water systems and groundwater is that subsurface properties are difficult and expensive to observe. The important properties of an aquifer are its *transmissivity*, or ability to transmit water, which in the case of the Chalk is largely governed by the degree of fissure flow in the chalk block, and its *storativity*, which represents the water released from the aquifer for a unit change in pressure. Local estimates of these can be obtained by controlled pumping of a well and observation of groundwater response in an adjacent borehole. However, in fractured rock systems, such as the Chalk, such estimates can be misleading (Butler *et al.*, 2009), and more generally a numerical model of an aquifer is needed to assimilate the available information on geology and aquifer properties, to simulate recharge (which depends on the balance of rainfall and actual evaporation, subject to the effects of soil moisture, and properties of the unsaturated zone), and compare the results with observed river flows and groundwater levels. With such a model, a well-field can be designed, and environmental effects (such as impacts on river flows and water tables) can be predicted. It is commonly the case that, where a major groundwater development is implemented, the associated drawdown of groundwater levels will affect existing private wells, so that alternative arrangements for supply will need to be made to affected properties.

Because groundwater provides a naturally occurring storage of water, there are many possibilities for the active management of a groundwater system. One possibility is known as aquifer storage and recovery (ASR), where surplus surface water (typically in winter) is recharged to an aquifer, for later abstraction (in the summer). Artificial recharge can occur by injection of water into a well system, or through creation of infiltration ponds. Critical issues concern the protection of groundwater quality and the prevention of clogging or fouling of the facilities. Hence, a long-running ASR project in the Lee Valley, north of London, injects water treated to potable standards into a well-field (Hawnt *et al.*, 1981; O'Shea *et al.*, 1995).

4.2.3 Resource characteristics and conjunctive use

Surface and groundwater resources have contrasting economic characteristics. Construction of a reservoir is a highly capital-intensive one-off investment, but with relatively low running costs. Groundwater can be developed well-by-well, at low capital cost, but then carries the associated costs of pumping.

In terms of physical characteristics, surface water resources are mainly vulnerable to summer low flows, and hence summer drought. Groundwater recharge predominantly occurs in winter, once soils have rewetted from summer drying (and the associated development of soil moisture deficits), and are therefore particularly susceptible to dry winters (and especially sequences of dry winters).

Clearly, in any given water resource system, there is the potential to optimise the combined use of surface and groundwater resources, depending on their availability, with respect to the economics of investment and operational performance, particularly drought security. The Severn (see Figure 4.5) provides a useful illustration. Although surface water resources predominate, there is a limited area of Triassic Sandstone which has been developed as the Shropshire Groundwater Scheme to provide an additional resource to be used in the case of severe surface water drought conditions, when the regulating reservoir (Llyn Clywedog) can no longer provide sufficient water to maintain downstream flows. The Shropshire scheme can be used to augment river flows, and to provide a source of supply to local users, thus relieving pressure on the overall system.

4.2.4   Environmental effects of abstraction

As noted in Section 4.1 above, human use of water must be balanced with the needs of the environment, and in the EU the Water Framework Directive includes the protection of ecological quality as a major requirement for water management. The water environment supports a myriad of species of flora and fauna which represent 'ecological diversity'. Each river or wetland water body has diversity characteristics which depend on local indigenous species and the prevailing hydrological conditions such as flow, water levels and topography. The populations of these species may vary with seasonal and annual weather patterns, but the rate of change is often different for different species, so that the pattern of variation can be specific to a particular weather sequence. During extreme hydrological events such as droughts and floods the population of indigenous species may be affected to such an extent that they are significantly harmed. Populations are also affected by local abstractions, whether directly from the wetland or watercourse or indirectly from groundwater if the reduction in groundwater levels and/or pressures around an abstraction (the cone of depression) affects the hydraulic conditions of the water body.

These complex relationships are studied by monitoring changes in hydrological conditions and comparison with indicator species, and are described in detail by Magurran (2004) and Southwood and Henderson (2000). The design of a monitoring programme depends on the characteristic of each water body. The hydrological regime may be assessed using a combination of flow measurement (e.g. using temporary river gauging or permanent weir or flume installations) and water level monitoring (e.g. using gauge boards or pressure transducers) and dip tubes or piezometers to measure groundwater levels. The population of indicator species may be monitored using river corridor survey techniques, and quantified using scoring systems. There are a number of models and proprietary software available for collation and assessment of results such as the Biological Monitoring Working Party Score System (Armitage *et al.*, 1983; Hawkes, 1997), which may be used to relate the ecological population with the quantity or quality of water available. A particularly useful tool is the Lotic-invertebrate Index for Flow Evaluation (LIFE): this is a method for linking macro-invertebrate populations to prevailing flow regimes, and is widely used for British waters (Extence *et al.*, 1999).

Detailed site-specific modelling techniques have been developed to investigate the relationship of taxalogical and hydrological data. For British waters, PHABSIM (Spence and Hickley, 2000) is becoming widely used for modelling river and wetland flow regimes and forecasting the effects of future change, which then provides valuable guidance for the management of water taken from the environment, discussed below.

4.2.5   Water resource system performance

We all need water to drink, and a reliable water supply is essential for a modern society. Water is taken from the environment to meet that need, but it is essential that this is done in such a way as to prevent harm to the natural environment, if we are to preserve the long-term equilibrium. So, we need to assess how much water can be taken from our rivers and groundwater.

4.2.5.1 Exploitation of water resources

A key component of water resources planning for industrial use or public water supply is the assessment of the *yield* or *deployable output* of a particular river or groundwater water body. Surface water systems are assessed by considering the long-term river hydrograph and empirical evidence of low flow in drought conditions (UKWIR, 2000). Where data are not available for appropriate historical drought conditions, estimates are made by the extrapolation of the data set or statistical modelling.

The methodology for assessing the yield of groundwater systems considers the empirical relationship between historical abstractions, either from a particular borehole or a set of boreholes where these are linked hydraulically (UKWIR, 1995) and the level of water in the aquifer. In this case it is necessary to have a reliable record of pumping water levels and rest levels to compare directly with volumes abstracted. For groundwater, the system response is much slower than surface water, as groundwater level changes as a result of rainfall may be days or months later due to the time taken for percolation through the overlying strata. The yield of a borehole is also significantly influenced by the character-istics of the surrounding aquifer. The nature and extent of fractures and fissures over the operational depth of a borehole means that flow into it may not be linear, and the presence of significant flow horizons may cause a step in yield at a particular depth. Performance data are plotted on a 'source reliable output summary diagram', as shown in Figure 4.8, and the yield of the borehole is defined by the boundary condition for stable operation of the well.

Groundwater supply systems often comprise a number of borehole elements over a geographical area. Separation in this way reduces the interference between boreholes and distributes the drawdown of overall groundwater levels in the aquifer. Where interference between boreholes is seen, or they are linked by a common factor such as proximity to a local river, then yield is assessed for the group of boreholes. Equally, when boreholes are operated across a wider area or a groundwater unit, the aggregate yield should be assessed for the purposes of maintaining equilibrium with overall effective rainfall percolating to the aquifer. Complex hydrogeological behaviour for multiple abstraction sites may be modelled using proprietary software such as MODFLOW (Harbaugh, 2005).

In both surface and groundwater systems, it is necessary to consider the sustainable yield over different durations so that this can be compared with the range of operational pumping conditions that are to be employed to meet the requirements of the user.

For public water supply, the boundary conditions used to determine yield also deter-mine the 'level of service' for operational purposes. 'Level of service' in this context refers to the frequency of restrictions placed on supplies to customers by a water supplier, and this acts in such a way as to limit the maximum resources needed to meet a given demand. This means that where yield has been assessed for historical conditions as corresponding, for example, to a ban on hosepipe use at a rate of 1 year in 10, then this condition must be placed on the future operation of that system and

Figure 4.8 Source reliable output summary diagram. Note that water levels may be given as metres below ground (m BGL), as metres relative to ordnance datum (m AOD) or as metres relative to another stated datum

Reproduced with permission from UKWIR (1995). © UKWIR

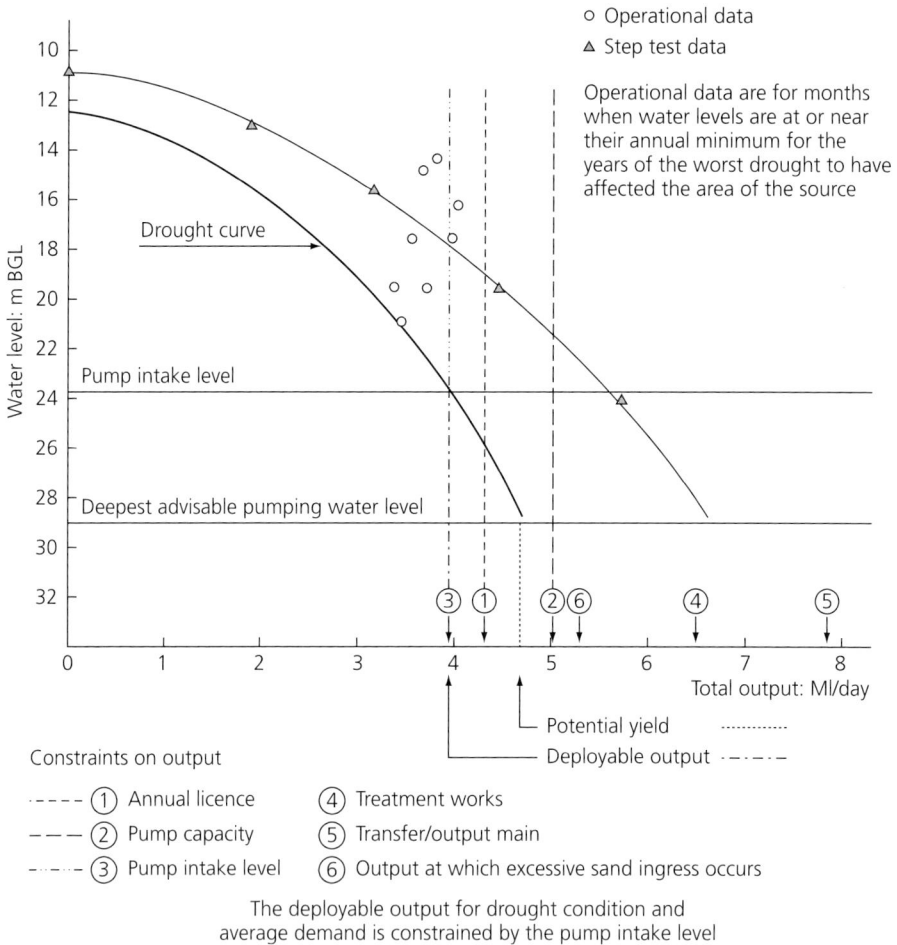

Constraints on output

---- ① Annual licence   ④ Treatment works
--- ② Pump capacity   ⑤ Transfer/output main
--·- ③ Pump intake level   ⑥ Output at which excessive sand ingress occurs

The deployable output for drought condition and average demand is constrained by the pump intake level

thus on the service provided to customers, unless account is taken of differences in future operating conditions. If an improved level of service is offered to customers, then demand is likely to be higher during the critical design period, and a shorter yield duration may have to be used to ensure that a higher level of abstraction can be sustained to meet demand.

Until recently, resource performance has been evaluated under the assumption of climate stationarity, i.e. that the past climate is a reasonable guide to the future. Current concerns for climate change mean that risks to future resource availability need to be assessed, including the effect of future changes in weather patterns and climate (see Kundzewicz *et al.* (2007) for information on water aspects of the IPCC fourth report). This is a major challenge, as while models of global climate have considerable utility in representing large-scale features such as temperature, precipitation is poorly represented, particularly at the spatial scales of individual catchments required for water resources assessment. For this reason, statistical downscaling methods are widely used to transform global climate model outputs into the time series of precipitation and potential evaporation needed for the assessment of climate change effects on water resources (Chun *et al.*, 2009, 2010). A nationally applicable weather generator model for future climate has been developed as part of the UK Climate Impacts Programme UKCIP09 (Defra, 2010). It must, however, be understood that the climate models have major limitations that cannot be wholly overcome by statistical methods, and there are concerns that the effects of the persistence of rainfall anomalies on drought are under-represented by downscaling methods. Recent developments are reported by Chun *et al.* (2009a,b). Implications of climate change for England and Wales are discussed in a recent report by the Environment Agency (2009).

Returning to the issues of environmental impact assessment for water abstractions, we note that natural water systems are variable in character, and techniques for assessing species populations are often limited to the sampling of representative or critical elements of a water body. Therefore, it is important to also assess the variance and uncertainty of the methods employed in any particular regime by statistical analysis of the data and to ensure that data collection takes place over sufficient time to capture the full range of seasonal and annual weather patterns. This is particularly so when considering future behaviour so that decisions taken to manage the water environment are reliable. A simple illustration of this is to consider the effect of a 5% change in river flow where the uncertainty in assessment of population has a standard error of 10%. Clearly, any conclusion on the quantum of future populations would be unreliable in this case; however, if the change in population over time and under a particular set of hydrological conditions has a standard error of 2%, then a reliable forecast of change may be made.

4.2.5.2 Augmenting water resources

When abstraction takes place, the water body or watercourse loses the benefits of that water. Conversely, there may be environment benefits at a point of discharge to a water body, provided the quality of water is consistent with the receiving waters. So, the net effect of water abstracted to supply a community can be mitigated by discharging treated wastewaters as close as possible to the point of abstraction. A water body that has abstractions equal to discharges may be said to be *water neutral*.

The quantum of water available in the natural environment may be supplemented by recharge mechanisms. The natural water cycle can be enhanced by artificial recharge. For example, where winter groundwater abstraction depletes flow from springs to a river, flows in the watercourse may be maintained by pumping water from alternative water bodies with a surfeit of resources. Groundwater may also be recharged by infiltration. The most common mechanism for this is through soakaways and infiltration basins linked to highway drainage, but this would also apply to rainwater-harvesting systems and direct aquifer recharge, as discussed above. This latter technique can be used to store surplus winter water for summer use, provided that the flow out of the groundwater body between the time of recharge and re-abstraction is not large in comparison with the recharged volume.

4.2.6   Regulatory control of abstraction

As abstracting water from the environment may limit availability for downstream users or harm in-stream or wetland ecology, it is necessary to limit the amount removed. This is achieved by licensing (currently required in the UK for any abstractions greater than $20\,m^3/day$). Abstraction licences in England and Wales are issued by the Environment Agency, and in Scotland by the Scottish Environmental Protection Agency (SEPA), and relate to water taken from any water body. Licences may have conditions that limit the amount taken at any point in time (e.g. varying seasonally from winter to summer) and over any period of time, from hourly to annually. These constraints may also relate to minimum flow requirements, often referred to as *hands-off flow* conditions, and which may be a particular proportion of the long-term flow hydrograph. The *Lower Thames Operating Agreement* is an example of a minimum flow requirement. In this case, abstraction from the river may only take place when flow over Teddington Lock (the point of transition from non-tidal to tidal reaches) exceeds $300\,m^3/h$.

Water bodies may be designated in accordance with the scarcity and critical nature of the indigenous species. Sites that are of European importance are designated under the Habitat Directive as *Special Areas of Conservation* or *Special Protection Areas*, and protection of these habitats is mandatory. Sites may also be designated as nature reserves of national, regional or local importance or Sites of Special Scientific Interest. In these cases, it may be necessary to apply licence conditions to control the water regime to protect particular species or habitats.

In order to assess the environmental requirement for a particular water body or collection of water bodies, the Environment Agency examines the prevailing conditions under its programme of catchment management strategies (CAMSs). These studies are also fundamental to the preparation of *river basin management plans*, which are required by the *Water Framework Directive* to determine actions necessary to achieve targets for good ecological and chemical status by the long-stop date of 2027. River basin management plans include the Environment Agency's National Environment Programme, which is a range of investigations and options appraisals to determine the cost of changing licensed abstraction compared with the environmental benefits that are derived. Licences that are required to be reduced are known as *sustainability reductions*.

Licences can take a number of forms. *Licences of right* are perpetual, and were granted when licensing first started, following the Water Resources Act 1963. In recent years, only *time-limited licences* have been granted, and as these are set for a fixed period the Environment Agency has the discretion to renew, terminate or amend the terms of the licence when an application is made on expiry of the licence. This means that the Environment Agency is able to adjust the abstraction regime in a catchment to maintain ecology. Licences may also be granted for specific purposes such as agricultural irrigation or cooling for energy generation. Licence charges are set in accordance with purpose and the proportion returned to the environment. The licensing system in England and Wales was re-examined in the Cave Review of competition and innovation in water markets (Cave, 2009).

4.3   Water treatment
4.3.1   Introduction to water quality
We turn now from the provision and management of water resources to the treatment and distribution of potable water. The primary objective of potable water treatment is to provide water that is of appropriate quality for the intended use at a reasonable cost. Potable water should be microbiologically and chemically safe and aesthetically acceptable to the consumer. In addition, the water quality may be adjusted to avoid or minimise undesirable effects such as plumbosolvency, corrosion or limescale deposits. The quality of water is defined in terms of individual physical, chemical and biological parameters, and these parameters are used for both operational and regulatory purposes. For the latter, they form the basis of legal standards that define the health risk and palatability of drinking water, and this will be discussed in detail in the next section. Most water quality parameters are unambiguous in that they refer to specific chemicals (e.g. nitrates, lead, manganese), particular micro-organisms (e.g. *Escherichia coli*) or physical properties (e.g. temperature, electrical conductivity), but others are surrogate values which give a general indication of water quality that can be conveniently measured manually on-site or by on-line instrumentation (e.g. turbidity, colour, pH).

Among the very large number of parameters that define water quality (>100), several key parameters are universally applied in the context of drinking water: these include turbidity, colour, taste, odour, hardness, residual chlorine, and indicator bacteria. In the context of water distribution systems, additional parameters may be important, such as pH, dissolved oxygen, redox potential, conductivity, Langelier index, lead, dissolved organic carbon, chloramines and disinfection by-products. The presence of particulate contaminants is usually quantified by turbidity, which is an indirect measure determined by nephelometry (relative light scattering). Turbidity is expressed in NTU (nephelometric turbidity units) by comparing the degree of light scattering with standard solutions. Impurities that cause water to have a visible colour are also determined optically, either by manual comparison with standard colour solutions (units: degrees Hazen), or spectrophotometrically (units: light absorbance at 400 nm per metre). Residual chlorine in drinking water after disinfection will exist either as free chlorine, in the form of hypochlorous acid and the hypochlorite ion, or combined chlorine, where the chlorine is bound to organic substances, or as chloramines (e.g. monochloramine, NH_2Cl); total chlorine is the sum of free and combined chlorine. While combined chlorine has less disinfection power than free chlorine, it persists longer, and this is believed to be advantageous in large distribution networks. Routine monitoring of micro-organisms involves the enumeration of a limited number of indicator organisms. These comprise total coliform bacteria, faecal coliforms, faecal streptococci and colony/plate counts. The absence of the aforementioned three organisms (per 100 ml), and no significant change in colony/plate counts, is assumed to signify the microbiological safety (absence of pathogens) of the water.

Apart from turbidity, colour, free and total chlorine, pH, temperature and conductivity, which are normally measured at water treatment plants by on-line instrumentation, the large majority of other water quality parameters are determined off-line in the laboratory from grab samples. Full descriptions of all water quality parameters and standard methods of determination can be found elsewhere (APHA, 2005).

Typical sources of water for drinking water supply have been mentioned previously in this chapter. As in most countries, the proportion of drinking water in the UK arising from groundwater or surface water sources varies region by region depending on the local topography and rainfall patterns. In very general terms, the north, north-west and west of the UK are upland areas that receive above-average rainfall, and thus surface water is the predominant source of drinking water, usually via direct supply reservoirs. In contrast, the east and south-east of the UK are much drier, lowland areas utilising impounded reservoirs and groundwater from greensand or limestone/ chalk aquifers. Consequently, current treatment process streams vary widely, depending on the nature of the source water. Upland waters are often rich in organic content, low in dissolved solids and salts, and acidic. In contrast, lowland waters are nutrient-rich, high in dissolved and suspended solids (e.g. phytoplankton) and salts (highly buffered), and contain substantial quantities of micro-organisms, arising from wastewater effluents, urban drainage and run-off. Groundwaters are a preferred source, owing to the low level of contaminants as a consequence of natural filtration and microbiological processes. However, specific water quality problems may exist owing to the presence of natural contaminants (e.g. arsenic, iron, manganese) or the accumulation of pollutants from surface sources (e.g. nitrates, pesticides, solvents).

The minimum quality of raw waters that can be used for drinking water supplies has been defined by the EU in relation to the degree of treatment that will be required to achieve the potable water standards. In addition, water service providers (WSPs) may have the flexibility to blend either raw or treated waters in order to meet the required standards in the most cost-effective way. While source waters are, and have been, adequately protected by well-established pollution control measures and legislation for many years, there still remain a number of challenges to the WSPs in the developed world in terms of fully complying with prevailing water quality regulations. Many of these relate to greater expectations from consumers plus increased scientific knowledge of the long-term health impacts of certain pollutants (e.g. disinfection by-products). However, deterioration of source waters can also be caused by the effects of changing land use and climate (e.g. organic colour, algal blooms), intensified agricultural practices (e.g. protozoan cysts, fertilisers, pesticides) and urban run-off/wastewater discharges (e.g. nitrates, phosphates, trace organic compounds, pharmaceutical/healthcare products). In the UK and in other countries, specific problems currently include the prevention of *Cryptosporidium* (a protozoan cyst) and pesticide breakthroughs, taste and odour incidents, plus reducing the leaching of iron, particulates and micro-contaminants from, for example, pipe-lining materials (e.g. PAHs – polynuclear aromatic hydrocarbons) within water distribution networks.

4.3.2 Water standards, monitoring and regulation

The required quality of drinking water is defined in terms of guideline and/or legal values for the various physical, chemical and biological parameters that represent no significant risk to health or undesirable aesthetic impact, sometimes described as 'wholesomeness'. For virtually all these parameters, the specified values are maximum concentrations, but in the case of water that is softened during treatment there is a minimum limit for the hardness and alkalinity concentrations. Internationally, the quality of water suitable for public water supplies is based on the World Health Organisation's *Guidelines for Drinking-water Quality*, which provides extensive reference material for all aspects of drinking water quality; this is summarised in report form (WHO, 2006) or via the WHO website. In the UK, water companies must comply with legal standards defined in The Water Supply (Water Quality) Regulations 2000, and associated amendments, which incorporated, and added to, those of the 1998 European Union Directive (European Council Directive 98/83/EC on the quality of water intended for human consumption). In the USA, drinking water standards are set by the Environmental Protection Agency. The UK Regulations set out the maximum (and minimum) values at particular locations in the water supply system (mostly at the consumer's tap), the designation of monitoring areas ('zones'), the number, frequency and location of samples, and specifications for the collection and analysis of samples. Each year, the water company must designate the names and areas within its area of supply that will be the water supply zones for the year, and the population served by these zones must not exceed 100 000 per zone. The annual sampling frequency per zone, and from treatment works/supply points, depends on the population and flow rate, and on the particular parameter. For the majority of water quality parameters, the location of compliance is at the consumers' taps, as indicated in Table 4.1. Water utilities also monitor many parameters at the water treatment plant and in the water distribution network in order to detect changes in water quality through the water supply system to the consumer's tap. Some of the regulated parameters that might change as water passes through pipe networks are shown in Table 4.1. The effects of continuing chemical reactions in the bulk water and long contact times with pipe surfaces may lead to reductions in parameter concentrations, such as residual free and combined chlorine and dissolved oxygen, while other parameters may increase due simply to corrosion and biofilm effects, such as iron, manganese, PAHs and turbidity.

In the UK, specifically England and Wales, the regulatory organisation responsible for ensuring drinking water quality complies with the standards is the Drinking Water Inspectorate (DWI). The DWI is an autonomous government body which is part of the Water Directorate of the Department of Environment, Food and Rural Affairs (Defra). It monitors all aspects of the drinking water supply that are managed by the private water companies in England and Wales, including the collection and reporting of compliance samples, assessment of water quality incidents, technical audits, European standardisation, research on emerging issues and hazards, and approval of the use of

Table 4.1 Some water quality standards relevant to potential changes in water distribution systems

Parameter	WHO guideline[a]	UK Regulations[b]	
		Value	Monitoring point
Turbidity	0.1 NTU	4 NTU (1 NTU)	Consumers' taps (WTWs[d])
Colour	<15 TCU[c]	20 mg/l Pt/Co	Consumers' taps
Coliform bacteria		0/100 ml	Service reservoirs and WTWs
Escherichia coli	0/100 ml	0/100 ml	Service reservoirs and WTWs, and consumers' taps
Nitrate[e]	50 mg NO_3/l	50 mg NO_3/l	Consumers' taps
Nitrite[e]	3 mg NO_2/l	0.5 mg NO_2/l	Consumers' taps
Lead	10 µg Pb/l	25 µg Pb/l[f]	Consumers' taps
Iron	0.3 mg Fe/l	0.2 mg Fe/l	Consumers' taps
Manganese	0.1 mg Mn/l	50 µg Mn/l	Consumers' taps
PAHs	0.7 µg/l[g]	0.1 µg/l[g]	Consumers' taps
Total trihalomethanes[g]	Sum ≤1[j]	100 µg/l	Consumers' taps
Chloroform	0.3 mg/l		
Bromoform	0.1 mg/l		
Dibromochloromethane	0.1 mg/l		
Bromodichloromethane	0.06 mg/l		

[a] WHO (2006)
[b] England and Wales, 2000
[c] True colour units
[d] Water treatment works
[e] $[NO_3]/50 + [NO_2]/3 < 1$
[f] 10 µg Pb/l after 25 December 2013
[g] Benzo[a]pyrene
[h] Benzo[b]fluoranthene, benzo[k]fluoranthene, benzo[ghi]perylene, indeno[1,2,3-cd]pyrene
[i] Chloroform, bromoform, dibromochloromethane, bromodichloromethane
[j] Sum of the ratio of the concentration of each compound to its respective guideline value

products and substances. Each year, over 2 million compliance samples are taken, and compliance typically exceeds 99%; details of sample compliance and other aspects of water company performance can be obtained from the DWI annual reports. Among the most common reasons for compliance failure are discolouration, microbiological contamination, disinfection/treatment failures, source contamination and loss of supply.

In recent years, the DWI has adopted the WHO recommendation of the creation of a holistic risk management approach to the provision of drinking water in the form of drinking water safety plans (DWSPs). This applies risk management to all stages of the water supply chain, from the raw water catchment, through the treatment and distribution, and finally to the consumer. Water companies are expected to establish

DWSPs for each of their supplies, and these have three key components: a comprehensive system assessment; operational monitoring; and documentation of management arrangements. At the heart of the DWSP procedure is the identification of hazards and assessment of the risks posed by each hazard, often via a risk-scoring matrix. The DWI provides independent verification of DWSPs, and expects all significant risks to be appropriately identified and assessed, and clear roles and accountabilities to be established.

4.3.3   Typical treatment methods

The treatment of raw water to meet drinking water standards involves a number of individual steps, or unit processes, where the type and precise number depend on the nature of the raw water. The processes may be grouped into broad categories, as shown in Table 4.2.

Conventional water treatment processes that have been used extensively for many decades, such as coagulation/flocculation, sedimentation, filtration and chlorination, continue to be generally successful in achieving water quality objectives. However, the rising demand for water by society and increasing impacts on water quality by human activities have lead to new challenges for the water treatment industry. Furthermore, advances in analytical chemistry methods have allowed the detection of contaminants whose presence in water was previously unknown. These new pressures and improved knowledge base have driven the development of more stringent regulations and treatment goals. This has also led to innovation in terms of the implementation of advanced treatment technologies and efforts to optimise existing processes to achieve even higher levels of treatment. This chapter summarises some of the current and future water quality challenges and trends and the state-of-the-art in potable water treatment processes.

At the simplest level in terms of water treatment, a high-quality groundwater may only require the application of a disinfection residual, such as a few milligrams per litre of chlorine. However, it is more usual that groundwaters will contain low levels of undesirable chemicals such as free and bound metals (iron, manganese and arsenic), anions (fluoride and nitrate) and organic compounds (solvents, fuel additives and pesticides). In order to achieve compliance with water quality standards, such waters will need additional treatment, depending on the contaminant, but typically involving one or

Table 4.2 Water treatment processes

Preliminary treatment	Raw water storage, screening, oxidation (aeration and/or chemical addition), pH adjustment
Primary treatment	Coagulation, flocculation, sedimentation, flotation
Secondary treatment	Rapid filtration, slow sand filtration
Advanced treatment	Pre- and intermediate ozonation, powdered and granular activated carbon, softening, denitrification
Final treatment	pH stabilisation, fluoridation, phosphate (lead stabilisation), disinfection (e.g. chlorination or chloramination)

more processes such as oxidation (aeration, chlorination), ion exchange or filtration (granular media or membrane), and adsorption. Surface waters are normally abstracted either directly from a river with sufficient flow reliability, a natural lake, or from an impounded reservoir constructed to provide adequate capacity for continuous supply. Direct supply from a river is advantageous because of the avoidance of substantial costs in constructing a reservoir, but the variability in flow associated with seasonal changes leads to marked changes in source water quality, which can present difficulties in providing a consistent treatment. Lakes and raw water reservoirs provide a more stable water quality, in general, but excessive algal activity can arise in nutrient-rich eutrophic waters during spring through to autumn in temperate climates. In dealing with high-algal waters, which can cause high-solids loadings on works, taste and odour issues, plus the release of algal toxins, various technologies have been applied, including double filtration (rapid-rate granular media or micro-strainers, followed by slow sand filtration), pre-oxidation (chlorine or ozone) and DAF (dissolved air flotation). Where lake or reservoir water is of relatively unchanging and good quality, the primary and secondary treatment steps (see Table 4.2) can be combined in a simplified process called 'direct filtration', in which coagulated raw water is applied directly to a rapid-rate filter without any clarification (sedimentation or flotation) stage. For moderately contaminated surface waters, such as a lowland urban river source, a typical treatment train would comprise pre-oxidation, chemically assisted clarification (by sedimentation or flotation), rapid-rate dual-layer filtration, intermediate ozonation, granular activated carbon (GAC) and final chlorination. For several major European cities (e.g. London, Paris, Amsterdam, Stockholm and Zurich) slow sand filtration (SSF) was the basis, historically, of the water treatment process, and this filter type remains in operation today, owing to its ability to achieve a high treatment performance for both particulate and microbiological contaminants. The enhanced performance is partly due to the intense biological surface layer ('*schmutzdecke*') that develops with time (Figure 4.9a), which is a unique feature of SSF, assisted usually by

Figure 4.9 Typical water filtration processes: (a) biological surface layer in a slow sand filter; (b) rapid sand filtration

(a)   (b)

an upstream rapid-filter stage to lessen the solids loading on the SSF (Figure 4.9b). With SSF, either an incorporated GAC layer is used in the filter or a separate GAC adsorption stage is needed to facilitate the removal of micro-pollutants such as pesticides. In certain circumstances, pre-treatment of river water may be achieved by drawing the water through sufficiently permeable river bank strata via infiltration wells, and the treatment performance of such systems is receiving considerable attention at present (Partinoudi and Collins, 2007). Finally, both groundwater and surface water can contain undesirably high concentrations of calcium and magnesium (e.g. >400 mg/l as calcium carbonate), requiring a softening step involving either precipitation or ion exchange.

In the USA and Europe there is continuing concern about the presence of potentially harmful organic micro-pollutants in drinking water, and particularly those arising as by-product compounds from disinfection. Since disinfection must be carried out rigorously and without compromising micro-organism inactivation, emphasis is being placed on improving the removal of by-product precursor materials prior to disinfection. In many treatment plants, chemical coagulation is the principal unit process for removing natural organic substances, and methods of improving the efficiency of this (enhanced coagulation) continue to be of interest and the subject of research investigation. New types of coagulant chemicals offer greater performance and/or lower cost, and much development work in recent years has concentrated on low basicity, highly charged polymeric aluminium and ferric salts, and the combination of metal salt coagulants with organic polyelectrolytes. For example, laboratory trials have demonstrated the superior treatment performance of particular forms of poly-ferric sulphate (Jiang and Graham, 1998a) and poly-alumino-ferric sulphate (Jiang and Graham, 1998b) in terms of the removal of natural organic matter (including disinfection by-product precursors), lower sensitivity to pH and temperature, and reduced sludge production; these improvements need to be replicated at pilot and full scale. The beneficial combination of new inorganic coagulants and organic polyelectrolytes is also being actively considered for particular types of water quality.

At present, coagulation and floc separation are widely achieved through the use of sludge blanket clarification or dissolved air flotation (DAF). Both processes have their advantages and disadvantages in regard to cost, treatment performance and operational requirements. In recent years, various innovations have been applied to improve their performance, such as the addition of fine sand suspensions to increase the density of chemical flocs (e.g. the Veolia 'Actiflo' process) and the combination of DAF and rapid filtration in one unit (CoCoDAFF). A new approach nearing full-scale application at the moment is that of using an electro-coagulation–flotation (ECF) cell to generate the coagulant and bring about floc separation by flotation (Cerisier and Smit, 1996; Jiang et al., 2002). This process system offers many potential advantages, particularly in providing a more compact process configuration, telescoping into one unit the three conventional stages of coagulant mixing, mechanical flocculation and flotation separation. In addition, ECF avoids the need for storage and dosing of chemical solutions, and it is possible that ECF may provide other benefits in terms of lower energy demand, more-efficient treatment and some microbiological inactivation. Various aspects of the technology are still under investigation, such as the avoidance of anode passivation and the optimal electrode arrangement (e.g. mono-polar or bipolar).

Another treatment process innovation which is receiving considerable interest is that of magnetic ion exchange (MIEX). The MIEX process (Orica Watercare, Wigan) employs a re-usable suspension of fine magnetic particles coated with an anionic resin to remove natural organic matter (NOM). As will be described later, NOM is a major concern in water treatment because of its role as a disinfection by-product precursor material and as a nutrient for biological activity in water distribution pipe networks. As mentioned earlier, conventional water treatment is relatively poor at removing NOM, and this has led to the need for enhanced coagulation. MIEX has been studied as a pretreatment for conventional coagulation processes, and generally has been found to achieve the higher organics removal expected (e.g. Shorrock and Drage, 2006). The economic advantages of MIEX over other methods of achieving enhanced organics removal appear to be case-specific, and this necessitates the need for pilot scale testing to justify its application.

In recent years there has been a widespread interest in, and application of, oxidant chemicals for a range of water treatment purposes. Such oxidants can be applied at different stages of treatment, depending on their purpose, as shown in Figure 4.10. Often this has been done to either replace, or reduce, the use of chlorine because of the concern over the formation of halogenated by-product compounds. While the pre-dominant oxidant applied so far has been ozone, there have been successful applications of chlorine dioxide and potassium permanganate in the treatment of surface waters (Ma *et al.*, 1997). Ozone has been of particular interest because of its ability to degrade pesticide compounds and other organic micro-pollutants. This normally occurs as an intermediate treatment step, after the conventional processes (clarification and filtra-tion), in combination with GAC (Figure 4.10). However, typical water treatment conditions limit the effectiveness of ozone treatment by minimising the generation from ozone of highly reactive radical species. Research interest is currently focused on methods of enhancing radical formation, including combinations of ozone with hydrogen peroxide (Lambert *et al.*, 1996), ultraviolet (UV) irradiation, metal catalysts

Figure 4.10 Application of oxidants in water treatment (BAC, biological activated carbon; FBC, flat bottomed clarifier; RG, rapid gravity; UV, ultraviolet irradiation)

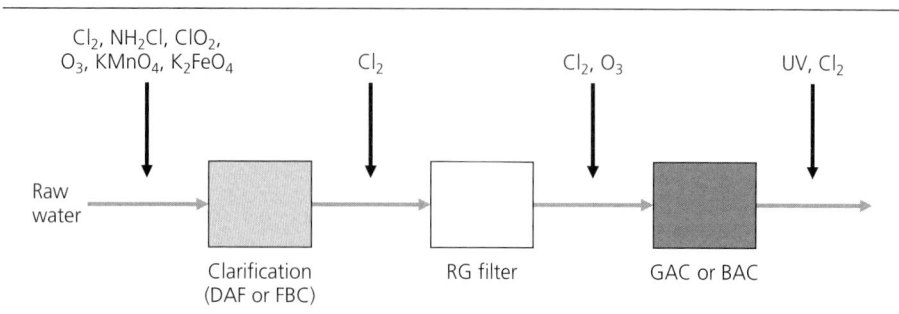

(Ma and Graham, 1999) or activated carbon (Jans and Hoigné, 1998). For example, studies of the ozonation of the herbicide atrazine have shown that the presence of a small concentration of manganese ($\sim$0.5 mg/l) can catalyse the degradation of atrazine (Ma and Graham, 1999). A possible application of this phenomenon is the ozone treatment of contaminated surface water and groundwater that also contain low levels of manganese.

Recently, there has been a change in the type of micro-pollutants being observed in raw waters to more environmentally sustainable and lower human toxicity chemicals, especially regarding pesticides. These new chemicals tend to be more polar in nature and therefore are not as easily removed by the current levels of ozone and GAC treatment. It has been shown that increased concentrations of ozone, including hydroxyl radical enhancement, are required to break down these compounds. However, increased ozone can produce elevated concentrations of bromate, which is a reaction product of ozone (with background bromide) and has a legislated maximum limit of 10 µg/l. Therefore, alternative highly oxidative treatment and adsorptive processes are being explored to achieve the most cost-effective treatment: an example of the former is ultraviolet irradiation combined with hydrogen peroxide (UV/H$_2$O$_2$), which has been successfully used at full scale in the Netherlands (Kruithoff et al., 2007).

Other treatment chemicals that combine oxidation and coagulation–precipitation capabilities are also of interest and undergoing evaluation. While permanganate, as mentioned above, can also assist coagulation through the precipitation of solid phase manganese dioxide (Ma et al., 1997), Fenton's reagent (ferrous ions and hydrogen peroxide) and ferrate (FeO$_4^{2-}$) are also able to produce iron-coagulating species as a result of powerful oxidation reactions. While the application of Fenton's reagent is limited to low-pH waters, the optimal pH conditions for ferrate are typically above neutral, which corresponds to a strong oxidation potential and adequate chemical stability (Jiang et al., 2004). Recently, the potential combination of ferrate and photocatalysis (UV/TiO$_2$) has been studied (Yuan et al., 2008), since there may be an advantageous synergy arising from the ferrate scavenging of conductance band electrons.

Membrane treatment continues to provide an important alternative to other established technologies for particular raw water qualities and to deal with specific contaminants where it is cost-effective to do so (Taylor and Wiesner, 1999). Membranes can be classified in simple terms according to the nominal size of the pores and the membrane material (ceramic, polymer) and form (spiral wound, hollow fibre); the most common membranes based on pore sizes are microfiltration (MF) (0.1–1 µm), ultrafiltration (UF) (0.01–0.1 µm), nanofiltration (NF) (1–10 nm) and reverse osmosis (RO) (<1 nm). Currently, the principal applications of membrane technology to water treatment are clarification and disinfection (MF and UF); NOM and trace organics removal (NF); nitrate, hardness and pesticide removal (RO); and brackish and seawater desalination (RO). Two examples demonstrating the versatility of combining membrane processes with conventional treatment have been reported recently. First, the use of UF after clarification for the treatment of low-alkalinity and highly coloured upland water (Hillis, 2006), and, secondly, the conversion of a conventional sludge blanket clarifier to a UF unit to treat heavily laden algal waters (Redhead, 2007). The use of membrane technology is expected to continue to increase in the future, as more-efficient membrane materials are developed and the causes of membrane fouling are understood and methods of mitigation applied. However, before adopting a membrane solution, careful consideration needs to be made of pre-treatment and waste stream management, as they are critical to the process, especially for NF and RO applications.

A significant development for the UK water industry was the change of wording in the UK Water Regulations in January 2008 regarding protozoa such as *Cryptosporidium*. The previous legislation required the removal of the oocysts independent of their viability, and the main process adopted to achieve compliance at high-risk sites was UF/MF membranes. The change in regulations has allowed other technologies such as UV irradiation to be considered to inactivate the organism and provide a more cost-effective solution. The UV doses required for *Cryptosporidium* are also adequate to inactivate other pathogenic bacteria. This potentially provides an opportunity to simplify the final disinfection stage on some very high-quality waters and good distribution network systems to UV and marginal chlorination, thereby reducing infrastructure and chemical requirements.

While current shortcomings in drinking water quality can be overcome to a large extent by new investment in conventional treatment technologies, and by renewal of the capital infrastructure, new technologies and methodologies are likely to offer advantages of greater process efficiency and reliability, capital and operational cost savings, more compact plant, and a lower dependency on energy and chemicals. In addition, future developments will be required to provide higher standards of treated water quality and a greater degree of sustainability, principally in terms of the reuse and recycling of chemicals and materials, waste minimisation and less energy consumption. These developments will be evaluated within the new approach to managing drinking water quality based on the establishment of DWSPs, as described earlier. Thus, conventional and new water treatment processes will be subjected to detailed analysis increasingly in the future within DWSPs to identify whether improvements are required or existing processes should be replaced with superior technologies.

Currently, the level of process monitoring and direct computer control at water treatment works is fairly extensive. In contrast, real-time process simulation and optimal operation remain undeveloped due to the complexity and inadequacy of unit process models. While considerable effort has been invested over the last 15 years in the particular aspect of in-line coagulation control (e.g. via streaming current detectors), the success of full-scale systems has been site-specific and partial. At present, a very limited number of water treatment process simulation models exist, notably OTTER, developed by WRc plc (UK) (Head *et al.*, 2002), Stimela, developed by TU-Delft/DHV (the Netherlands) (van der Helm and Rietveld, 2002), and Metrex, developed by the University of Duisburg/IWW (Germany) (Mälzer and Nahrstedt, 2002). These are being evaluated for their ability to simulate actual process performance, and as a predictive tool for future scenarios. These models are generally deterministic in nature, and suffer from the lack of a sufficient range of on-line water quality data to employ them. Thus, they are principally of value as a design aid, or planning tool, but have limited application to real-time operation and control. Currently, an EU research project, Techneau, is developing a European platform for modelling drinking water treatment processes, based on integrating the OTTER, Stimela and Metrex models. While developments and improvements in these models will continue in the future, there is also complementary and growing interest in the use of artificial neural network, 'black box'-type models of treatment processes, which are inherently more capable of dealing with complex systems, such as water treatment processes, but are site-specific, requiring considerable effort to train the model to the local conditions (e.g. Adgar *et al.*, 2000; Maier *et al.*, 2004; AWWARF, 2006).

Among the techniques of computer modelling being applied increasingly to drinking water treatment processes is computational fluid dynamics (CFD) modelling. This approach has matured considerably in the last 5–8 years with the increasing availability of affordable commercial software. CFD is useful for quantifying flow patterns, mixing regimes, and contact times. Modelling allows the consideration of alternative designs and extremes of conditions that may not always be possible to test on site using physical tests, such as tracer tests. CFD modelling has been previously applied to the analysis of the hydrodynamics in chlorine contact chambers (e.g. Templeton *et al.*, 2006) and UV disinfection reactors (e.g. Valade *et al.*, 2003).

4.3.4   Management of residuals

Associated with the water treatment processes is the production of waste flows ('residuals' or 'sludges') containing solid matter, principally arising from the particle separation processes. These processes include screens, sedimentation and flotation clarifiers, granular media filters and activated carbon adsorbers, membrane processes, and softening processes. The large majority (70–80%) of sludges arise from the coagulant solids separated in the clarifiers, and most of the remainder (15–20%) is from softening. Although such waste flows may be small compared with the flow treated (e.g. 1–2% by volume), they cannot be discharged directly to the environment, and therefore represent a significant cost and operational burden to water companies and utilities in terms of handling and final disposal. A survey of waterworks sludge management in the UK in 1997–1998 estimated the annual total production of sludge in the UK to be 131 000 tonnes dry solids (tDS), and indicated that approximately 50% of sludges (in tDS/per year) was disposed offsite in landfills and a further 25% was discharged to wastewater sewers (Simpson *et al.*, 2002). With the rising cost of disposal to landfill (~£80/tonne in 2009 for sludge from water treatment works), water utilities are actively investigating novel, more-sustainable options for sludge disposal, such as incorporation into building materials, land reclamation and application to agricultural land (Owen, 2002). Since waste flows are initially very low in solids content (<5%), some degree of thickening and dewatering is normally carried out at the treatment plant to substantially reduce the flow volume. Depending on the final disposal method, the sludge may receive thickening by sedimentation (solids content up to 10%), and dewatering by mechanical plant, such as a filter plate press or centrifuge (solids content up to 30%). Transport costs also have to be taken to account, as any benefit from recycling may be negated if a local application cannot be adopted.

4.4   Water distribution systems

This section provides a brief introduction to water distribution systems, since subsequent chapters will review in much greater detail the key elements of such systems, their design and modelling, their operation, maintenance and performance, and their long-term management.

4.4.1   Introduction to water distribution systems

Once water has been treated to the required standards at the water treatment plant, it is transported to the consumer via a water distribution system. For the consumer, the water must arrive at an acceptable pressure and in sufficient quantity, and the water quality must retain its wholesomeness and aesthetic acceptability during its passage through the pipe network. In the UK, the requirements for pressure and quantity are defined by the Office of Water Services (Ofwat), which specifies for domestic consumers an indicative minimum water pressure of 10 m at the boundary of the consumer's property, and a legal minimum of 7 m, at a flow rate of 9 l/min. To achieve the pressure and flow requirements, the water is conveyed from the water treatment plant through a large network of pipes by means of energy provided by pumping plant and/or elevation head (potential energy). Since water treatment is best achieved under relatively constant flow rate conditions, and the demand for water by the consumer is highly variable, there is a need to balance the two flow patterns via storage. Thus, the major components of water distributions systems are typically the following: storage tanks (water towers) or (service) reservoirs, pumping plant, and a network of pipes arranged in an optimal layout (Figure 4.11). The layout invariably involves interconnections of pipes in closed loops in order to provide security of supply in case of failure of a pipe element, and to minimise pressure loss.

The overall supply network is divided into operational zones defined on the basis of service reservoirs, pumping stations, pressure zones or other operational considerations. In the UK, these operational zones are further divided into district metered areas (DMAs), the boundaries of which are defined by isolation valves, in order that leakage can be monitored and managed by metering the quantities of water entering and leaving the DMA. The subsequent analysis of flow and pressure, especially at night when a high proportion of users are inactive, enables leakage specialists to calculate the level of leaks in the district. DMAs were first introduced to the UK at the start of the 1980s, by the UK water industry, and typically cover a part of the supply network corresponding to 500–3500 connections. The supply network includes three principal types of pipe made from various materials, such as iron, steel, asbestos cement and plastic (uPVC and MDPE). Trunk mains are the largest in size (~300 to >1200 mm diameter), and carry large flows to/from treatment and to the reservoirs and towers in the distribution network area. Distribution mains are smaller (>90 mm diameter), and take the flow from the trunk mains to the streets/roads outside the consumers' properties, from which service pipes (~20 mm) provide the connection to the domestic property. Operation of the distribution network is facilitated by the

Figure 4.11 Water distribution systems

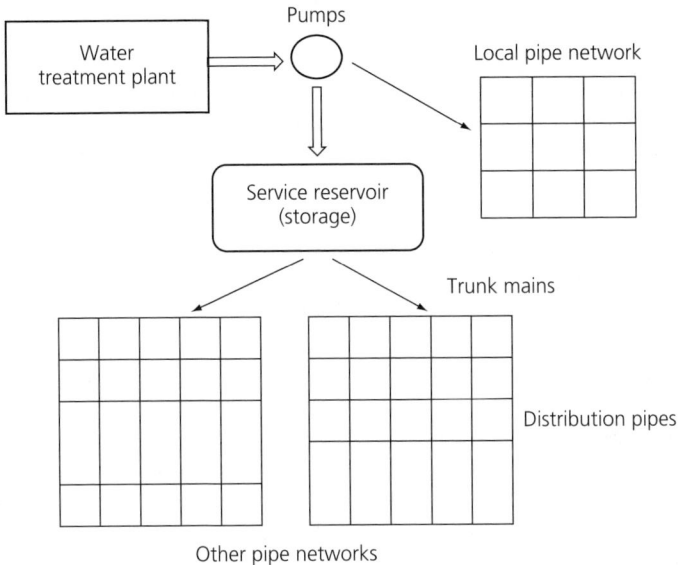

inclusion of hydraulic devices in appropriate locations, and these include control valves, air valves, washout/flushing chambers, pressure-reducing valves, flow and pressure meters, and fire hydrants: these are described in later chapters. Developments in such equipment in the future will improve operation of the networks and reduce pressure, with direct beneficial impacts on energy, leakage and bursts.

Innovative tools and techniques to assess and improve the integrity of the existing distribution network are also being explored, to provide greater resilience and minimise disruption. These include novel pipe-relining materials (e.g. Aqualiner) and non-intrusive mains assessment methods (e.g. Smartball), which are potential solutions to the current challenge of ageing assets in the network system.

Some water utilities are in the process of developing a more-integrated approach between the water treatment works and the distribution network systems by utilising modern data management systems in real time. If successful, this could provide, when completed, the most cost-effective water supply for the company and its customers.

4.4.2 Water quality

The quality of potable water is likely to incur minor change as it passes from the treatment plant to the consumer, since the flow may take many hours, or even several days in large networks, to reach the consumer. These changes are inherently undesirable, since the quality invariably deteriorates with flow time owing to the reactivity, and loss, of the residual disinfectant and contamination from internal pipe surfaces. The latter has various components, which include, principally, sediments that have accumulated during low-flow conditions, corrosion products from pipes and fittings, and biofilms that have developed owing to the presence of opportunistic micro-organisms and biodegradable organic substances in the bulk flow. Although the water leaving treatment plants complies with the necessary drinking water standards, low levels of particulates (<0.5 NTU) and dissolved organic substances (<2 mg C/l) are sufficient precursors for the development of sediments and biological activity in pipe systems over long time periods. Under normal conditions, increasing water age leads to a reduction in disinfectant residual (i.e. free and combined chlorine), an increase in trihalomethane compounds, and a reduction in dissolved oxygen, which must be managed to ensure the determinands are maintained within the DWI's prescribed concentration values.

The choice of disinfectant can be treatment-works- and network-specific. If the treatment works is sited in a highly populated conurbation, there will be a preference not to use gaseous chlorine as the prime disinfectant due to the inherent health and safety issues. In these cases, on-site electrolytic chlorination (OSEC) or sodium hypochlorite solution may sometimes be utilised. Outside these issues, the impact on water quality and aesthetic implications will be reflected in the choice of disinfectant. In the UK, the normal choice of disinfectants is based around chloramine or chlorine chemistries. Chloramine, in general, is used for larger distribution networks to ensure that the water receives adequate disinfection at the furthest point of the distribution system and/or to minimise disinfection by-product formation. It also has the benefit of having no residual taste and odour at the customer tap. This can be compared with chlorine, which is a more cost-effective and efficient disinfectant for distribution systems normally encountered in the UK. The concentration of chlorine leaving a water treatment works is controlled again to provide adequate disinfection to the furthest point of the distribution network. However, consideration is needed of the amount of chlorine that can be applied at the works and its implications on disinfection by-products (e.g. trihalomethanes and haloacetic acids) and aesthetic parameters. To minimise these effects, the concentration of the free chlorine leaving the works is balanced. Control is needed to not create taste and odour complaints from customers receiving water at the first take-off point away from the works while maintaining disinfection further along the distribution mains. To manage the issue and minimise disinfection by-products, booster chlorination may be considered. These installations will have a corresponding operational impact due to potentially remote location and corresponding costs, and hence are only installed where necessary.

It has to be recognised that, together with discolouration, taste and odour complaints relating to chlorination are the main customer contacts with water companies. The aesthetic effects of any changes are easily detected, and, as such, form a priority focus for operating companies. The worst-case scenario would be adopting a mixture of chloraminated supply with that of a chlorinated system either via a treatment works or booster chlorination, as the inevitable change of disinfectant would be readily detected at the customer tap.

Other countries, including those in continental Europe and America, sometimes utilise chlorine dioxide as their chosen disinfectant, and this has few disinfection by-products and little taste and odour potential if managed correctly. However, it has not been adopted in the UK due to its production costs, increased oxidative power and, therefore, shorter life, especially in older and longer distribution networks, plus the potential for the formation of chlorate in supply.

In some networks, intermittent problems with water quality arise, often associated with sudden changes in flow rate caused by bursts, pumping or valve operations, which lead to unacceptable discolouration, indicated by high turbidity and colour concentrations. In these situations, the water companies will carry out repairs and flushing of the relevant distribution mains to remove accumulated sediments, corrosion products and biofilms. When flushing occurs, the water companies will always try to achieve a higher flow rate than the calculated demand under normal operation, to reduce further discolouration events. Sometimes this is difficult to achieve due to the size of some of the larger mains and disposal routes for the flushing water.

Water quality in the distribution network is monitored by manual sampling and analysis, either in the field by portable test kits for some parameters (e.g. pH, conductivity, dissolved oxygen, residual chlorine), or in the laboratory if the parameters are stable or require more advanced instrumentation, such as pesticide concentrations. Currently, a limited number of commercial in-line sensors are beginning to be developed and are undergoing trials. These are able to detect up to 12 basic physicochemical parameters, including residual chlorine, turbidity, pressure, temperature and conductivity. This type of instrument could provide a valuable audit tool, and in the future may be further deployed to assist with the real-time management of the distribution network if incorporated with a suitable data management and control system.

In the future it is expected that more attention will be given by water companies to instigate a more-integrated management system for the supply of water to the customer. DWSPs and the associated monitoring and control of water quality from source to tap will provide a strong foundation. This will assist in the greater use and inter-mixing of alternative water sources to optimise water resources, plus changes in network operational practices to reduce chemical and energy consumption and improve the supply/demand balance.

REFERENCES

Adgar, A., Cox, C. S. and Böhme, T. J. (2000) Performance improvements at surface water treatment works using ANN-based automation schemes. *Transactions of the Institute of Chemical Engineers, Part A* 78: 1026–1039.

APHA (2005) *Standard Methods for the Examination of Water and Wastewater*, 21st edn. Baltimore, MA: American Public Health Association, American Water Works Association, Water Environment Federation.

Armitage, P. D., Moss, D., Wright, J. F. and Furse, M. T. (1983) The performance of a new biological water score system based on macroinvertebrates over a wide range of unpolluted running-water sites. *Water Research* 17(3): 333–347.

AWWARF (2006) *Real-time Artificial Intelligence Control and Optimisation of a Full-scale WTP*. Denver, OH: AWWA Research Foundation.

Beven, K. and Germann, P. (1982) Macropores and water flow in soils. *Water Resource Research* 18(5): 1311–1325.

Butler, A. P. (2010) Groundwater vulnerability and protection. In: Wheater, H. S., Mathias, S. P. and Li, X. (eds), *Groundwater Modelling in Arid and Semi-arid Areas.* Cambridge: Cambridge University Press.

Butler, A. P., Mathias, S. A., Gallagher, A. J., Peach, D. W. and Williams, A. T. (2009) Analysis of flow processes in fractured chalk under pumped and ambient conditions. *Hydrogeology Journal* 17(8): 1849–1858.

Carroll, Z. L., Bird, S. B., Emmett, B. A., Reynolds, B. and Sinclair, F. L. (2004) Can tree shelterbelts on agricultural land reduce flood risk? *Soil Use and Management* 20: 357–359.

Cave, M. (2009) *Cave Review of Competition and Innovation in Water Markets.* www.defra.gov.uk/enviornment/quality/water/industry/cavereview/ [accessed 01.09.10].

Cerisier, S. D. M. and Smit, J. J. (1996) The electrochemical generation of ferric ions in cooling water as an alternative for ferric chloride dosing to effect flocculation. *Water SA* 22(4): 327–330.

Chun, K. P., Wheater, H. S. and Onof, C. J. (2009) Streamflow estimation for six UK catchments under future climate scenarios. *Hydrology Research* 40(2–3): 96–112.

Chun, K. P., Wheater, H. S. and Onof, C. J. (2010) Projecting and hindcasting potential evaporation for the UK between 1950 and 2099. Unpublished.

Defra (2010) *UK Climate Projections – Reports and Guidance: About the Weather Generator.* http://ukclimateprojections.defra.gov.uk/content/view/858/500/ [accessed 01.09.10].

EC (2009) *Water Framework Directive. Groundwater.* http://ec.europa.eu/environment/water/water-framework/groundwater.html [accessed 01.09.10].

Environment Agency (2009) *Water for People and the Environment: Water Resources Strategy for England and Wales.* London: Environment Agency.

Extence, C. A., Balbi, D. M. and Chadd, R. P. (1999) River flow indexing using British benthic macroinvertebrates: a framework for setting hydroecoglocial objectives. *Regulated Rivers: Research and Management* 15(6): 545–574.

Gash, J. H. C., Wright, I. R. and Lloyd, C. R. (1980) Comparative estimates of interception loss from three coniferous forests in Great Britain. *Journal of Hydrology* 48: 89–106.

Groundwater Forum (1998) *Groundwater – Our Hidden Asset.* www.groundwateruk.org.

Harbaugh, A. W. (2005) MODFLOW-2005, the U.S. Geological Survey modular groundwater mode – the groundwater flow process. *Techniques and Methods 6,* A16. Reston, VA: US Geological Survey.

Hawkes, H. A. (1997) Origin and development of the Biological Monitoring Working Party (BMWP) Score System. *Water Research* 32(3): 964–968.

Hawnt, R. J. E., Joseph, J. B. and Flavin, R. J. (1981) Experience with borehole recharge in the Lee Valley. *Journal Institution of Water Engineers and Scientists* 35(5): 437–451.

Head, R., Shepherd, D., Butt, G. and Buck, G. (2002) OTTER mathematical process simulation of potable water treatment. *Water Science and Technology: Water Supply* 2(1): 95–101.

Hillis, P. (2006) Enhanced coagulation, flocculation and immersed ultrafiltration for treatment of low alkalinity and highly coloured upland water. *Journal of Water Supply: Research and Technology – Aqua* 55(7–8): 549–558.

Holman, I. P., Hollis, J. M., Bramley, M. E. and Thompson, T. R. E. (2003) The contribution of soil structural degradation to catchment flooding: a preliminary investigation of the 2000 floods in England and Wales. *Hydrology and Earth System Sciences* 7: 754–765.

Ireson, A. M., Wheater, H. S., Butler, A. P., Mathias, S. A., Finch, J. and Cooper, J. D. (2006) Hydrological processes in the Chalk unsaturated zone – insights from an intensive field monitoring programme. *Journal of Hydrology* 330: 29–43.

Ireson, A. M., Mathias, S. A., Wheater, H. S., Butler, A. P. and Finch, J. (2009) A model for flow in the Chalk unsaturated zone incorporating progressive weathering. *Journal of Hydrology* 365(3–4): 244–260.

Jans, U. and Hoigné, J. (1998) Activated carbon and carbon black catalyzed transformation of aqueous ozone into OH-radicals. *Ozone Science and Engineering* 20(1): 67–90.

Jiang, C., Li, X.-Z., Graham, N., Jiang, J.-Q. and Ma, J. (2004) The influence of pH on the degradation of phenol and chlorophenols by potassium ferrate. *Chemosphere* 56(10): 946–956.

Jiang, J. and Graham, N. J. D. (1998a) Preparation and characterisation of an optimal polyferric sulphate (PFS) as a coagulant for water treatment. *Journal of Chemical Technology and Biotechnology* 73(4): 351–358.

Jiang, J. and Graham, N. J. D. (1998b) Evaluation of poly-alumino-iron sulphate (PAFS) as a coagulant for water treatment. In: *8th International Gothenburg Symposium on Chemical Treatment*. Prague.

Jiang, J.-Q., Graham, N., André, C., Kelsall, G. H. and Brandon, N. (2002) Laboratory study of electro-coagulation–flotation for water treatment. *Water Research* 36: 4064–4078.

Jones, M. (2010) *Rising Groundwater in Central London. UK Groundwater Forum.* www.groundwateruk.org/Rising_Groundwater_in_Central_London.aspx [accessed 01.09.10].

Kirkby, M. J. (ed.) (1978) *Hillslope Hydrology*. Chichester: Wiley.

Kruithof, J. C., Kamp, P. C. and Martijn, B. J. (2007) UV/H_2O_2 treatment: a practical solution for organic contaminant control and primary disinfection. *Ozone Science and Engineering* 29(4): 273–280.

Kundzewicz, Z. W., Mata, L. J., Arnell, N. W., Döll, P., Kabat, P., Jiménez, B., Miller, K. A., Oki, T., Sen, Z. and Shiklomanov, I. A. (2007) Freshwater resources and their management. In: Parry, M. L., Canziani, O. F., Palutikof, J. P., van der Linden, P. J. and Hanson, C. E. (eds), *Climate Change 2007: Impacts, Adaptation and Vulnerability. Contribution of Working Group II to the Fourth Assessment Report of the Intergovernmental Panel on Climate Change*, pp. 173–210. Cambridge: Cambridge University Press.

Lambert, S. D., Graham, N. J. D. and Croll, B. T. (1996) Degradation of selected herbicides in a lowland surface water by ozone and ozone-hydrogen peroxide. *Ozone Science and Engineering* 18(3): 251–269.

Ma, J. and Graham, N. J. D. (1999) Degradation of atrazine by manganese-catalysed ozonation: influence of humic substances. *Water Research* 33(3): 785–793.

Ma, J., Graham, N. and Li, G. (1997) Effectiveness of permanganate pre-oxidation in enhancing the coagulation of surface waters – laboratory case studies. *Journal of Water Supply: Research and Technology – Aqua* 46(1): 1–10.

Magurran, A. E. (2004) *Measuring Ecological Diversity*. Oxford: Blackwell.

Maier, H. R., Morgan, N. and Chow, C. W. K. (2004) Use of artificial neural networks for predicting optimal alum doses and treated water quality parameters. *Environmental Modelling and Software* 19: 485–494.

Mälzer, H.-J. and Nahrstedt, A. (2002) *Modellierung Mehrstufiger Trinkwasseraufbereitungsanlagen Mittels eines Expetensystem-basierten Simulationsmodells (Metrex) am Beispiel von Oberflächenwasser*. Mülheim: IWW Water Centre.

Marc, V. and Robinson, M. (2007) The long-term water balance (1972–2004) of upland forestry and grassland at Plynlimon, mid-Wales. *Hydrology and Earth System Science* 11(1): 44–60.

O'Shea, M. J., Baxter, K. M. and Charalambous, A. N. (1995) The hydrogeology of the Enfield–Haringey artificial recharge scheme, north London. *Quarterly Journal of Engineering Geology and Hydrogeology* 28(2): S115–S129.

Owen, P. G. (2002) Water-treatment works' sludge management. *Water and Environment Journal* 16(4): 282–285.

Partinoudi, V. and Collins, M. R. (2007) Assessing riverbank filtration removal mechanisms. *Journal of the American Water Works Association* 99(12): 61.

Perry, M. and Hollis, D. (2005) The development of a new set of long-term climate averages for the UK. *International Journal of Climatology* 25(8): 1023–1039.

Redhead, A. (2007) Membrane technology at St. Saviours water treatment works, Guernsey, Channel Islands. *Water and Environment Journal* 22(2): 75–80.

Shorrock, K. and Drage, B. (2006) A pilot plant evaluation of the magnetic ion exchange process for the removal of dissolved organic carbon at Draycote water treatment works. *Water and Environment Journal* 20: 65–70.

Simmons, C. T., Bauer-Gottwein, P., Graf, T., Kinzelbach, W., Kooi, H., Li, L., Post, V., Prommer, H., Therrien, R., Voss, C., Ward, J. and Werner, A. (2010) Variable density flow: from modelling to applications. In: Wheater, H. S., Mathias, S. P. and Li, X. (eds), *Groundwater Modelling in Arid and Semi-arid Areas*. Cambridge: Cambridge University Press.

Simpson, A., Burgess, P. and Coleman, S. J. (2002) The management of potable water treatment sludge: present situation in the UK. *Water and Environment Journal* 16(4): 260–263.

Southwood, T. R. E. and Henderson, P. A. (2000) *Ecological Methods*, 3rd edn. Oxford: Blackwell Science.

Spence, R. and Hickley, P. (2000) The use of PHABSIM in the management of water resources and fisheries in England and Wales. *Ecological Engineering* 16(1): 153–158.

Sumbler, M. G. (1996) *London and the Thames Valley*. London: HMSO.

Taylor, J. S. and Wiesner, M. (1999) Membranes. In: *Water Quality and Treatment: A Handbook of Community Water Supplies*, Ch. 11. Denver, CO: American Water Works Association.

Templeton, M. R., Hofmann, R. and Andrews R. C. (2006) Case study comparisons of computational fluid dynamics (CFD) modeling versus tracer testing for determining clearwell residence times in drinking water treatment. *Journal of Environmental Engineering and Science* 5: 529–536.

UKWIR (1995) *A Methodology for the Determination of Outputs of Groundwater Sites*, 95/WR/01/2. Cambridge: Groundwater Development Consultants.

UKWIR (2000) *A Unified Methodology for the Determination of Deployable Output from Water Sources*, 00/WR/18/1. London: Halcrow.

UNESCO (2009) *Water in a Changing World*. www.unesco.org/water/wwap/wwdr/wwdr3/ [accessed 01.09.10].

Valade, M., Rokjer, D., Peters, R., Keesler, D. and Boryskowsky, M. (2003) Development of alternate means for validation of UV reactors. In: *Proceedings of the American Water Works Association Water Quality Technology Conference*. Philadelphia, PA.

van der Helm, A. W. C. and Rietveld, L. C. (2002) Modelling of drinking water treatment processes within the Stimela environment. *Water Science and Technology: Water Supply* 2(1): 87–93.

WHO (2006) *Guidelines for Drinking Water Quality*, 3rd edn, Vol. 1: *Recommendations*. Geneva: World Health Organisation.

Yuan, B.-L., Li, X.-Z. and Graham, N. (2008) Aqueous oxidation of dimethyl phthalate in a ferrate–TiO_2–UV reaction system. *Water Research* 42: 1413–1420.

REFERENCED LEGISLATION

The Drinking Water Directive. Council Directive 98/83/EC of 3 November 1998 on the quality of water intended for human consumption. *Official Journal of the European Communities* L330: 32–54.

The Water Framework Directive. Directive 2000/60/EC of the European Parliament and of the Council of 23 October 2000 establishing a framework for community action in the field of water policy. *Official Journal of the European Communities* L327: 1–73.

The Water Resources Act 1963. London: HMSO.

The Water Supply (Water Quality) Regulations 2000, SI 2000, No. 3184. London: TSO.

Water Distribution Systems
ISBN: 978-0-7277-4112-7

ICE Publishing: All rights reserved
doi: 10.1680/wds.41127.111

ice
Institution of Civil Engineers

publishing

Chapter 5
Distribution network elements

Tiku Tanyimboh University of Strathclyde, UK
Myles Key South West Water Ltd, Exeter, UK

5.1   Introduction

The water distribution network is an essential element of the service provision to water-using customers. It is too simplistic to think of this service as simply processes (treatment) and pipelines. Beyond the water treatment works the distribution network is a complex system that stores, transports and reconditions the water both from a quality and pressure perspective. The largest asset base of potable water service provider (WSP) is the pipeline network. This chapter is concerned with the main elements of water distribution networks, including pipelines, pumps, valves and service reservoirs. The chapter begins with an overview of the history of water transportation and explains how pipelines are designed and maintained. The importance of valves and fittings for managing pressure and controlling water as it is transported around the distribution network is explained. To complement this, some practical aspects of surge control are also covered. Types of pump and pump selection are explained. WSPs must cope with a wide variety of normal and abnormal operating conditions, such as very large flows for fire-fighting, routine repairs and maintenance and the inherent variability in water consumption. This chapter also includes service reservoirs that play a key role in this respect.

5.2   Pipes
5.2.1   A brief history of water networks and pipe materials

To satisfy our early desires for water it was sufficient for water to be found and used at its source. The Romans were the first civilisation to consider transporting water over distances from the water's source to the populated cities. Ancient Rome built aqueducts to deliver water to the populace, but the aqueducts were open channels exposed to the elements, and the quality of the water was poor by today's standards.

As civilisations developed and the population boomed, the resourcefulness of the human race combined with improved understanding of material properties and advancements in technology saw water transportation systems move from open channels to purpose-made pipelines.

Pipe construction moved through various stages, from the development and use of lead piping in the Roman Empire to cast iron pipes as early as the 14th century to asbestos, steel and plastic products in the 20th century (discussed in greater detail in Chapter 1).

The present chapter discusses pipeline design and material selection in modern water distribution networks.

5.2.2 Pipeline design and material selection

Pipeline material selection, until recently, was relatively straightforward: iron or steel for high-pressure applications, highways, rivers, and other difficult environmental conditions, and plastic pipes for lower-pressure applications, rural networks and housing developments. However, advances in modern plastic pipes with improved pressure ratings and the ability to create continuous plastic pipelines has provided much greater choice and, at the same time, a selection headache for pipeline designers.

The main considerations for a water engineer designing for a new distribution pipeline are:

■ the environment (ground conditions) in which the pipe will be installed
■ the volume of water the new pipeline needs to carry, considering immediate and future demands
■ the static pressure and pressure variation that the pipeline will be subjected to.

5.2.3 Interaction of pipelines with the local environment

When selecting the pipe material, consideration must be given to the natural environment in which the pipe will be laid. Pipes are generally laid at a minimum depth of 750 mm of cover, although some engineers insist on 900 mm of cover being provided. This will generally protect the pipeline from damage by frost penetration and third-party activity (e.g. from a plough blade).

Pipes laid in unmade ground are subjected to forces exerted by the natural movement of the ground. Typically, shrinkage and expansion of the ground will occur in the UK as a consequence of the temperate climate; therefore, cold winter weather can lead to ground movement through rain penetration followed by freezing temperatures, which causes the water to expand as it turns to ice. At the opposite end of the scale, the warmer, summer months cause further ground movement as the soil dries out and shrinks.

Pipelines need to be able to absorb the stresses created by ground movement, otherwise stress fractures will occur. As a consequence, WSPs will incur costs through the leakage detection effort expended in locating the leaks and in dealing with customer contacts and resultant compensation payments due to the supply interruptions caused while the repairs are carried out.

Many pipelines are laid in the highway, which creates different stresses to the pipes due to traffic loadings. Highways, of course, are designed to withstand traffic loadings, and are less influenced by weather-related movement, as the surface/wearing course is designed to carry water away from the sub-structure; however, surface cracking and wear and tear on the road surface can lead to localised water penetration, which will permit expansion and shrinkage to occur.

Pipes laid at normal depth in narrow trenches are afforded support by the surrounding undisturbed ground so that traffic loadings are effectively dispersed. However, if pipes are laid too close to the surface or if extra wide trenches have been excavated, then traffic (surcharge) loading may be significant and could have a detrimental effect on the pipe. The selected pipe bedding material and the effectiveness of the compaction technique used for the pipe layer are significant factors in affording adequate support to any pipeline, but particularly where adverse conditions prevail.

When considering the effects of loadings applied to pipelines, pipes are generally classed as rigid or flexible. With rigid pipes the weight of the soil and any imposed load is transmitted around the pipe wall from the crown to the invert, which, by its nature, resists deformation. In flexible pipes, any imposed load that is not absorbed by the pipe surround can cause the pipe to deform, pushing the pipe wall out horizontally into the surrounding bedding. All pipes have an acceptable degree of deformation, buckling pressure or ovalisation which they are able to withstand before joint leaks, stress fractures or lining cracks occur.

In non-trafficked areas, the weight imposed on a pipeline (dead load) is fairly constant, varying only through changes in moisture or air content of the soil, and can largely be ignored. In trafficked areas, it is important to understand the weight of traffic that can be applied to a buried pipeline and to ensure that the applied traffic weight together with the overburden weight of the soil does not exceed the capacity of the pipeline to resist the resultant wall deformation.

The following formula adapted by Saint Gobain Pipelines,[a] is modified from Spangler's Iowa formula, and can be used to calculate the deformation (ovalisation) percentage for ductile pipes in order to ensure the design capacity of the pipe material is not exceeded. The acceptable percentage of ovalisation can be determined for a given pipe material and size by reference to the manufacturers' technical details:

$$\Delta = \frac{100K(\mathrm{Pe} + \mathrm{Pt})}{8S + 0.061E^1/\mathrm{DLF}} \qquad (5.1)$$

where Δ = the ovalisation (%)
K = the bedding coefficient
Pe = the earth load ($\mathrm{kN/m^2}$)
Pt = the traffic load ($\mathrm{kN/m^2}$)
S = the pipe diametral stiffness ($\mathrm{kN/m^2}$)
E^1 = the modulus of the soil reaction ($\mathrm{kN/m^2}$)
DLF = the deflection lag factor

The bedding coefficient K reflects the angle of support at the pipe invert and the quality of the bedding and sidefill material used in the pipe trench. The DLF is influenced by the

---

[a] Reproduced with permission.

nature of the native soil, side-fill and the working pressure of the pipeline; it is assumed that for pressure pipelines the DLF = 1.

In addition to stresses through ground movement, material selection also needs to be considered in light of the natural soil type, which will influence the longevity of the pipeline through chemical or electrolytic reaction with the pipe wall. This consideration extends beyond the natural environment, as, increasingly, pipelines are being laid to support new housing and industrial developments in land which has been subjected to previous use, typically referred to as brownfield sites. In circumstances where the previous use has been for heavy industry, fuel stations and other instances where petroleum products have been used, the ground is often contaminated, and pipelines need to be able to provide a barrier against penetration from the ground contaminant.

There is no single product solution for any of the conditions described in the preceding paragraphs. The pipeline designer needs to ensure that the selected product is able to provide the expected service life considering the external stresses created by the environment.

5.2.4 Pipe selection (size and pressure rating)

The next consideration is the size (diameter) of the pipe that is selected. Distribution networks are typically categorised into trunk mains, distribution mains and service pipes.

5.2.4.1 Trunk main

A trunk main is classified as a large-diameter main (nominal $\geqslant 300$ mm ID) which is used for transferring bulk supplies of water usually between treatment works and service reservoirs but also between one service reservoir and another.

5.2.4.2 Distribution main

Distribution mains generally supply water into communities and are supplied mainly from service reservoirs but are sometimes linked directly from trunk mains. Distribution mains typically range in size from 80 mm to 300 mm in diameter, but regional WSPs will have examples of distribution mains that fall outside of this typical size banding.

5.2.4.3 Service pipes

Finally, service pipes connect customers to distribution mains, and range in size from 15 mm to 50 mm, but, as with distribution mains, there will be examples where supply pipe diameters extend beyond the 50 mm size range.

It is important to remember that demands on networks will often change through the lifetime of the pipeline. This is particularly true for the larger-diameter pipes simply because there is greater scope for influence due to the variety of networks that are supported downstream. Pipeline designers must therefore consider potential future growth and allow headroom in their design to enable the downstream network demands to grow before the pipeline needs to be reinforced. The headroom allowance could be based on known future growth, which can be established through discussion with local planning authorities. Pipeline design is discussed in greater detail in Chapter 7, 'Design of water distribution systems'.

5.2.5   Protection systems

The ground conditions in which a pipeline is laid influence the longevity of the pipe, as the soil might be aggressive and cause external damage to the pipe if it is not adequately protected. Similarly, the water flowing within the pipe will have the potential to cause internal degradation. Knowledge of the previous occupancy of brownfield sites is also essential, as this can influence a pipeline designer's material selection, since a barrier product might be required to avoid any potential permeation of ground contaminants through the pipe wall.

5.2.5.1   External protection

With increasing awareness of our carbon footprint and with landfill costs rising, WSPs are encouraged to re-use excavated material as backfill wherever possible. In addition to the landfill cost, importing selective backfill material is also expensive, and even though non-aggressive material has been utilised to surround the pipe, waterlogging of the ground will still create the potential for corrosion to occur to iron or steel pipelines. Designers are therefore left with the option to choose the best pipe to suit the prevailing environmental conditions.

Historically, grey iron pipes in either the spun or cast form were laid with little or no external protection, and relied mainly on the wall thickness to provide longevity. In corrosive soils with low resistivity, pipes would often fail due to localised pitting causing pinhole-sized breaks to occur, which, if not located and repaired quickly, would erode further to create larger holes and consequential leakage losses.

When ductile iron pipes were introduced in the UK water industry in the 1960s, a crude form of loose polythene wrapping was employed to protect the pipeline. Although the theory was sound, in reality the wrapping was easily damaged and provided minimal protection to the pipeline. Close-fitting polythene wrapping was added later to the pipes, but they still suffered from damage during transportation, handling and installation activities.

External protection systems on ductile pipes are now highly developed, with the addition of zinc, bitumen and epoxy coatings in various combinations to suit the ground conditions in which the pipeline is laid.

Steel pipes from their inception in the UK water industry have invariably been protected by way of cathodic protection systems. Metal structures are prone to damage through electrolytic action from the surrounding soil and water, which will interact with the steel pipe through imperfections in the external protective coating, commonly referred to as 'holidays'. Even though every precaution is taken by the installers to avoid damage to the protective coating, sharp objects in the pipe surround or 'holidays' that occur during the manufacturing process mean that the pipeline requires additional protection in order to achieve its anticipated design life.

115

Cathodic protection works by creating a sacrificial anode either by using an impressed current (rectifier) or natural anodes such as magnesium or zinc. Current flows from the protection system (anode) through the natural electrolyte (soil) to the steel pipeline. This must be sufficient to reduce the electrical potential of the metal in the pipe below its corrosion potential. Cathodic protection systems need to provide protection to all parts of the pipeline, and must be regularly checked and maintained to ensure sufficient potential remains to protect the pipeline.

Theoretically, cathodic protection systems could be used to protect iron pipelines; however, individual mechanical joints would need to be bonded across to ensure the continuity of the current along the pipeline. This would add significant cost to the pipe-laying project, and would therefore render this process prohibitive in most cases.

Steel pipe manufacturers, like their ductile iron competitors, have refined and developed their protective coating systems to an extent where some now claim that additional external protection is no longer necessary. However, the potential for damage to occur to the coating during the pipe-laying process would persuade most pipeline designers to include the cost of a cathodic protection system in the scheme budget.

5.2.5.2 Joint and fittings wrapping
The majority of pipelines will require that fittings such as tees and valves are added to provide the necessary control and management of the pipeline for the asset manager. Inevitably, the addition of a fitting requires the pipe to be cut, which breaks the continuity of the pipeline and interferes with the protective coating. In order to protect the integrity of the external protection system, joints created through this process need to be wrapped using an oil-based hemp or polyethylene shrink wrap in the case of ductile iron and/or over-bonded in the case of steel. Failure to do this adequately can create corrosion points around the joints, which, over time, can result in a reduction in the mechanical strength of the pipeline, leading to failure through bursting.

5.2.5.3 Barrier systems
As development land becomes increasingly sparse in the UK, developers inevitably look towards brownfield sites to meet the housing needs for a burgeoning population. In many cases, the previous use of the land has left a legacy of 'contaminated' ground; this is particularly true of former garage forecourts and industrial sites.

Developers are presented with the option of 'remediating' the land to remove the contaminant, which is often a prohibitively expensive option, or asking WSPs to install barrier pipe products to eliminate the risk of the contaminant affecting water supplies to the new houses. Invariably, housing developments are serviced using smaller-diameter polyethylene (PE) or polyvinyl chloride (PVC) water pipes for both mains and services. Chemicals contained in the contaminant may have the potential to permeate through the wall of a plastic pipe, imparting a taste to the water and affecting the life expectancy of the pipe.

Plastic pipe manufacturers have risen to the challenge, and many now produce a barrier pipe system. Pipes need to provide a barrier to prevent the potential contaminant from reaching the pipe bore in order to protect the water supply. This is achieved by incorporating an aluminium layer; additionally, an outer protection layer is required to maintain the long-term strength of the pipe.

In order to maintain the integrity of the barrier system, joints need to be protected either by incorporating barrier fittings or by wrapping with overlapped aluminium tape, which is then wrapped again with an oil-based waterproof tape.

5.2.5.4 Internal protection

Iron and steel pipelines were historically laid with either crude (at best) or non-existent internal linings. Many WSPs have seen significant elements of their funding programmes expended in relining grey iron mains that have corroded internally, causing dramatic reductions in internal bore and discoloured water problems.

Developments in lining technologies have given rehabilitation engineers a range of products to consider for developing a solution to extend the life of pipelines and for providing improved water quality to customers. A spin-off benefit of mains rehabilitation is the recovery of headroom in the network and the resultant hydraulic improvements. There are many examples of water-pumping stations being rendered redundant as a consequence of mains rehabilitation projects.

Epoxy resin, PE, including high-build PE, and close-fit non-structural and semi-structural linings are some of the alternatives that have been used to rehabilitate water mains. High-build and semi-structural products can be used where the host main has lost some of its mechanical integrity, and therefore provide an alternative to mains replacement in a limited number of cases.

The decision to reline rather than renew is complex, and is covered in greater detail in Chapter 10, 'Finance and project appraisal'. Relining is much less disruptive to customers and to the environment than replacement and is far less expensive.

Modern ductile iron and steel pipes are supplied from the manufacturer with applied internal linings to protect the pipeline from internal corrosion. Cement mortar linings were used from the 1970s, with later additions of epoxy coatings to control pH reaction and extend residence times of water in transit.

5.2.6 Pipe restraint

In an earlier section pipe material selection was discussed with due consideration to ground movement. In pipelines where socket and spigot joints are utilised, it is necessary to consider the need to provide thrust restraint at the points where bends, blank ends, tees and, to a lesser extent, tapers are incorporated into the pipeline. This is required to prevent the pipe from moving due to the pressure exerted on the pipe at these points of deviation by both static and dynamic thrusts acting within the pipeline.

Traditionally, this is achieved by the provision of a thrust or anchor block, which is designed to provide a restraining force to oppose the direction in which the thrust is acting on the pipe. The size and shape of the block required to overcome the thrust is dependent on the pressure acting on the pipe, the direction in which the pressure is applied and the bearing pressure of the ground in which the block will be constructed.

Designers should also consider the possibility that flow can be reversed in certain pipelines, for example a flow and return main between a pump station and a service reservoir. In such cases, the direction of thrust may change, and the thrust block must be designed to restrain the thrust forces in both directions. Consideration must also be given to the accessibility of the joint, so that repairs can be made if necessary without the need to destroy the anchor block.

There are alternative methods of restraining bends such as tie-bars, self-anchoring joints, stub flanges and end-restrained fittings, but these forms of anchorage often serve to move the potential point of separation to another nearby joint. They should therefore be considered as a complementary form of restraint rather than a substitute for a well-designed thrust block.

There are two types of thrust exerted on a pipeline: static and dynamic. Both types act in the same direction within a pipeline, but dynamic thrust, which occurs under high velocity, can create an additional burden at direction changes. The following formula, adapted by Saint Gobain Pipelines,[b] can be used to calculate the combined static and dynamic thrust at bends:

$$T = (P + 0.01v^2)10^2 A_e \times 2 \sin \theta/2 \qquad (5.2)$$

where T = the thrust force acting on bend (kN)
P = the internal pressure (bar)
v = the velocity of water (m/s)
A_e = the cross-sectional area of the pipe from its external diameter (m^2)
θ = the angle of the bend (°)

Once the thrust acting on the bend is known, then a thrust block can be designed to restrain the force (Figure 5.1). Thrust block design is complex, and expert advice should be sought in order to ensure that pipelines are properly restrained. Disregarding this need can result in catastrophic failures of pipelines, disruption to customers and expensive compensation and repair bills to WSPs.

5.2.7 Pipe jointing

Most pipe materials have a selection of jointing techniques that can be applied. This short section describes the techniques that can be used for joining pipes and pipe fittings on the most commonly used pipe materials.

---

[b] Reproduced with permission.

Figure 5.1 The direction of thrust acting on a selection of bends and fittings and the typical location of thrust blocks installed to resist the thrust. Arrows represent the direction of thrust

Reproduced with permission. © Saint Gobain PAM UK, www.saint-gobain-pam.co.uk

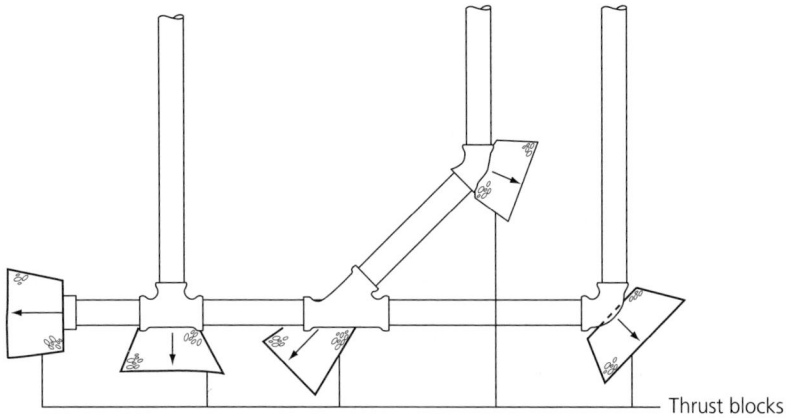

Thrust blocks

5.2.7.1 Ductile iron pipes

Socket and spigot push-fit joint – a male (spigot) to female (socket) flexible joint which uses a rubber ring gasket to provide the leak-tight seal. The joint provides (typically) 5° angular deflection to accommodate some ground movement and to aid installation where consecutive pipes are not perfectly aligned. It can be joined with an anchor gasket, to provide a restrained joint.

Restrained flexible weld bead joint – a socket and spigot joint but with the addition of a bead welded to the spigot end and a locking ring which locates adjacent to the bead, to stop the joint from pulling apart. It is typically used for high-pressure applications or where the use of thrust blocks is impractical.

Tied joint – a socket and spigot joint with the addition of a flange that is welded to the spigot end of the pipe. A second loose flange is located behind the socket and tie bars between the two flanges, and ties the joint together to prevent separation (Figure 5.2).

Figure 5.2 Example of a tied socket and spigot joint

Reproduced with permission. © Electrosteel Castings (UK) Ltd, www.electrosteel.co.uk

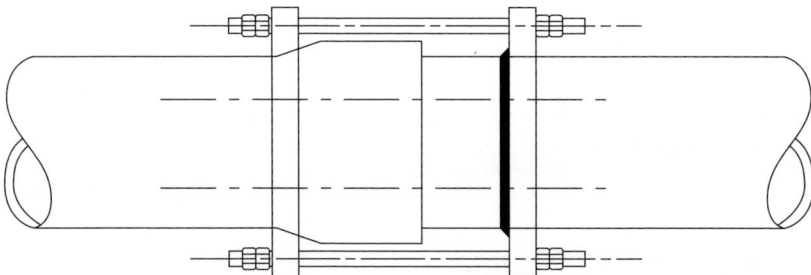

Flanged joint – generally used for joining pipes to fittings, this is a fully rigid joint connecting flange to flange with nuts, bolts and flange gaskets to provide the water-tight seal. The flanges carry a range of pressure ratings, typically 16, 25 or 40 bar.

General – historically, iron pipes have been joined using other techniques that are largely redundant in modern distribution networks but will still be present for old spun and cast grey iron pipelines. These include run lead, bolted gland, victaulic coupling and dresser coupling joints.

5.2.7.2   PE pipes

Butt-fused joint – butt fusion uses an electrically heated plate against which two pipe ends are drawn together simultaneously. When the pipe ends are sufficiently heated, the plate is withdrawn and the pipe ends are brought together at a predetermined pressure until the PE has cooled to form a continuous pipe. The resultant 'joint' will be end-load resistant and possess similar properties to the parent pipes.

Electrofused joint – pipe ends are pushed together within an electrofusion coupler after appropriate preparation of the pipe ends and pipe wall. The coupler and the pipe ends are then fused together by use of an electric current, to form a continuous pipe. A correctly fused joint has an equivalent pressure rating to the parent pipes that have been fused together. Electrofusion has an added advantage over butt fusion in that pipes of different diameters can be fused together.

Puddle flange – where end loads will be exerted on a PE pipeline, such as at line valve installations, puddle flanges are often used to absorb the force exerted on the valve gate when the valve is closed. The fitting is installed in the line, connected to the valve. The puddle flange is usually encased in concrete, so that forces created by virtue of closing the valve are transferred through the flange to the concrete thrust block and the surrounding soil (Figure 5.3).

Stub flange joints – an alternative method of joining PE pipes to metal fittings, the stub flange is a drilled flange which is formed on one end of the PE pipe. Used with a backing

Figure 5.3 Example of a puddle flange

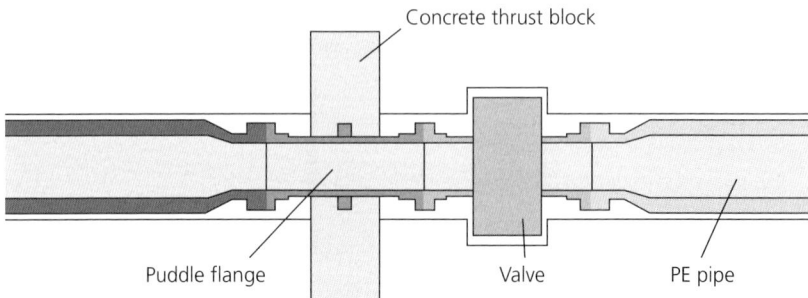

Concrete thrust block

Puddle flange   Valve   PE pipe

ring, the flanged end of the pipe is joined to the flanged fitting using standard flange nuts and bolts.

General – standard mechanical compression fittings such as couplings, adaptors and ferrule straps can also be used with PE pipes. PE is therefore a versatile product in terms of the jointing solutions that can be considered for use with this pipeline material.

5.2.7.3   PVC pipes

Push-fit mechanical joint – a male (spigot) to female (socket) flexible joint which uses a rubber ring gasket to provide the leak-tight seal. The joint provides a small degree of angular deflection to accommodate some ground movement and to aid installation where consecutive pipes are not perfectly aligned. This is the most commonly used joint for PVC pipe systems.

Straight coupler – a mechanical compression fitting that is designed to join the plain or cut ends of pipes together using an internal sealing gasket, to provide a leak-tight seal.

Stepped coupler – based on the same sealing principles as the straight coupler but designed to provide transition between pipes of different outside diameters and materials, and therefore permits the connection of PVC to ductile iron, cast iron or asbestos cement.

Flange adaptor – based on the same sealing principles as the straight coupler but will join the plain end of a PVC pipe to flanged fittings such as tees and sluice valves. It therefore enables PVC pipelines to simply incorporate control devices for network management.

5.2.8   Modes of pipeline failure

If all of the preceding considerations are properly accounted for in pipeline design and installation, then network managers should have every confidence that their network is fit for purpose. However, inevitably faults will occur due to poorly laid pipes, insufficient thrust restraint or inadequate external protection systems or as a result of third-party activities. The resulting damage will need to be located and repaired.

Failures can occur in a number of ways, and the following sections describe modes of failure and repair techniques.

5.2.8.1   Joint failure

Flanged joints and socketed joints on push-fit pipes both require a rubber gasket to provide a water-tight seal. It is important that the correct gasket is used for the application and that the gasket is fitted correctly. Equally important is the making of the joint, so that bolts are tightened to the correct torque and in the right sequence in the case of a flange, and that pipe ends (spigot) are pushed home correctly in the case of a push-fit joint.

Failure to do this correctly can result in the gasket being forced out of the joint and a leak occurring at this location. If this type of failure is identified quickly, the repair may be simply made by breaking and remaking the joint. However, if the leak is left undetected for a period of time, then material can be lost from the pipe wall or the face of the flange, which may require that an encapsulation fitting is needed to completely surround the joint. Alternatively, the joint might have to be cut out and replaced using flange adaptors and couplings combined with a cut section of pipe or using welded couplers in the case of plastic pipes.

5.2.8.2 Circumferential pipe breaks

A common mode of failure, particularly in iron pipes, is a fracture around the circumference of the pipe. Often caused by ground movement, the pipe wall is subjected to stresses which cause the pipe to fracture.

In most cases, this can be rectified by using a repair collar or coupling, but if the fracture requires significant cut-back to create an even face on both sections of pipe, then two couplings and a section of cut pipe may be required to remake the joint. If the fractured pipe has deflected significantly through the course of the fracture, then repair gangs must always check the integrity of any nearby joints as part of their repair works.

Early (1980s) medium-density PE (MDPE), or PE80, pipes were prone to rapid crack propagation (RCP), particularly when the crack initiated on the external wall of the pipe. In instances where external impact damage was caused to an MDPE pipe under pressure, the initial damage would propagate axially at extremely high speed. Larger-diameter MDPE pipes were particularly prone to RCP, and, as a consequence, MDPE pipes, DN250 and greater, were de-rated. The PE pipe industry has, to a large extent, resolved the problem in its modern PE80 pipes and a Water Industry Standard (WIS) specification has been introduced to enable manufacturers to demonstrate that their modern PE pipes meet the necessary toughness and fatigue standards. In addition, a high-performance PE (HPPE), or PE100, pipe has been introduced to the water sector which is not de-rated for RCP.

A UK Water Industry Information and Guidance Note, IGN 4-32-18, provides guidance on the choice of pressure classifications or ratings for buried and above-ground PE pipe systems (UK Water Industry, 2003).

5.2.8.3 Longitudinal pipe breaks

Longitudinal fractures occur along the length of a pipe, and can be caused by a number of factors, including ground movement, pressure surges and excessive mains pressure. By their nature they are expensive to repair, since more ground has to be excavated to expose the extents of the fracture, and in the case of uPVC (rigid PVC) and asbestos cement pipes usually result in a full pipe length being replaced from joint to joint.

Repairs are invariably completed by using a new length of pipe and suitable couplings, although it is often necessary to cut off the socket end of the adjacent section of main in order to make the repair.

Figure 5.4 Example of pinholes

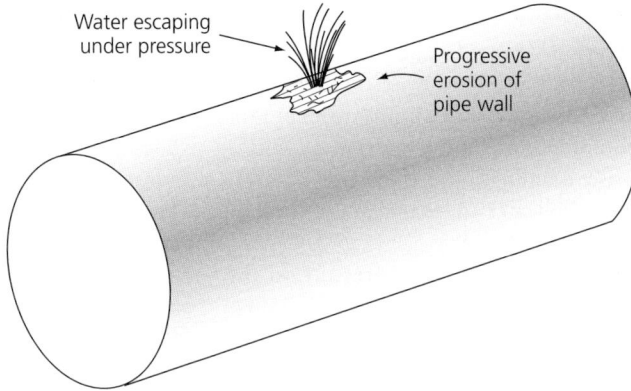

5.2.8.4 Pinholes

As the term suggests, pinholes are small breaks in a pipe wall, usually caused when pipe walls are damaged or pitted due to corrosion so that the mechanical strength of the pipe is reduced. Pinholes can develop into larger breaks if they are not located and repaired quickly, as the escaping water will quickly erode the pipe wall around the hole as it escapes under pressure (Figure 5.4).

Repairs to pinholes are usually accomplished by fitting a split repair collar around the pipe to cover the hole. Historically, experienced repair crews would often plug the hole first by using a hazel twig or section of broom handle to stop the leak, in order that the collar can be fitted.

5.2.8.5 Complications

Water main breaks will not always occur in convenient locations; therefore, repair techniques for the types of failures covered previously may have to change because of the local environment or due to inaccessible sites. Water mains laid at significant depth or buried under buildings, in river beds or under major highways, breaks on encased bends or special fabricated fittings may test the inventiveness of the maintenance engineer. In many of these examples, it is often easier to re-lay the main in order to bypass the break, but, of course, that is not always possible.

It is wise for a duplicated main to be incorporated in the design of a water network that needs to cross a major road or river, or for pipes to be laid in ducts or tunnels so that pipe work can be withdrawn and replaced in the event of failure. In the case of special fittings, often the only solution is to weld the fitting in order to carry out a repair or to replace it like for like.

5.3   Pumps

Pumps play a vital role in water supply systems. They provide or maintain the energy needed to carry water from the treatment works to the service reservoir within the distribution system and to the distribution mains, and ensure the required amount of water can be delivered with sufficient pressure. By law, WSPs in England and Wales must ensure that there is enough pressure to supply the top floor of a house under conditions of normal demand. The stipulated criteria or reference level of service – known as DG2 – are a pressure head of 10 m of water at the boundary stop tap, at a flow of 9 l/min. WSPs are also required to maintain a minimum static pressure of 7 m of water in the communication pipe that supplies a property. These criteria do not apply to high-rise buildings, for there are different requirements; for example, it is common to pump the water to a tank in the roof, which then supplies the building.

Pumps operate primarily by adding pressure to the water to achieve a static lift and overcome the head losses that occur in the system. The main cause of head loss in distribution networks is pipe friction. However, other, minor, losses can occur at pipe junctions, bends and other locations where there are fittings such as valves or discontinuities. The major types of pump include the rotodynamic and positive displacement types. Positive displacement pumps are not normally used for water supply purposes these days. They are often of the reciprocating type, and can be found at water and wastewater treatment works. Water is drawn into a cylinder through an inlet valve and released on the forward stroke of the piston through an outlet valve.

A rotodynamic pump imparts energy to the flow from the motion of a rotating component called an impeller at the centre of the pump. The volute is the casing that surrounds the impeller. The impeller blades are mounted on a shaft that is typically driven by an electric motor. Water enters the pump axially. The discharge from the pump can be tangential at the circumference of the casing or axial along the shaft. The pumps are thus referred to as centrifugal and axial flow, respectively. In mixed-flow pumps, the discharge direction is intermediate between axial flow and centrifugal.

In a centrifugal pump, the water entering the pump is forced outwards by the rotating impeller. Therefore, the water gains speed and moves into the volute, where it slows down and so acquires extra pressure. The volute is snail shaped, and collects and directs the flow to the outlet of the pump. Centrifugal pumps are versatile in that impeller design variations enable them to operate over a wide range in terms of the pressure head added and the volumetric flow rate, and they are thus widely used. Centrifugal pumps are suitable for pumping relatively small volumes of water against high heads. Multi-stage pumps, an arrangement with multiple identical impellers in series on a single shaft and driven by the same motor and in which water flows through the impellers sequentially, can be used to achieve even greater lifts.

The impeller of an axial flow pump operates in a similar fashion to a propeller, in that the rotation of the impeller drives the water forward in the axial direction. Some of the increase in kinetic energy thus achieved is converted into a pressure increase. Axial flow pumps have a limited suction lift, and should be submerged. They are utilised in situations in which the volumetric flow is large and the head required of the pump is low.

Mixed-flow pumps combine both radial and axial flow. Their range of operation in terms of flows and heads is intermediate between centrifugal and axial flow pumps.

The efficiency of a pump is the ratio of the power generated by the pump to the power supplied to the pump. The power generated is given by

$$P = \rho g Q H \tag{5.3}$$

where P = the power
ρ = the density of the fluid
g = the acceleration due to gravity
Q = the volume flow rate
H = the difference in total head between the inlet and outlet of the pump

5.3.1 Performance characteristics
The head generated by a pump depends on the flow rate, and, for a given pump, the relationship between the head and the flow defines its performance characteristic, which can be shown graphically as a plot of the head against the discharge. Along with the head and discharge, the power required to operate a pump and its efficiency are primary considerations when selecting a pump to perform a particular duty. Self-evidently, it is desirable to operate a pump at its maximum efficiency, and as these parameters are all related to the discharge, it is advantageous to superimpose the plots, showing their variations, as illustrated in Figure 5.5. The graphs in Figure 5.5 show the performance characteristics of the pump, and are thus called performance or

Figure 5.5 Characteristics curves for a centrifugal pump

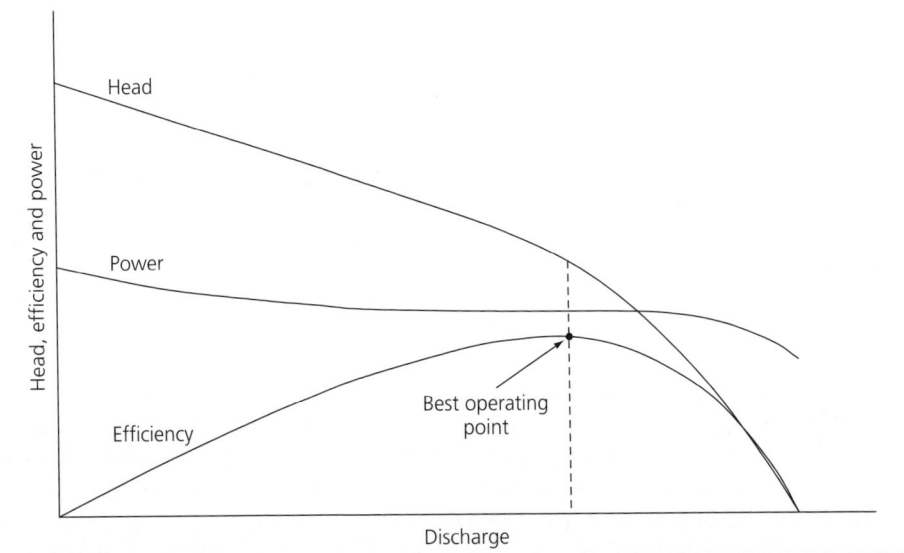

characteristics curves. The discharge corresponding to the maximum efficiency is referred to as the rated discharge. This is also called the design point, often quoted as the head and discharge at the point of maximum efficiency. It should be noted that the performance characteristics depend on the speed of the pump (as shown in the next section). Also, the shapes of the curves depend on the type of pump (e.g. see Potter and Wiggert, 2010).

For a pipeline or pipe network, the head to be added by a pump increases as the flow rate increases, because the pipe friction and other losses increase with the flow rate. The operating point is the point of intersection between the head–discharge curve of the pump and that of the pipe network (the system curve), as explained in the next section. In a design situation, a key objective when selecting a pump is to make sure that the operating point is as close as possible to the maximum efficiency or rated discharge of the pump.

Pumps may be operated in series or in parallel. When operated in series, the water flows through each of the pumps in turn, and thus the head generated by the pumps collectively is the sum of the heads developed by each of the pumps. The head–discharge curve for such a system is obtained by summing the heads for a given discharge. As the flow through the pumps is the same, pumps that operate in series should be identical or similar, to reduce the possibility of one or more pumps operating outside the normal operating range. For pumps that operate in parallel, the heads added by the pumps are identical, while the overall discharge is the sum of the individual discharges. Accordingly, the head–discharge curve for the system is obtained by adding together the discharges for a given head. As the head across the pumps must be the same, pumps operated in parallel should be identical or similar, to reduce the possibility of one or more pumps operating outside the normal operating range.

Cavitation is an important design consideration for pumps. It is characterised by the formation and collapse of bubbles, and causes vibration and noise, reduces the efficiency of operation and damages impeller blades by erosion. Cavitation occurs when the pressure is less than the vapour pressure, causing air bubbles to form, typically on the suction side of the impeller, where the pressure is lowest. The bubbles then collapse as they encounter a region of high pressure within the pump. Cavitation can be prevented by ensuring that the pressure within the pump is kept above the vapour pressure.

The pressure above which cavitation will not occur is termed the required net positive suction head (NPSH), and corresponds to the minimum pressure that will raise water from the sump to the impeller. It depends on the flow rate and speed of the pump, and is provided by the manufacturer. The NPSH is given by

$$\text{NPSH} = p/\rho g + v^2/2g - p_v/\rho g \tag{5.4}$$

where p = the pressure at the pump inlet
v = the velocity at the pump inlet
p_v = the vapour pressure

In terms of heads,

$$p/\rho g = p_{atm}/\rho g - H_s - h_l \qquad (5.5)$$

where H_s = the static lift in the suction pipe

h_l = the head loss (i.e. pipe friction and minor losses) in the suction pipe

p_{atm} = the atmospheric pressure

To avoid cavitation, the actual NPSH should exceed the required NPSH.

5.3.2   Pump selection

When selecting a pump, one of the tasks accomplished early on in the process is to identify the type of pump that would be most suitable. This can be done with the help of the specific speed. The specific speed is defined as the rotational speed (rpm) required to develop a unit head (1 m) for a unit discharge (1 m^3/s), i.e.

$$N_s = NQ^{1/2}/(gH)^{3/4} \qquad (5.6)$$

where N_s = the specific speed (rpm)

N = the impeller speed (rpm)

Q = the discharge at the point of maximum efficiency: the duty point (m^3/s)

H = the head at the duty point (m)

Great care is required with regard to the specific speed, as g in the denominator is often omitted and consistent units are not always adhered to.

Centrifugal pumps work best for specific speeds between approximately 2 and 16 rpm; for mixed-flow pumps the range is approximately 7–29 rpm; and for axial flow the range is about 27–76 rpm (Twort et al., 2000).

For a given pump, the effect of changing the rotational speed can be investigated using the following relationships, often referred to as affinity laws. The affinity laws can also be used to compare two geometrically similar pumps.

$$Q_1/Q_2 = N_1/N_2$$
$$H_1/H_2 = (N_1/N_2)^2 \qquad (5.7)$$
$$P_1/P_2 = (N_1/N_2)^3$$

where Q = the discharge

H = the head developed

P = the power required

Pump selection and sizing go hand in hand with the design of the delivery pipe, as the head H to be generated by the pump consists of the total static lift (i.e. the difference between the delivery and suction levels) and the pipe friction and minor head losses:

$$H = H_s + h_l \tag{5.8}$$

$$h_l = (\lambda L/D + \sum K)(v^2/2g) \tag{5.9}$$

where H_s = the total static lift
  h_l = the head loss
  λ = the friction factor
  L = the pipe length
  D = the pipe diameter
  K = the minor loss coefficient: the summation includes all the minor losses
  including bends, fittings, etc.
  v = the velocity

A plot of H against the discharge yields the performance curve for the piping system, which, when superimposed on the performance characteristics curves for the pump, identifies the operating point as the intersection of the head–discharge curves of the system and pump (Figure 5.6). The objective is to match the rated discharge (the duty point) and the operating point as closely as possible, to ensure that the pump operates at the highest efficiency possible and to minimise the power required.

Alternative solutions based on different pump and delivery pipe systems may be arrived at in this way, and the one with the smallest overall cost (i.e. capital and operating) selected.

Figure 5.6 Pump characteristics and system curves. Option 1 is the preferred solution for the pump shown

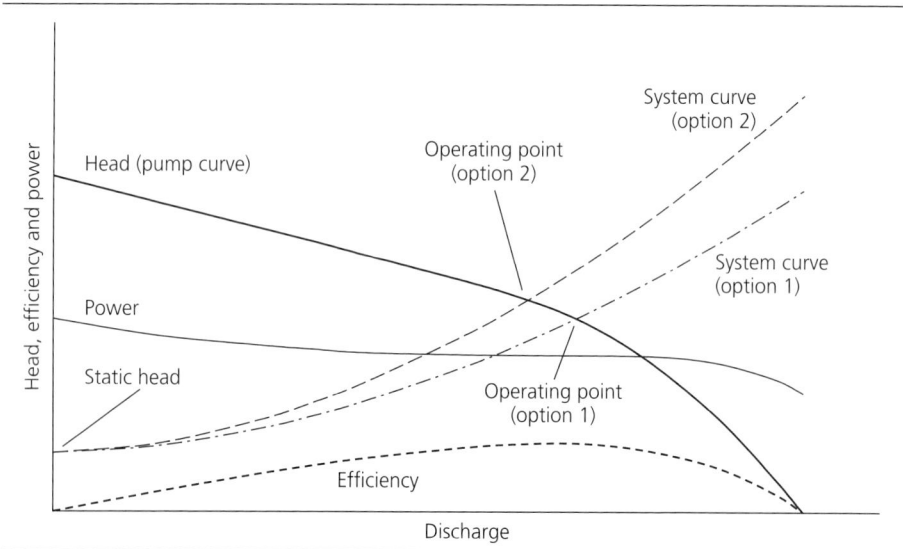

Pumps in series or parallel may be adopted instead of a single pump. This may be because the design constraints cannot be met by a single pump or due to operational reasons. Pumps in series can achieve a higher head and offer greater flexibility in terms of the head developed. Pumps in parallel can achieve a larger discharge and offer flexibility in terms of the discharge because the number of pumps deployed can be varied, since each pump can operate independently. In any case, adequate standby capacity must be provided to ensure continuity of supply in the event of a malfunction and to allow for routine maintenance.

5.4 Valves
5.4.1 Control and operability of distribution networks
Earlier sections of this chapter included information on pipe selection and the importance of installing the right pipe in terms of size and material. In essence, the role of pipes (water mains) is to ensure that the volume of water within the distribution network is suitable to meet the needs of the customers that will use it. Having established that the network is capable of distributing the correct volume of water, it is then necessary to provide the means by which to manage the network for both planned and unplanned maintenance and to take account of the types of conditions that will occur naturally within a hydraulic network.

Modern control rooms are able to interact with the distribution system to open and close valves and to start and stop pumps to balance demands across the network. To do this with confidence, key installations are enabled to feed back information through supervisory control and data acquisition (SCADA) networks, so flow, pressure and reservoir level data are visible 24 hours a day. SCADA and systems operations are discussed in greater detail in Chapter 8, 'Operation, maintenance and performance'.

This section will deal with the control valves that are present in a distribution network that provide its functionality and operability. It will also cover valves that are inserted into the network to permit distribution engineers and managers to interact with the network to create changes to the operating conditions.

5.4.2 Line valves
'Line valve' is a generic term that covers a variety of valves which have a primary purpose of changing the flow volume within a pipeline. This can vary from the fullest extents of operation, i.e. fully open to fully closed, or to a controlled, sometimes referred to as a 'throttled', position somewhere between fully open and fully closed.

The following sections describe the various types of valve in common use in the water industry, together with an explanation of their functionality.

5.4.2.1 Gate or sluice valve
The purpose of the gate valve is merely to stop the flow of water in a pipeline, usually to permit maintenance activities to be undertaken. They are also used to separate areas of different pressure or to create water quality zones and meter areas. The valve consists of a gate that moves vertically within a guide rail by operation of a screw thread. The screw

Figure 5.7 Gate valve

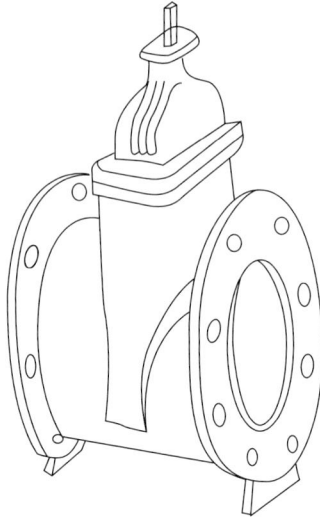

thread connects to a stem which can be specified as either rising or non-rising, depending on its intended use. Typically, gate valves are offered with either hard or soft faces (sometimes referred to as resilient faced) to suit the preference of the water undertaking or to suit specific conditions. A gate valve is shown in Figure 5.7.

In high-pressure applications on larger-diameter mains, gate valves are often difficult to open from a fully closed position because of the pressure differential across the gate. To assist with this difficulty, larger gate valves can be purchased with a smaller-diameter integral bypass which incorporates a small gate valve that can be operated to equalise pressure on either side of the larger gate, to permit the main valve to be opened.

5.4.2.2   Butterfly valves

The butterfly valve comprises a disc that operates on a horizontal pivot around the centre (or offset from the centre) of the valve bore (Figure 5.8). The valve operates to any position between fully closed, perpendicular to the water flow, and fully open, parallel with the water flow, through a quarter turn. Like the gate valve, it is offered by most manufacturers with a metal-faced or resilient-faced option. Butterfly valves are typically used as flow control devices, and can be operated by hand or by motorised actuation.

Butterfly valves are often used in water treatment works where flows need to be regularly changed, and can be used as service reservoir inlet control valves linked with telemetry networks. While the butterfly valve offers excellent flow controllability, it is sometimes dismissed by distribution engineers because the pivot and gate occupy the pipeline even in the fully open position. This inevitably creates a head loss within the pipeline and restricts the passage of a swab during mains cleaning operations.

Figure 5.8 Butterfly valve

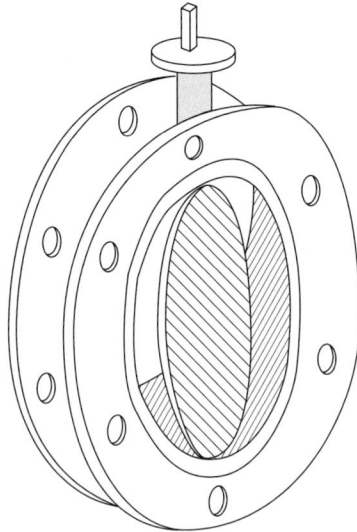

5.4.2.3   Eccentric plug valves

These valves are designed specifically for flow control. The control element is a profiled plug which pivots around a vertical stem. The plug rotates away from the seat as it begins to open, reducing wear on the seat and exposing an increasing amount of the pipe bore as it moves towards its fully open position. The valve therefore provides high flow capacity with virtually a full bore when open.

Used increasingly for service reservoir inlet control and strategic network control, the valve can be operated manually by way of a hand wheel, but mostly are configured with motorised actuators for remote operations via SCADA networks.

5.4.3   Air valves

When water mains burst or are fully/partially drained down for maintenance purposes, the water column is replaced by air. On refilling the main after the maintenance or repair operation, the system must be capable of exhausting the air from the pipeline in order to avoid air locks or aerated water.

In the event of a planned or unplanned evacuation of water from a pipeline, air needs to be able to enter the pipe at a similar rate to the water escaping from it. If the network is not capable of achieving this, then there is a risk that a vacuum can be created within the network, which can lead to the main imploding or at best to the creation of a potential weakness that can lead to a future failure of the pipeline. When a distribution pipeline is sucking in air, other unwanted detritus can be drawn into the network: it is vital, therefore, that air valve chambers are free draining, so that water and associated debris cannot accumulate within the chamber.

131

Sudden starting or stopping of flows in a pipeline due to, for example, the rapid closure of a valve or sudden stopping of a pump can lead to a separation of the water column within the network. This in turn can lead to momentary negative pressures occurring in the pipeline, followed by equally short-lived high pressures, referred to as transients. These undesirable events can create high stresses within a pipeline.

Air transfer in and out of pipelines can be managed in several ways, but it is essential that pipelines are designed with appropriately sized and strategically placed air valves to manage the air movement to and from the distribution network. The importance of correct air valve placement and sizing cannot be over-emphasised.

Air valves should be located at high points on the distribution network, where air will naturally accumulate. In addition, consideration should be given to installing air valves at gradient changes, both upwards and downwards, and at regular intervals along constant gradients.

There are a number of options to consider when selecting the most appropriate air valve for the location and dynamic network events that the valve will need to be able to accommodate. Pipeline designers must consider these carefully before selecting a valve for each location.

An example of an air valve is shown in Figure 5.9.

5.4.4   Automatic control valves

In the context of this chapter, the term 'automatic control valve' refers to a valve that has been added to the distribution network to change flow or pressure automatically, i.e. without manual or remote intervention.

Figure 5.9 Single-orifice air valve

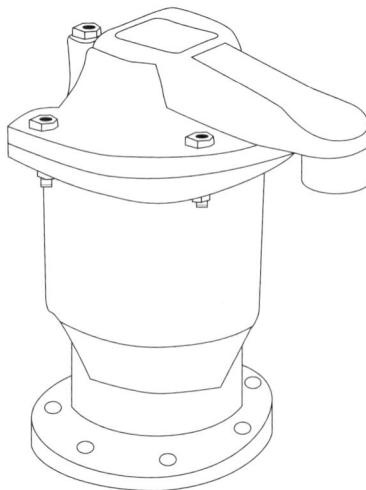

There are many different control variations that can be deployed through the automatic control valve, limited only by the imagination of the end user. Additionally, the pilot rails and regulators can be designed to provide multiple control functionality, such as pressure reduction and pressure sustaining within the same valve.

The following sections describe the principal control functions of the automatic control valve, although it should be noted that this is not an exhaustive list of control functions that can be performed by the valve.

5.4.4.1   Pressure-reducing valve (PRV)

The purpose of the PRV is to maintain a constant pressure in the water distribution network downstream of the valve. PRVs are used extensively in networks characterised by hilly topography with gravity sources at high elevation, and are installed to support pressure management initiatives for WSPs. The pilot regulator is sensing pressure variations in the downstream section of the valve created by changes in demand. This causes the pressure to change in the control space on the top of the main valve. The stem and disc then alter position relative to the seat, in order to restore the downstream pressure to the desired level (Figure 5.10).

Many WSPs employ modulating devices with PRVs to alter (modulate) the pressure control set point to suit significant variations in demand within the downstream network due to, for example, industrial parks with large commercial water users. Electronic modulators can be deployed using timed set point changes or remote pressure-sensing feedback. Alternatively, hydraulic modulators, using an orifice plate to

Figure 5.10 A cut-away view of a Cla-Val globe-style automatic control valve

Reproduced with permission. © Cla-Val UK Ltd, www.cla-val.co.uk

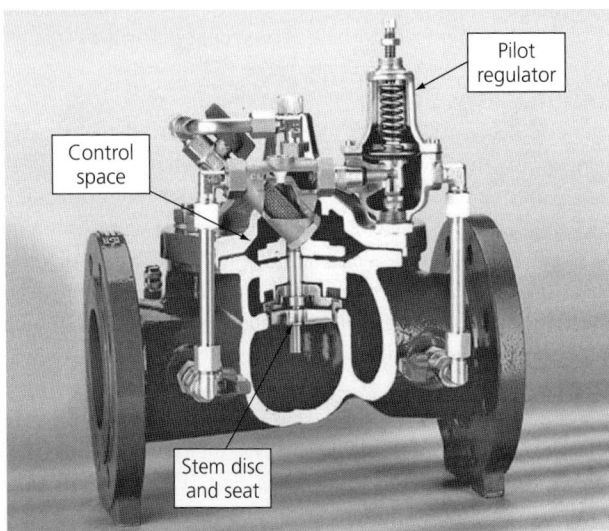

sense flow rate changes and to modulate the outlet set point, are becoming increasingly popular with many engineers.

5.4.4.2   Pressure-sustaining valve (PSV)

The PSV works to stop the pressure in the network upstream of the valve from dropping below a prescribed minimum level. A typical use for a PSV is on the inlet to a service reservoir, where it is configured to prevent the demand into the reservoir from drawing down the pressure on the water main providing the feed, perhaps where there are properties connected to the inlet main at high elevation. The pilot regulator senses the pressure in the upstream section of the valve. As the upstream pressure reduces towards its minimum setting, the regulator begins to close, causing increased pressure in the control space, which causes the main valve to restrict the volume of water that passes through it.

5.4.4.3   Burst main valve (BMV)

BMVs are used where properties or sensitive environments are at risk of flood damage as a consequence of a burst main. The valve also serves to protect the network upstream of the valve from excessive drain down or de-pressurising due to the burst. The BMV is configured to sense a minimum downstream pressure that would be characterised by a burst in the network supplied through the valve. The regulator is configured to either fully or partially fill the control space, to cause the main valve to fully or partially close, to restrict the volume of water feeding the burst.

5.4.4.4   Altitude control valve

Altitude control valves are used at service reservoir inlets. The regulator is configured to sense the maximum and minimum control levels (static back pressure) within the reservoir and operates the main valve at a linear or graduated rate, to increase flow when the reservoir is low or reduce it as it approaches the top water level (Figure 5.11). The altitude control valve has replaced the equilibrium ball valve at many sites, as it is considered by many distribution engineers to be a more reliable form of inlet control.

5.4.5   Non-return valves (NRV)

The NRV is a generic description for a valve that is referred to by various alternative names: reflux, check, swing check, one-way, recoil. The purpose of the NRV is to permit water to flow in one direction and to prevent water from flowing in the opposite direction (backwards) through the valve. Most prevalent at water-pumping stations, the NRV is installed to prevent water passing back through the pumps when they are in their standby configuration. The designer needs to carefully consider several key design criteria when specifying NRVs, not least the installed angle of the valve, which may influence the correct operation of the valve, particularly where the gate relies on gravity to return it to its closed position.

There are numerous alternative designs for NRVs, including ball check, butterfly disc, dual plate, and spring-assisted cone to name just a few. It is vital that the design engineer determines the critical parameters, including the pressure drop across the valve, the flowing velocity of the water through the valve and the rate of deceleration of the

Figure 5.11 A typical control configuration for altitude control valves (two-way flow)
Reproduced with permission. © Cla-Val UK Ltd, www.cla-val.co.uk

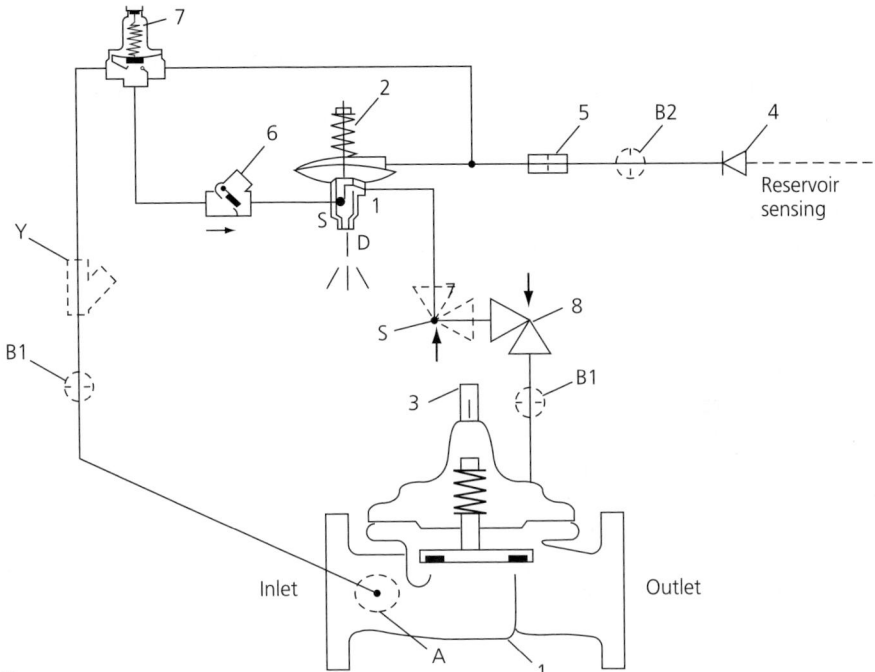

Key:
1, main valve; 2, altitude control; 3, valve position indicator; 4, bell reducer; 5, restriction assembly;
6, swing check control; 7, flow check control; 8, flow control (closing); A, strainer; B1 and B2,
isolation valves; S, flow control (opening); D, check valve with isolation; Y, 'Y' strainer.

water after the pump has stopped (e.g. see Thorley, 1991). Deceleration is not easy to calculate, so simulation of the network using a computer model with surge analysis software would be advised, to ensure that appropriate valve selection is made for the prevailing conditions. An example of an NRV (swing check) is shown in Figure 5.12.

Figure 5.12 Swing check valve

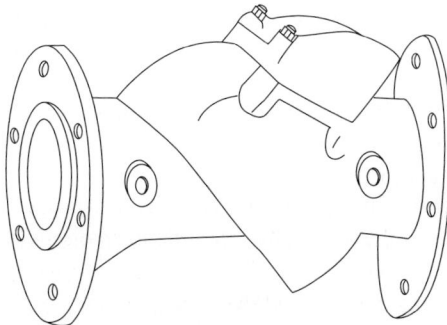

5.4.6   Ball float valves

We have already stated earlier that modern, automatic control valves are increasingly being deployed in water distribution networks to replace some of the traditional control methodologies. The ball float valve is one such device that has been relied upon for many years to manage flows into service reservoirs, tanks and water towers, and while it may have a more attractive alternative in the altitude control valve, it continues to perform its function in many water distribution networks globally.

5.5   Service reservoirs

Water undertakings in England and Wales have a legal duty to maintain a wholesome, continuous and adequate supply of water to their customers. Therefore, the supplies must be uninterrupted and achieve the required standards with regard to quality, quantity and pressure. Self-evidently, the components (pumping stations and treatment works included) of water supply and distribution systems require regular maintenance, and, to carry it out, some parts of the system have to be taken out of service. A similar situation with some elements of the system being unavailable arises when repairs are carried out. Superimposed on these 'abnormal' operating conditions are the inherent variability and uncertainty associated with water consumption from one day to the next, as discussed in Chapter 3, 'Water demand: estimation, forecasting and management'. This includes the diurnal variations as well as any underlying weekly pattern. Water undertakings deal with these and many other eventualities such as large flows for fire-fighting using a variety of strategies, a major component of which is the storage of treated water in reservoirs close to the demand centre. Surface or underground tanks within the distribution system are commonly referred to as service reservoirs, whereas elevated tanks are often called water towers. A portion of the water in service reservoirs is held in reserve to deal with contingencies, with the remainder being the variable storage that balances hourly fluctuations in demand. The contingencies may include major fires and source failures.

The principal function of service reservoirs is thus to improve the security and continuity of supply by providing additional storage. There are additional benefits that may be summarised as follows. The variable storage balances hourly fluctuations in demand and smoothes out the peaks. This has the advantage that it reduces the capacities of the water treatment and transmission (as opposed to distribution) facilities, including pumps and trunk mains. Thus, the treatment and transmission facilities may be sized on the basis of the consumption that occurs on the maximum day, whereas the distribution system would be sized according to the maximum hourly consumption. Service reservoirs also simplify the operation and improve the efficiencies of treatment works and pumping stations. The treatment works can produce water at a constant rate (for longer periods). Similarly, the pumping need not track the daily consumption pattern closely but instead can be optimised to achieve a more steady operation and take advantage of any periods during which electricity is cheap. Service reservoirs also improve the pressures within the distribution system, as discussed later in this section.

The common practice is to locate service reservoirs on high ground close to the demand centre, to provide adequate pressure for the system served. Elevated tanks or water towers are often used as an alternative. A relatively central location will help avoid excessive pressure near the service reservoir and low-pressure problems in the outlying areas. With respect to the main supply source, the service reservoir should be located beyond the demand centre, to ensure the service reservoir and main source supply water from different directions to more easily cope with the peaks in demand. This arrangement should yield a more economical design compared with a service reservoir ahead of the demand centre. The adoption of two major supply directions in this way also enhances the reliability of the distribution system considerably. Other factors such as the aesthetics of the structure come into play, and the final choice from among potential locations requires an economic and technical appraisal, including hydraulic and water quality analyses.

Service reservoirs are normally reinforced concrete structures. Steel tanks and pre-stressed concrete are also used sometimes, but the interested reader should refer to more specialised texts on the structural design and constructional aspects, e.g. the BS EN 1992-3:2006 standard. A minimum of two compartments are required to help maintain continuity of supply, as one compartment will remain operational if the other is taken out of service for maintenance and repairs, etc. For this reason, service reservoirs are usually rectangular in plan. They are normally covered, to prevent contamination of the water. The Water Supply (Water Quality) Regulations 2000 are the statutory instruments set up to safeguard the quality of water in service reservoirs in England and Wales. They require samples to be taken weekly from each service reservoir. In the UK, the Reservoirs Act 1975 embodies the legal framework for the safety (design and construction included) of reservoirs that hold $25\,000\,m^3$ or more above the adjoining natural ground level. In England and Wales, under the Act, the Environment Agency is entrusted with the responsibility to ensure that owners and operators of service reservoirs observe the Act.

5.5.1 Balancing storage

Given the daily water consumption pattern and the water production rate, the storage needed to cope with the hourly fluctuations can be calculated using a mass balance approach on an hourly basis, as summarised in the following procedure. The calculations may be carried out using the average daily consumption based on the peak week; different water undertakings apply peaking factors in slightly different ways.

1. Initial conditions:
   volume required $= 0$
   cumulative depletion $= 0$; i.e. the reservoir is assumed full.
2. Surplus $=$ average hourly demand – an hour's demand
3. If 'surplus' is positive:
   cumulative depletion $=$ max$\{0,$ (cumulative depletion – surplus)$\}$
   If 'surplus' is negative, i.e. there is a supply shortfall:
   cumulative depletion $=$ cumulative depletion – surplus
   volume required $=$ max$\{$volume required, cumulative depletion$\}$
4. If the analysis is complete (24 hours):
   balancing storage $=$ volume required
   Otherwise, return to step 2.

Table 5.1 Balancing storage calculations

Time: h	Demand: m³/h	Surplus flow: m³/h	Cumulative depletion: m³
08:00	176	−16	16
09:00	198	−38	54
10:00	208	−48	102
11:00	236	−76	178
12:00	226	−66	244
13:00	227	−67	311
14:00	214	−54	365
15:00	189	−29	394
16:00	163	−3	397
17:00	163	−3	400
18:00	172	−12	412
19:00	183	−23	435
20:00	181	−21	456
21:00	168	−8	464
22:00	162	−2	**466**
23:00	155	5	461
00:00	134	26	435
01:00	105	55	380
02:00	89	71	309
03:00	85	75	234
04:00	88	72	162
05:00	92	68	94
06:00	96	64	30
07:00	130	30	0

Average demand $= 160\,\text{m}^3/\text{h}$
Supply $=$ average demand (assumed)
Surplus flow $=$ supply − demand
Balancing storage required $=$ maximum cumulative depletion $= 466\,\text{m}^3$

Table 5.1 illustrates the calculations and the data used in the calculations were provided by South West Water.

It is worth restating that there is a lot of uncertainty and randomness associated with water demands. These and other relevant factors (e.g. any reserve capacity at the treatment works), including the need for operational flexibility, should be allowed for. A balancing storage of about 25% of the average daily consumption is generally considered reasonable in the UK.

5.5.2 Contingency storage

The estimation of contingency storage is not straightforward, and is highly system-specific, as it is governed by the relevant risk factors mentioned earlier and any available additional contingency measures (e.g. alternative sources/service reservoirs from which some water could potentially be rerouted; any standby power generation capacity; the capacity of any duplicate major supply mains; etc.). The monitoring and operational control practices of the water undertaking and the structure of the electricity tariff may also influence the overall amount of storage (balancing and contingency) provided. An overall capacity of up to 24 hours supply is generally considered reasonable in the UK.

Water quality deteriorates as the residence time increases (Grayman and Clark, 1993; Ghebremichael *et al.*, 2008) and, along with the location and operating policy, should also be considered when determining the overall storage volume. The overall age of water throughout a distribution system is very sensitive to the locations of service reservoirs and their operating policies, as characterised by the diurnal water level fluctuations (Grayman *et al.*, 1991). Other relevant factors are the shape and inlet and outlet arrangements. Mixing inside the service reservoir is very important, and should be addressed at the design stage. Dead zones with limited water exchange with the main bulk of the water in the rest of the reservoir should be avoided. Short circuiting between the inlet and outlet should be avoided also. These issues are best assessed with the help of detailed hydraulic and water quality models of the reservoir and distribution system. Chapter 6, 'Network modelling', and Chapter 8, 'Operation, maintenance and performance', contain more information on water quality.

The UK Water Supply (Water Quality) Regulations 2000 require disinfection of all water from a treatment works. Chlorine is the most common disinfectant used to render pathogenic organisms in water harmless. A sufficient residual chlorine concentration should be maintained throughout the distribution system. However, chlorine reacts with organic and inorganic substances in water, to produce a wide range of undesirable chlorinated compounds that are called disinfection by-products (DBPs). Therefore, the residual chlorine concentration decreases while, concomitantly, the concentrations of the DBPs increase with time. DBPs can cause taste and odour problems, and are thought to constitute a potential health risk under certain circumstances (Twort *et al.*, 2000; Clark and Sivaganesan, 2002). Therefore, with respect to both DBPs and the chlorine residual, the quality of the water deteriorates with time. Self-evidently, the overall age of the water in the system as a whole increases with the amount of storage provided. Water quality deteriorates with time, and thus excessive storage beyond that required to safeguard the continuity of supply is best avoided.

5.6 System integration
5.6.1 Regulation and monitoring
The EU Drinking Water Directive 98/83/EC established obligations for EU member states in respect of water intended for human consumption. To discharge this responsibility, the UK government requires that all statutory water undertakers and licensed water suppliers in the UK operate under strict regulatory conditions. The conditions require that accurate and auditable reports are produced in order that regulators can monitor WSPs for compliance against their prescribed standards.

The Drinking Water Inspectorate (DWI) regulates water supplies in England and Wales, and the Water Supply (Water Quality) Regulations 2000 (Amendment) Regulations 2007 are the statutory instruments which dictate how water quality must be monitored. WSPs divide their regions into 'water into supply' (WIS) zones, within which a mandatory sampling programme is undertaken to monitor and report their compliance levels. Results are provided to the DWI and are also reported to the Office of Water Services (Ofwat) as part of the June return reporting programme.

Ofwat is the economic regulator for the water industry in England and Wales. It is responsible for ensuring that WSPs provide good service and value for money to their customers. Part of its remit is to monitor the performance and serviceability of the water supply infrastructure, including leakage.

In order for WSPs to report their performance against the serviceability standards, district metered areas (DMAs) are established as a subset of WIS zones, primarily to aid the reporting on consumption and leakage levels in the water distribution network.

A further definition is ascribed to water networks in order to monitor distribution system pressures. Pressure managed areas (PMAs) are subsections of DMAs, and provide the means by which average zone night pressures (AZNPs) are monitored and reported to Ofwat (Figure 5.13).

Figure 5.13 Example of a WIS zone (BV, boundary value; DMA, district meter area; M, meter; PMA, pressure managed area; PRV, pressure-reducing valve; SR, service reservoir)

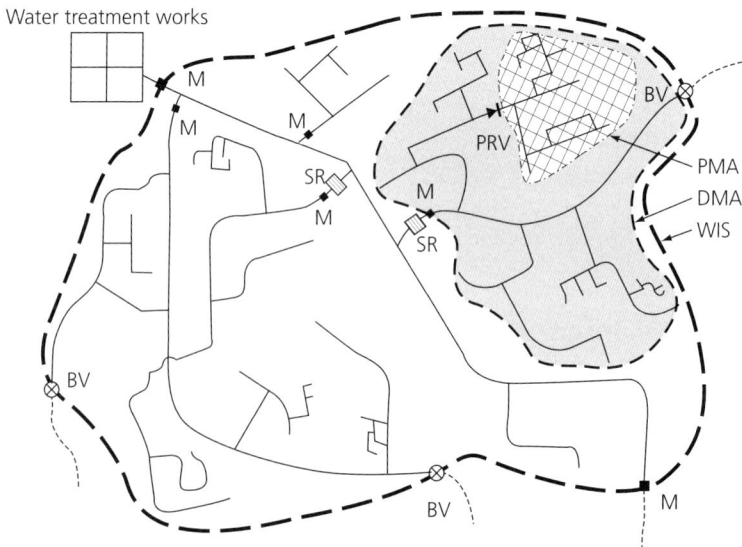

5.7   Surge control

Rapid changes in the flow rate or boundary conditions (e.g. line valve setting) in a distribution network can cause unsteady flow conditions, often characterised by brief fluctuations in pressure that can be detrimental, as stated in Section 5.3. For example, when the flow is stopped abruptly by the rapid closure of a valve, the water immediately upstream is compressed by virtue of its momentum, and, as a result, the density of the water is raised, with a concomitant increase in pressure. As the forward motion of the water in the pipeline upstream of the valve is progressively retarded and halted, a positive-pressure wave is transmitted upstream at a speed or celerity that depends on the dimensions and material properties of the pipe. On reaching the upstream end of the pipe, depending on the boundary conditions, the wave is reflected back downstream as a positive- or negative-pressure wave, which, in turn, will be reflected as a wave of the same type at the closed valve. In the absence of friction, this cycle would continue indefinitely, but in reality the pressure waves die out quickly due to friction. Conversely, a drop in pressure can occur immediately downstream of a pump that stops abruptly or a valve that closes rapidly. Thus, a negative-pressure wave travels downstream to the end of the pipe. These unsteady flow conditions, known as transients or pressure surges, generate additional stresses that can cause damage to pipelines. Large increases in pressure can lead to bursts, while pressure drops can cause pipes to collapse or fail subsequently due to fatigue. Sub-atmospheric pressures are undesirable, as they may facilitate contamination of the water through open air valves.

Although transients are an inherent property of water distribution networks that cannot be eliminated entirely, they should be borne in mind and addressed at the design stage, with the aim of keeping the pressure surges within the permitted bounds (as specified in codes of practice and standards). The analysis of transients is complex, and requires special software. As mentioned previously, pressure waves are reflected or transmitted at the ends of pipe sections and where the material properties or dimensions of the pipes change. These interactions mean that the critical condition does not always correspond to the initial pressure surge. It can be shown (e.g. see Potter and Wiggert, 2010) that the pressure generated by a sudden velocity change of Δv is given by

$$\Delta p = \rho c \, \Delta v \qquad\qquad (5.10)$$

This is equivalent to

$$\Delta H = c \, \Delta v / g \qquad\qquad (5.11)$$

Here, g = the acceleration due to gravity
 ρ = the density of the fluid
 c = the speed of the wave
 Δp = the change in the pressure
 ΔH = the change in the head

The speed (celerity) of the wave c depends on the pipe and fluid properties, and is given by

$$c = [K/\rho(1 + DK/dE)]^{1/2} \qquad\qquad (5.12)$$

where $K=$ the bulk modulus of the fluid
$\quad D=$ the internal diameter of the pipe
$\quad d=$ the pipe wall thickness
$\quad E=$ the pipe material Young's modulus

The most important consideration regarding the attenuation of pressure transients is the speed with which changes in the flow take place. If a pressure wave is initiated at a valve, a time of $2L/c$ will elapse before the reflected wave returns to the valve, where L is the length of the pipe section or the distance from the valve to the boundary where the wave reflection occurs. The parameter $2L/c$ is important, and is referred to as the pipe period. Thus, a valve closure that is completed within one pipe period is said to be rapid, and the magnitude of the pressure transient it creates will be the maximum as given by equation (5.10) (known as the Joukowski equation). Conversely, the surge will be less severe if the valve closure is slow, meaning it takes longer than one pipe period. The vital role that air valves, non-return valves and pressure relief valves play in controlling surge is explained in Section 5.4. Other surge control devices and measures are described below, while a more fundamental characterisation is provided in Chapter 2, 'Basic hydraulic principles'.

The speed with which a pump stops is self-evidently important, and should be considered. Increasing the time the pump takes to stop, e.g. using a flywheel, will decrease the risk due to negative pressure. This, however, is not a straightforward decision to make. A *pump bypass* permits water from the sump to enter the delivery pipe if low pressure develops in the delivery pipe. As mentioned in Section 5.4, a non-return valve in the delivery pipe prevents backflow. The bypass consists of a small-diameter pipe with a valve that opens and then closes slowly.

In a similar way, a *feed tank*, with a non-return valve, releases water into a pipeline to restore the pressure if low pressure develops. A feed tank can be deployed downstream of a pump or at high points in a pipeline where accumulation of air and separation of the water column are likely. The feed tank should be designed to safeguard the quality of the water by preventing its contamination and stagnation and ensure sufficient water is available when required.

A *surge shaft* rises above the hydraulic grade line and dampens pressure fluctuations in the pipeline. It is connected to the pipeline, and the water level in the shaft rises when the velocity is decreasing. On the other hand, when the flow accelerates, the surge tank augments the water in the pipeline, and thus avoids negative pressures. The fact that a surge tank must rise above the hydraulic grade line is obviously a disadvantage, as the location and construction cost are dictated by topography.

Air vessels are part filled with air under pressure, and are used at pumping stations. They release water when the pipeline pressure drops, e.g. when a pump trips, to avoid negative surge. Water enters the vessels as a buffer against high positive pressures. *Accumulators* are a variant in which a flexible rubber membrane separates the air from the water, to reduce the amount of air absorbed by the water. The air is replenished periodically.

REFERENCES

Clark, R. M. and Sivaganesan, M. (2002) Predicting chlorine residuals in drinking water: second order model. *Journal of Water Resources Planning and Management* 128(2): 152–161.

Ghebremichael, K., Gebremeskel, A., Trifunovic, N. and Amy, G. (2008) Modeling disinfection by-products: coupling hydraulic and chemical models. *Water Science and Technology: Water Supply* 8(3): 289–295.

Grayman, W. M. and Clark, R. M. (1993) Using computers to determine the effect of storage on water quality. *Journal of the American Water Works Association* 85(7): 66–77.

Grayman, W. M., Clark, R. M. and Goodrich, J. A. (1991) The effects of operation, design and location of storage tanks on the water quality in a distribution system. In: *Water Quality Modeling in Distribution Systems Conference*. Denver, CO: American Water Works Association Research Foundation and American Water Works Association.

Potter, M. C. and Wiggert, D. C. (2010) *Mechanics of Fluids*, 3rd edn. Andover: Cengage Learning.

Thorley, A. R. D. (1991) *Fluid Transients in Pipeline Systems*. Hadley Wood: D&L George.

Twort, A. C., Ratnayaka, D. D. and Brandt, M. J. (2000) *Water Supply*, 5th edn. Arnold.

UK Water Industry (2003) *The Choice of Pressure Ratings for Polyethylene Pipe Systems for Water Supply and Sewerage details. Water Industry Information and Guidance*, IGN 4-32-18. Swindon: WRc. www.water.org.uk/home/member-services/wis-and-ign/current-documents-plastics-and-rubbers/ign-4-32-18c.pdf [accessed 01.09.10].

REFERENCED LEGISLATION AND STANDARDS

BS EN 1992-3:2006: Eurocode 2. Design of Concrete Structures. Liquid Retaining and Containing Structures. London: British Standards Institute.

The Drinking Water Directive. Council Directive 98/83/EC of 3 November 1998 on the quality of water intended for human consumption. *Official Journal of the European Communities* L330: 32–54.

The Reservoirs Act 1975. London: HMSO.

The Water Supply (Water Quality) Regulations 2000. London: TSO.

The Water Supply (Water Quality) Regulations 2000 (Amendment) Regulations 2007, No. 2734. London: TSO.

Water Distribution Systems
ISBN: 978-0-7277-4112-7

ICE Publishing: All rights reserved
doi: 10.1680/wds.41127.145

ice

Institution of Civil Engineers

publishing

Chapter 6
Network modelling

Dragan A. Savić University of Exeter, UK
Rob Casey Thames Water, Bourne End, UK
Zoran Kapelan University of Exeter, UK

6.1   Introduction

This chapter provides a brief review of principles used for creating a computer model of water distribution network behaviour. Computer modelling or network modelling (also known as *network analysis*) will be discussed, with different types of network analysis models covered, including steady-state, dynamic (extended period), water quality, transient, etc.

The chapter starts with an introduction to different types of models, before moving on to consider basic modelling principles, including complexity and uncertainty associated with network analysis. This is followed by the discussion of data sources for model building, focusing, in particular, on procedures for water consumption and demand assessment. The final part provides a discussion on model building for a particular purpose and the steps in model building, including model calibration and maintenance.

6.2   Models

According to the *Oxford Dictionary*, a model is 'a simplified mathematical description of a system or process, used to assist calculations and predictions'. As modern cities and their associated infrastructure, including water distribution systems (WDSs), have grown over time, network analysis has become a crucial planning, design and operational activity. However, network analysis, like any other type of computer modelling, suffers from the so-called 'garbage in, garbage out' syndrome. Often abbreviated to GIGO, this is a well-known computer axiom meaning that if invalid data are fed into a computer model, the output would also be invalid. Therefore, data for network modelling will also be discussed.

Although there are a number of possible classifications of mathematical models, this review will focus on *simulation models* (Box 6.1) and their use in WDS management.

Over the last 20 years, there has been a significant increase in the number of network analysis software packages, both in the commercial and in the public domains. Coupled with increases in hardware power, it is now possible to simulate very large systems as all-mains models under complex dynamic scenarios. This increase in computer power coupled with software development has also lead to a significant increase in the range of applications supported by modelling (Figure 6.1), moving

---

Box 6.1 Network simulation

A simulation is the process of using a network model that provides insight into the dynamics of a particular water distribution system and shows how it will behave over time and in various situations. For example, one might want to simulate variations of flow and pressure in a system over a 24 hour period to check for regulatory compliance. A model would be used, and simulation runs performed to assess whether pressures satisfy minimum requirements under different demand scenarios.

Simulation is not only useful for testing and analysis when prediction or evaluation of system behaviour and management options is needed but also for training of staff.

---

from strategic and steady-state models confined to planning and engineering design departments to all-mains extended-period simulation (EPS) models used by various utility departments.

In addition, commercial software packages, such as WaterGEMS (Bentley Systems), H₂ONET and InfoWorks WS (MWH Soft), incorporate advanced graphical user interface (GUI) and geographic information system (GIS) capabilities to provide an accurate geographical representation of a network, thus enabling simple communication of complex phenomena, even to non-specialists. Even the freely available EPANET

Figure 6.1 Model uses have expanded with increased computing power

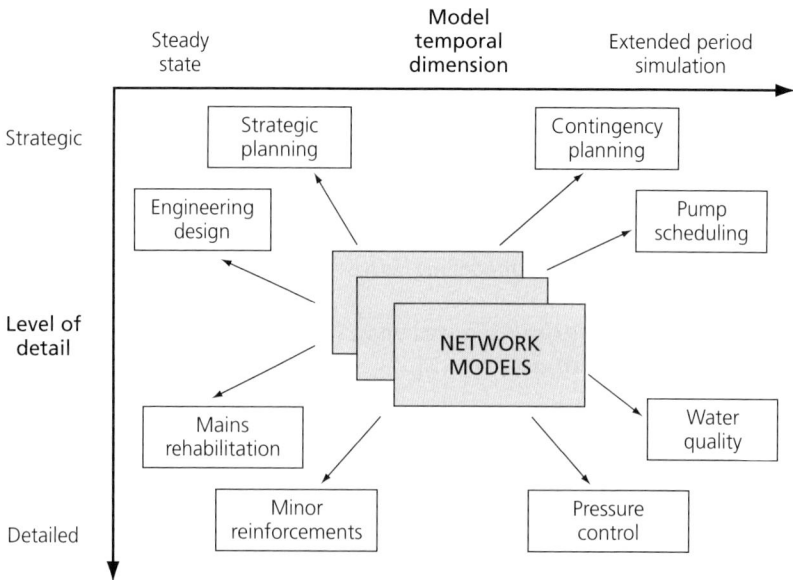

Figure 6.2 The EPANET GUI

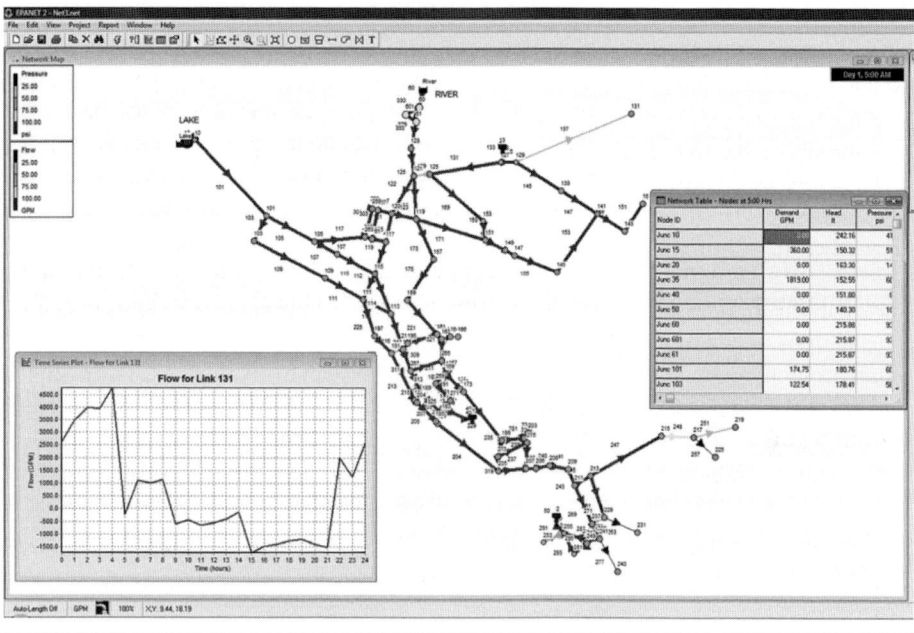

(Rossman, 2000) provides good modelling capabilities with a simple but effective GUI (Figure 6.2).

6.2.1 Steady-state modelling

Early network models simulated only steady-state (snapshot) hydraulic conditions, i.e. where all demand and operations are constant. Although conditions in a real WDS are always changing with time (e.g. demands, reservoir levels, etc.), these models are useful when particularly important instantaneous conditions within a network need to be investigated. For example, a steady-state simulation would be performed to analyse peak demand conditions for pipe sizing, or minimum demand conditions important for investigating sediment deposition potential. Many hydraulic problems in a WDS are solved by a steady-state simulation, most often those relating to infrastructure design problems. Steady-state modelling serves also as a building block for more advanced types of modelling (i.e. extended period simulation, water quality analysis and fire protection studies). Once it is well understood, it is easier to grasp the concepts of these advanced modelling approaches.

Input data for steady-state simulation include all static network data (e.g. pipe and node characteristics) and fixed operational conditions (e.g. reservoir levels, valve statuses, pump flow rates, etc.) for a particular time snapshot. The model calibration process often starts with considering a steady-state situation where nodes with reservoirs and pumps are removed from the model and flows are assigned to them as negative flows (inflow into the system). This allows for the flows through the system to be checked against measurements and errors corrected before calibration of the EPS model is attempted.

6.2.2   EPS modelling

In the 1970s, modelling capability was extended to include EPS models that could accommodate changing demand and operational conditions. These models are useful in determining the dynamic response of a network system, i.e. how it behaves under varying demands, how a stressed system reacts in emergencies or how, as in the case of the Anytown problem (see Section 6.5.1), the system could be strengthened most efficiently to provide sustainable service over a long period of time.

The established approach to performing EPS simulation computations involves steady-state analysis to determine pressure and flows in the system and a numerical integration technique to integrate reservoir flows in or out of the system to compute reservoir volume changes (see Chapter 2). This type of simulation is extremely important for complex systems where important dynamic interactions occur involving varying reservoir water levels, pumps turning on and off, and pressure and flow control valve operation between different parts of the network (e.g. district metered areas (DMAs) or pressure zones). An example of a dynamic output for pipe flow and node pressure is given in Figure 6.3.

Input data for EPS models require not only the static data used in steady-state modelling but additional data to describe time-varying demands, pump and valve characteristics (e.g. pump curves, valve losses, etc.) and reservoir levels. Furthermore, the EPS simulation requires information on the computational time step and the duration of the simulation run. The decision about the time step and simulation duration should be based on the intended purpose of the model, and should allow proper capture of equipment controls/functioning. For example, reservoir storage analysis should be performed over at least a 24 hour period to accurately capture the filling and emptying of the reservoir, with a time step that should not be longer than 1 hour, to avoid missing an important change in the status of valves or pumps, e.g. when the reservoir becomes full before the end of the computational time step. As for water quality modelling, the simulation is run for a sufficiently long period of time (e.g. several days) under a repeating pattern of source and demand inputs so that the initial water quality conditions, especially in storage reservoirs, do not influence the water quality predictions in the distribution system.

6.2.3   Water quality modelling

Around the early 1990s, hydraulic modelling software was enhanced to include water quality modelling (EPA, 2005). This software enabled factors such as chlorine decay, source water mixing, contamination spread and water age to be modelled. The output of a water quality model can include the temporal and spatial distribution of a variety of constituents within a distribution system, including the fraction of water originating from a source, the age of water, the concentration of tracers (e.g. non-reactive constituents such as chloride or fluoride), the concentration of a reactive compound (e.g. chlorine or chloramine) and the concentration of disinfection by-products (e.g. trihalomethanes).

Figure 6.3 Example output from an EPS model simulation run

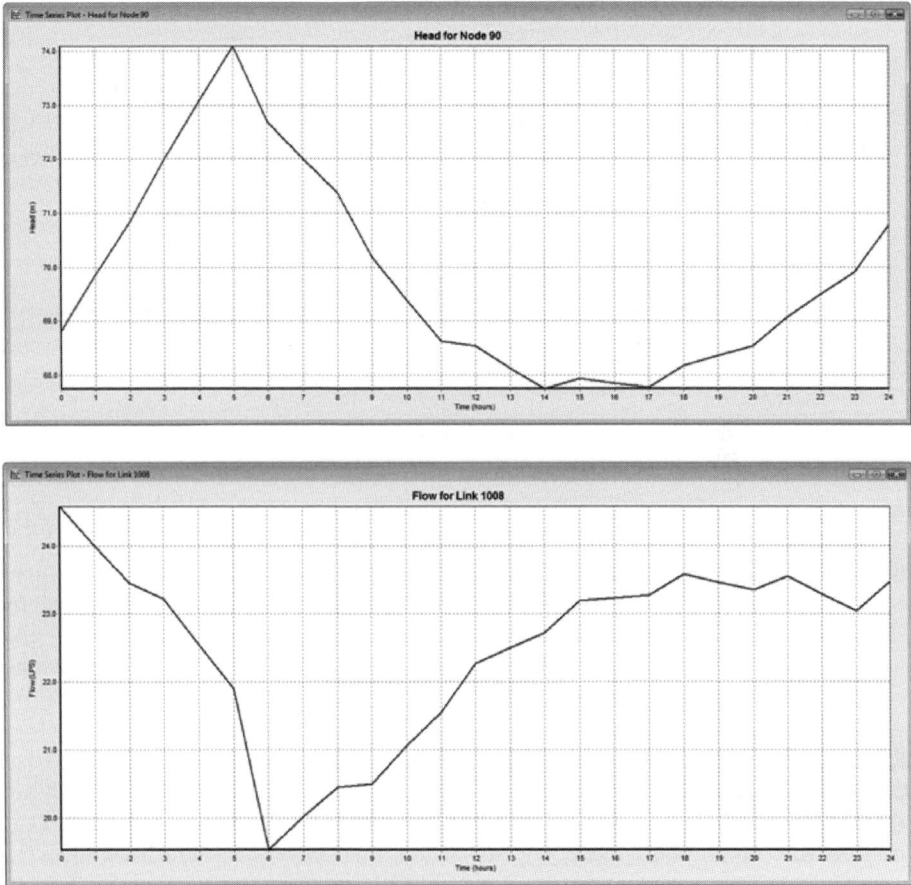

Although the modelling of chlorine decay and other chemical reactions is complex, key water quality parameters such as water age, source mixing and contamination spread can be successfully modelled using a standard hydraulic model which has been calibrated for flow and pressure. Water age models can often be used to gain some understanding of chlorine decay and bacterial growth issues within the distribution system, as high water age is often associated with these problems. With water quality models it is important to use all-mains models to accurately predict water age, mixing and contamination spread.

To successfully model chlorine and other chemical reactions in the distribution system requires these parameters to be sampled throughout the network, which is a lot more expensive and difficult than standard pressure measurements. Also, for chlorine decay, the chemical reactions are quite complex, with reactions occurring both within the bulk fluid and at the pipe wall. However, chlorine decay models have been successfully built, calibrated and used to resolve water quality issues in many different countries. An

Figure 6.4 Chlorine residual levels at a node over a 24 hour period

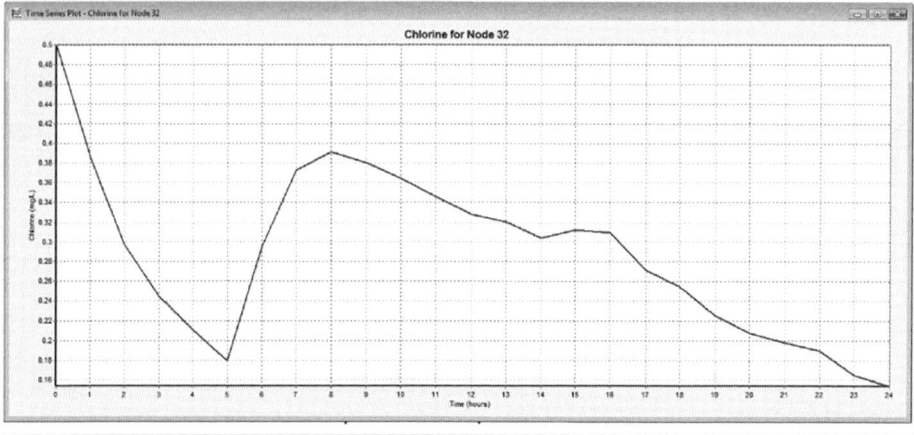

example of the chlorine residual level output for a particular node in a water distribution network is given in Figure 6.4.

Some modelling software suppliers have also included the capability to model sediment transport, to predict possible water quality issues such as brown water caused by significant velocity changes or flow reversals when sediment material is picked up and distributed around the network. These models can also be used to determine flushing routines to safely remove sediment.

However, once again, much can be done with a standard hydraulic model to predict sediment-based water quality issues by identifying pipes with low velocities which are significantly increased or reversed following valve operations. As a general rule, it is assumed that pipes with a peak velocity of less than 0.3–0.4 m/s are most likely to provide the highest risk in terms of sediment build-up. Furthermore, pipes with peak velocities above 0.6–0.8 m/s are most likely to be self-cleansing. This may vary from system to system and with different water sources, but local experience can be gained from flushing trials and turbidity monitoring.

With regard to flow reversals, the same rules given above for velocities will generally apply for sediment; however, where material is attached to the pipe wall (i.e. around the pipe as well as on the bottom), then the flow reversal may result in this material becoming detached into the bulk fluid.

6.2.4 Hydraulic transients

Pressure transients occur when there is a rapid change in the fluid velocity within a pipe-line. These pressure changes are often generated by the rapid opening/closure of valves or switching/failure of pumps within the pipe system.

Prior to the 1970s, before computers were available to the design office, solutions to the complex equations that describe unsteady or transient flow conditions (Wylie and

---

Box 6.2 Transient analysis

Transient analysis models are normally used to identify the extent of pressure surges generated by pump switching/failure or valve modulation within water networks. The model is then used to design surge relief measures such as air vessels, pump inertia and valve control.

Streeter, 1978) were provided by graphical methods of analysis, such as those developed by Schnyder and Bergeron (Bergeron, 1961). Analysis was usually confined to a single pipeline, which was sufficient to model surge issues on dedicated pumping mains.

The first computer programs, based on the method of characteristics, were only able to handle a few pipes, and hence a very simplified network. This, coupled with the fact that the transient flow equations are different to the steady-state or extended period simulations, and that the analysis time period is much shorter, led to the separate development of transient programs from the basic steady-state hydraulic models (Box 6.2).

However, the recent significant advances in available desktop computing power and analysis techniques have enabled surge analysis on much larger networks to be undertaken. This has led to some suppliers offering both extended period simulation and transient analysis facilities under a common GUI. This saves considerable time in model building, removes the need to transfer data from one package to another, and also enables the same modeller to undertake both the extended period simulation and the transient analysis.

The models can be calibrated using special transient pressure loggers that can record the pressure changes that occur over fractions of a second during a surge event.

6.3   Basic modelling principles
6.3.1   Complexity
A WDS may be represented as a set of nodes and links. As more and more water utilities are building their models from GISs, tens or even hundreds of thousands of such elements can be included in a single model. Nowadays, the tendency is to build all-mains models, which can then be simplified if necessary, e.g. for strategic decision-making (Box 6.3). A typical distribution system comprising urban and rural areas is shown in Figure 6.5. A close-up view of the smaller area, showing details such as streets and building, is provided in Figure 6.6.

It is not always feasible or necessary to include all elements and features in a model of a particular system, and a decision on the level of detail that the model has to be made to is often required, i.e. what elements and features to include and what to omit (Obradovic and Lonsdale, 1998). This largely depends on the intended use of the model, as, for example, a model used for master planning (i.e. long-term development and rehabilitation planning) will have much less detail compared with a model used for operational

---

Box 6.3 Occam's razor: complexity and parsimony

When building a model, it is advisable that modellers use the well-known simplicity principle when deciding on the level of model simplification for a particular purpose. Popularly interpreted, this principle, also called Occam's razor (William of Occam, *c.* 1288–*c.* 1348), states that *when there are two competing models that make exactly the same predictions [for a particular purpose], the simpler one is the better.* In other words, for the same level of predictive exactness (defined by the model intended use), the simplest model should be selected.

---

optimisation (e.g. energy use optimisation), where detailed pump and valve characteristics play a crucial role. Attempting to include each individual service connection, tap, valve and other component of a large system in a model could be an enormous undertaking that has no significant impact on the model results (Walski *et al.*, 2003).

Figure 6.5 A water distribution network

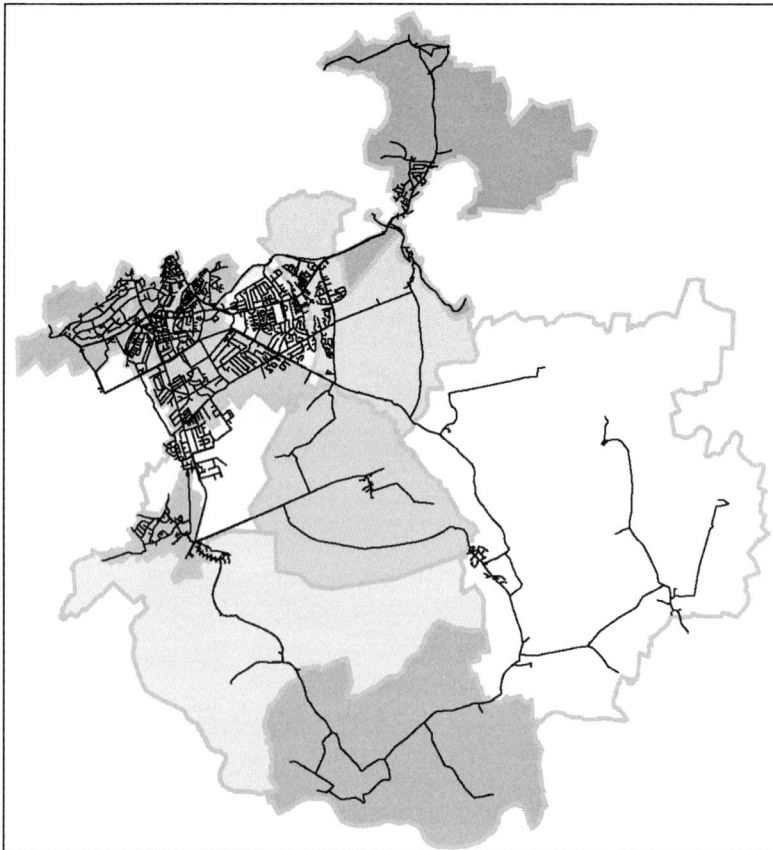

Figure 6.6 Detail from the water distribution network shown in Figure 6.5

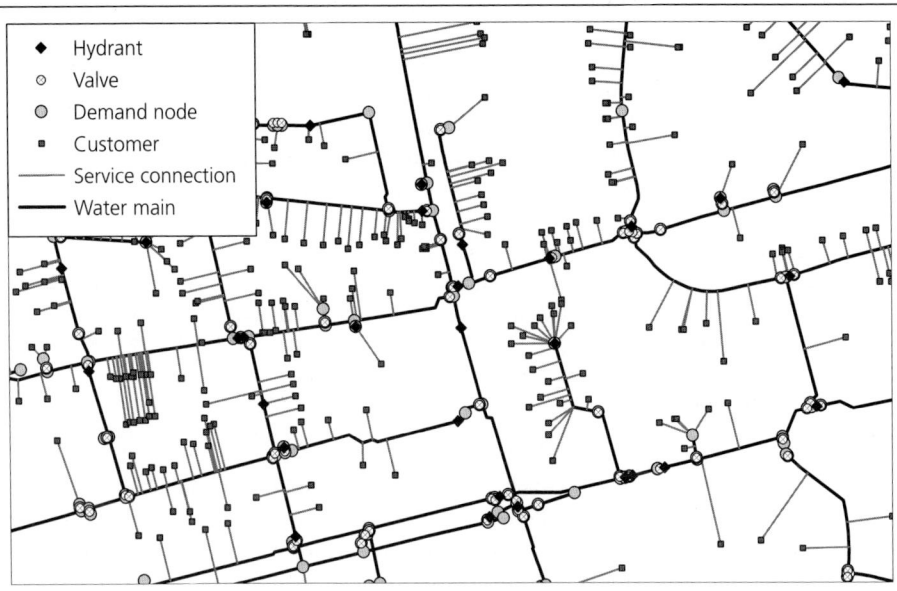

Legend:
- ◆ Hydrant
- ○ Valve
- ◎ Demand node
- ▪ Customer
- Service connection
- Water main

The process of model simplification in WDSs is called skeletonisation. It consists of selecting for inclusion in the model only those elements of the network that have a significant impact on the behaviour of the entire system. Choosing the degree of skeletonisation that is acceptable is one of the most difficult aspects of the modelling process, and should depend upon the ultimate use of the model and the sensitivity of the model results. Figure 6.7 show an all-mains model and its skeletonised counterpart obtained by removing small-diameter pipes.

6.3.2   Uncertainty

Like the vast majority of mathematical models in other areas of engineering, water distribution modelling relies on deterministic approaches to describe the behaviour of

Figure 6.7 (a) An all-mains model and (b) a corresponding skeletonised model

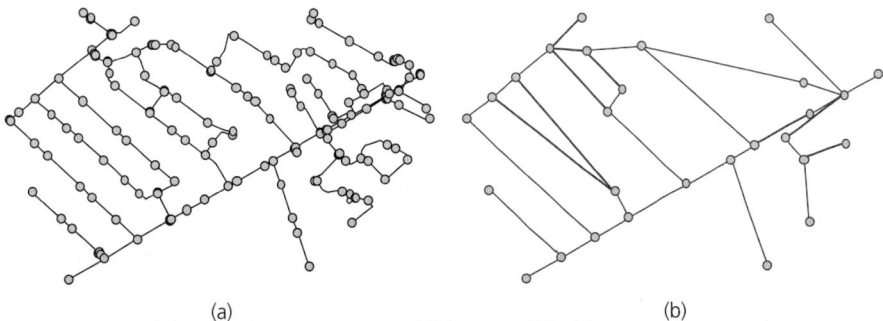

(a)   (b)

Box 6.4 Types of uncertainty

Two general types of uncertainty exist. The first is known as *reducible* (or *epistemic*) *uncertainty*. This uncertainty generally results from the lack of information about some aspect of the problem being analysed (e.g. the status of some valve in the WDS may not be known simply because that information is lacking; however, once the inspection is done, uncertainty can be reduced).

The second class of uncertainty is known as *irreducible* (or *aleatoric*) *uncertainty*. This type of uncertainty consists of fluctuations that are intrinsic to the problem being studied. Examples are the uncertainties associated with pressure and flow measurements.

---

a system. However, all real-life problems incorporate uncertainty in one way or another (Box 6.4). Such contradiction between 'mathematical determinism' and 'natural uncertainty' can seriously affect the reliability of the results of modelling. A large number of problems in the design, planning and management of engineering systems, including WDSs, require that decisions be made in the presence of various sources of uncertainty.

Different sources of uncertainties exist in WDS modelling (Filion and Karney, 2002). First, some of the input variables for WDS simulation models are uncertain. The most frequently analysed are the uncertainties associated with water consumption and pipe roughness coefficients (Xu and Goulter, 1998). However, many other sources of uncertainty exist in real-life WDSs, especially in systems that have been in use for many years (e.g. internal pipe diameter, status of some devices), basic characteristics of some devices (e.g. valve head loss curve or pump head flow curve), network connectivity (are the two pipes at a 'junction' linked or not?), etc. Secondly, WDS simulation models are not perfect, i.e. they are only an approximation of the complex real-life WDS and its behaviour. For instance, in a WDS hydraulic model, demands are allowed to occur at network nodes only, even though they are distributed along pipes in the real-life networks. Thirdly, uncertainties originate from the uncertain environment in which WDSs exist. This is especially true for the economic environment. For instance, the following uncertainties exist: uncertainties associated with the cost evaluation of different intervention (optimisation) options (e.g. the structure of different cost models, various unit costs, etc.), rehabilitation work budget uncertainties, macro economic uncertainties (e.g. the discount rate used to calculate the net present value of all costs), etc.

In order to be modelled, uncertainties need to be characterised first. Several general ways exist to do this:

- using probability theory, i.e. probability density functions (Press *et al.*, 1990)
- using possibility theory, i.e. fuzzy logic (Zadeh, 1965)
- using simple bounds on uncertain variables (Brdys and Chen, 1993).

Other, approaches exist too. Once characterised, uncertainties can be quantified by a number of different methods. These methods can be broadly classified into the analytical- and the sampling-type methods. Examples of analytical methods include (Press *et al.*, 1990) the first-order second-moment model (FOSM) and the second-order second-moment model (SOSM). The sampling-type methods include the Monte Carlo simulation (MCS) method and a large number of stratified sampling methods (e.g. the Latin hypercube method, etc.). As a general rule, analytical uncertainty quantification methods are computationally faster but normally work under certain assumptions only (e.g. model linearity, independent, normally distributed uncertain variables, etc.). The sampling-type methods tend to be more general, i.e. less restrictive (can handle non-linear models, etc.), but also much more computationally demanding.

6.4   Data for network modelling
An overview of the data collection, model building and calibration process is shown in Figure 6.8. The details behind this process are discussed in the following sections.

6.4.1   Water company data
Since the mid-19th century, asset records have been created and archived by water utilities in the UK and worldwide. Paper maps were generally used to save records, with cartographic maps used as a reference source (e.g. Ordinance Survey mapping in the UK). Since the mid-1980s, most water utilities have made significant progress towards digitisation of their data records. Nowadays, water utilities possess large

Figure 6.8 An overview of the model building and calibration process

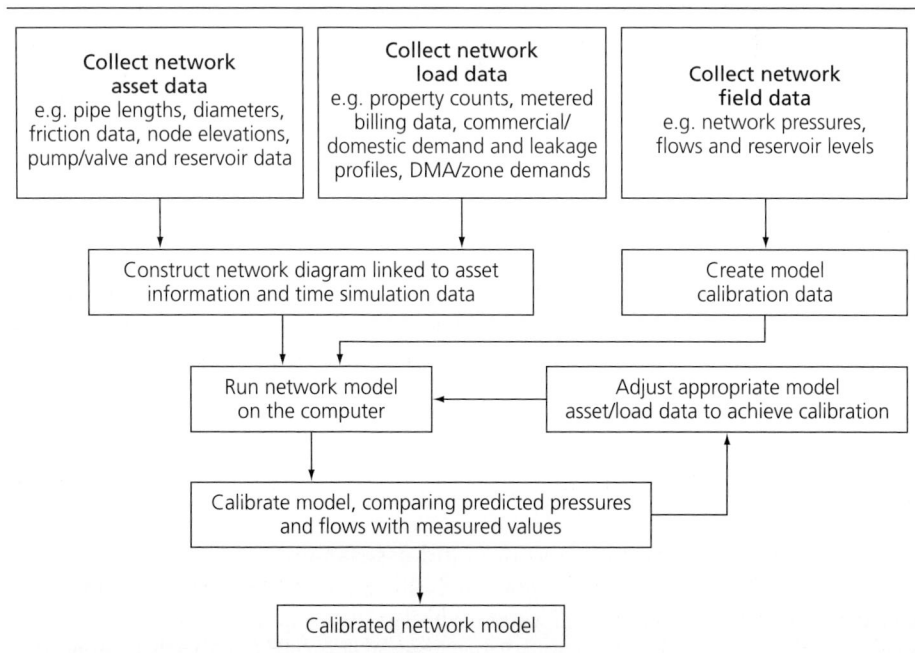

quantities of data derived from different sources (drawings, supervisory control and data acquisition (SCADA), asset and customer databases, work management systems, etc.).

In general, the following data sources may be available within a water utility:

- drawings (distribution network records, source and reservoir data, etc.)
- pressure and flow data (manually logged)
- SCADA/telemetry data (e.g. reservoir levels, pump stations, or pressure and flow data – continuously logged)
- work management data (e.g. job information based on work carried out as a result of mains bursts, service pipe repairs or work done at a pumping station)
- GIS data (information about above- and below-ground assets, e.g. treatment works, meter points, pipe age, pipe material, etc.)
- historic performance data (asset failure data, e.g. pipe bursts/leaks, discolouration events, interruptions to supply, etc.)
- customer contact data (customer phone calls made in relation to an issue/event in the system, e.g. no water, burst/leak, water quality, etc.)
- water quality data (collected as part of the water quality testing/sampling process)
- energy use data (pump schedules, energy consumption and cost at pumping stations)
- customer meter readings (e.g. via a billing system).

These digital records are embedded within many organisational functions, and are being used in a range of modelling and business scenarios, including network analysis. However, these data records are collected in different formats and often stored in different computerised systems (e.g. databases), requiring a number of different software packages and file formats for storing, editing, analysing and viewing data. For example, in a typical UK water company, different software systems would be used for leakage and pressure monitoring, static asset data, GIS asset data, real-time pressure and flow data, billing data, network model data, etc.

Although data quality is generally thought to have improved over time, much of the data held in water company records should be checked for consistency and accuracy, to avoid working with incorrect information, e.g. in terms of location, size of pipe or pipe material. The problem is that, historically, the record drawings (required to be maintained by law) were either the design drawings, which were never modified to show 'as-built' reality, or, in some cases, simply schematics showing pipe centrelines. Finally, there is always the problem of unrecorded 'operational' enhancements which were not included in the system drawings. The last several decades have shown considerable investment in data collection and improvements in quality of record drawings, with mobile IT allowing field workers to record what they actually encounter. However, caution is needed when this information is used for model building.

6.4.2   Water consumption and demand assessment

When speaking about water consumption and demand, it is commonly understood that consumption is the total amount of water used whereas demand is the immediate rate of that consumption. The usual measure of consumption is the amount supplied per head of population, e.g. in litres per capita per day (lcd), whereas the demand is measured in litres per second (l/s). Average per capita consumption across England and Wales in 2007–2008 was 148 lcd, but the water usage of households varies significantly from area to area. It is obvious that a good understanding of water consumption and demand estimates are a prerequisite for meaningful network analysis.

Metering is an essential component of water consumption assessment, as it could provide valuable information about how much water is used, by whom, where and when. The primary metering points in a water supply system include production (bulk) meters, meters in discrete zones within the WDS (e.g. DMAs) and customer meters. These meters are important for performing a water balance analysis aimed mainly at water loss management, but this information should also be used for network analysis. For example, water consumption in a supply zone (measured by a production meter) should equal the sum of demands in its DMAs. Similarly, the total consumption in an area should be equal to the sum of demands in its supply zones.

It is clear that metering is essential for the effective management of a WDS. However, it is not always possible to achieve complete metering coverage, as, for example, in the UK, where most residential customers do not have permanent meters. On the other hand, production (bulk) water meters, DMA/pressure zone meters and meters for large industrial/commercial customers are a must, and are almost universally available in all well-managed distribution systems. Information from these meters should be used for network analysis.

In order to use a model in the dynamic mode (i.e. for EPS), information on temporal variations in water usage over the period being modelled are required. A typical diurnal diagram of water usage (Figure 6.9) at a district level can be constructed by use of statistical analysis. These need to be developed considering different demand conditions, e.g. working versus weekend days, or summer versus winter days.

Figure 6.9 Typical diurnal demand curves

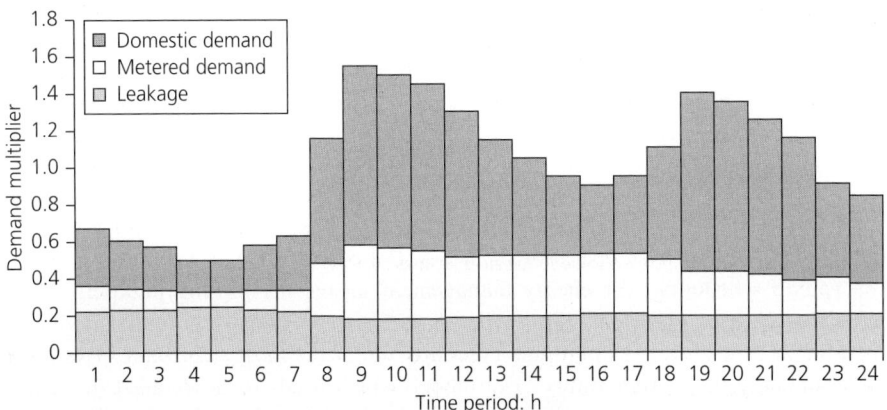

Diurnal demand curves should be developed for each major demand type or consumer class. For example, distinct diurnal curves might be developed for metered demand (commercial and industrial) and domestic demand (which is not always based on metered consumption). Leakage could represent another significant demand, as it is a portion of the water put into the distribution and should be accounted for in modelling. Due to possible heterogeneity of WDSs that have developed haphazardly over time, it may even be necessary to allocate different levels of leakage to different parts of the system.

When modelling a water pressure zone, metered demand profiles are normally developed from standard metered profiles by consumer type (and large consumer meter profiles), and the leakage profile is determined from nightline and pressure measurements. Thus, the domestic profile is often the remaining demand that makes up the total demand profile for the zone (see Figure 6.9). For zones where the total demand can be measured for each DMA, it is therefore possible to construct a separate domestic profile for each DMA. However, this makes factoring up the model for future demands cumbersome, and often a compromise is reached by using a common domestic profile, as long as the model DMA profile is within 10% of the measured values.

6.5   Model building
6.5.1   Model purpose
Models are built for various purposes, with the aim of aiding analysis, design, operation or maintenance of WDSs. The first step in model building is to define its purpose, which in turn will affect the level of detail contained in the model. Model applications include master (strategic) planning, energy management, water quality analysis, regulatory reporting, emergency planning, water age analysis, planning of local distribution improvements, leakage analysis, resolution of system anomalies, surge (water hammer) analysis, etc.

Master planning or *strategic planning* involves various analyses to support the selection of capital improvement projects necessary to ensure the required level of service for the future. It uses long-term demand projections to assess the capability of the network to adequately serve its customers under a variety of conditions. The baseline master planning applications could include sizing of piping improvements, analysis of system pressures, storage capacity analysis, fire flow analysis and pump operation analysis. Due to the uncertainty associated with future demand projections and population growth, it is not necessary to use a detailed all-mains model, and a strategic (skeletonised) model is often employed. This type of model is normally constructed from an all-mains model with certain pipe sizes removed (e.g. <200 mm diameter).

Operations management involves assessment of the current system operating conditions and planning of improved operations schemes. For example, pump operation optimisation (pump scheduling) for energy management means maximising pumping during periods of cheaper electricity. This type of analysis can identify potential savings in operational expenditure, as pumping costs are one of the most significant investments made annually by a water utility. The model, which needs to be detailed in treating existing pumps and control valves, can also be used to assess the efficiency of the pumps and any need for replacement/refurbishment and the potential reduction in running costs that might result. Other modelling applications related to operations management include pressure management (to design pressure management schemes that contribute to a reduction in system leakage and burst incidents, extending the life of the assets and cutting pressure-related demand), fire flow analysis (investigation of the effects of exceptional demands and fire flow), storage facilities optimisation (to prevent excessive retention times and improve water quality), surge (water hammer) analysis, etc. These types of models may include all pipes above a certain diameter size (e.g. >150 mm) and even some important pipes of smaller diameter, where they are significant carriers.

Water quality modelling can be used for assessing system chlorination requirements (e.g. chlorine residuals) and simulation of pollution incidents/scenarios. For example, a water service provider needs to ensure that adequate chlorine residuals are maintained throughout the entire system in order not only to achieve the required level of protection but also to avoid excessive levels of residual chlorine leaving the treatment plant that can cause taste problems, leading to customer complaints. A reasonable balance can be achieved by maintaining the chlorine levels leaving the treatment plant below the taste threshold and providing booster disinfection (i.e. adding disinfectant at some critical locations in the system) such that disinfectant residuals are maintained at a level greater than the minimum for public health. Accurate assessment of water quality through modelling can help the water utility reduce the probability of health incidents and customer complaints. A well-calibrated hydraulic model is a prerequisite for a good water quality analysis model. Hydraulic models are also used on their own (i.e. without the need for water quality modelling) to investigate water-quality-related issues, such as water age analysis, source blending (tracking of the hydraulic boundary between sources) or sediment modelling in a WDS. An example of such use is the simulation of mains flushing to facilitate the effective removal of the pollutant or sediment from the system, which helps to minimise the disruption to normal supplies or to identify customers affected by flushing.

To illustrate some of the issues with planning, design and/or operations of WDSs, a simplified design problem is given next (Figure 6.10). The network, also known as the 'Anytown' WDS, was originally created as a challenging design benchmark for optimisation models (Walski *et al.*, 1987). The problem is simplified, as it does not involve many of the features of real systems, such, as for example, multiple pressure zones, seasonal and local demand fluctuations, fiscal constraints, uncertainty of future demands, ageing pipe issues and complicated staging of construction. However, it has been demonstrated that the Anytown problem is complex and serves as an illustrative example exhibiting many real system features, such as locating and sizing new pumps and service reservoirs (Mays, 2000). The task is to determine the most economically effective design to reinforce the existing system to meet projected demands, taking into account pumping costs (OPEX) as well as capital expenditure (CAPEX). The town is formed around an old centre (solid lines representing pipes in this area) situated to the south-east of pipe 34 (see Figure 6.10), where excavations are more difficult to undertake and, consequently, are more expensive. There is a surrounding residential

Figure 6.10 A water distribution network design problem
Modified with permission from Farmani *et al.* (2006). © IWA Publishing

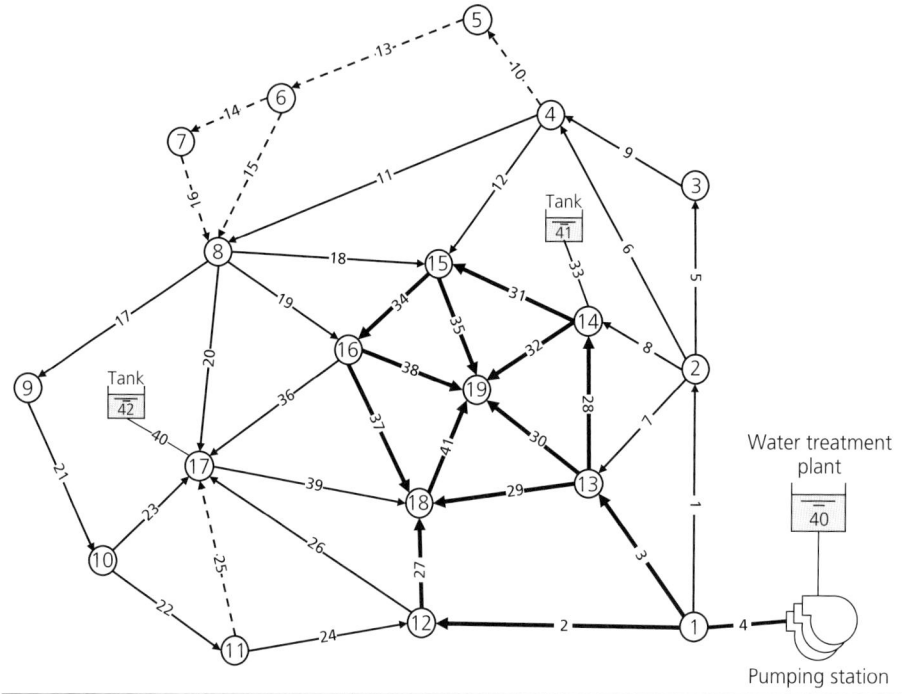

area (the pipes represented by thin solid lines), with some existing industries near node 17 and a projected new industrial park to be developed to the north. Options include duplication (in a range of possible diameters) of any pipe in the system, addition of new pipes (dashed lines), selecting the operation schedule of pumping stations and the provision of new reservoir storage at any location. Water is pumped into the system from a water treatment works by means of three identical pumps connected in parallel.

The design brief requirements are: (1) to maintain a minimum pressure of at least 28 m (or 40 psi, as in the original problem) at all nodes at average day flow as well as instantaneous peak flow (which is 1.8 times the average day flow); (2) to meet fire flow requirements with a minimum pressure of 14 m (20 psi); (3) to preserve the specified emergency volumes in reservoirs, which should empty and fill over their operational ranges during the specified average demand day; (4) to provide a certain level of reliability/redundancy in the network (e.g. by ensuring that each node is supplied by at lest two pipes); and (5) to achieve optimum OPEX and CAPEX for the solution. Additionally, one needs to consider water quality, as different design and operation plans will affect water quality in the network (which was not considered in the original problem statement). In order to address the design brief in a satisfactory manner, the designer has to consider installing new pipes, pumps and reservoirs or to consider existing pipes for cleaning and relining. Each of the 35 pipes could be considered for possible duplication or cleaning and relining, and six new pipes need to be sized (with 10 new diameters available). The existing pump station could be upgraded or a new one built. Furthermore, the location of elevated reservoirs, the overflow and the minimum normal day elevation, and the diameter and the bottom of the reservoir from the minimum normal day elevation should also be found. All of the above design options need to be individually costed, based on the type and size of the improvement and on the location of the particular network element (e.g. in the old centre or in the industrial zone) and the behaviour of the system tested against the above design requirements.

Even for this simplified problem, many alternative solutions are possible, each with its own advantages and disadvantages. Without network modelling, the task would be extremely difficult to complete, and it would not be possible to assess the behaviour of the system until the proposed improvements were implemented. Therefore, modelling can save some valuable time and resources. For example, if a model of the existing system is available, a number of alternative solutions can be examined using the updated model (i.e. with new pipes, pumps and reservoirs added) in a short period of time. To illustrate outputs and the benefits of modelling, two different solutions to the Anytown problem, with piping improvements and the reservoir locations and capacities, are given in Figure 6.11.

Both solutions satisfy the minimum pressure requirements and use the full reservoir operating range over the 24 hour simulation. However, the solution in Figure 6.11a has a total cost of about $13 million, a maximum water age value of about 35 hours (a surrogate for water quality: the higher the age, the worse the water quality) and a resilience index value of 0.18 (the larger the value, the better is reliability of the solution). The solution shown in Figure 6.11b has a total cost of about $17.3 million, a maximum water age of about 34.5 hours and a resilience index value of about 0.20. On closer inspection, the minimum cost solution needs one new reservoir at node 15, whereas the solution with the better reliability, but more expensive, needs one new reservoir at node 19 and relies on more pipes being duplicated, thus providing more capacity to supply demand even in the case of pipe failure in the system. The EPS also provides additional information on how the two solutions behave: for example, they both have high fluctuation in reservoir levels (i.e. high turnover, thus contributing to better water quality), but the solution in Figure 6.11b achieves a higher level of reliability, with one of the reservoirs staying full for most of the operating period (which, on the other hand, could contribute to water quality problems due to little turnover). Obviously, the ideal solution that satisfies all criteria to the fullest does not exist, but at least a compromise solution can be found based on detailed analysis afforded by modelling.

6.5.2 Data collection

Significant data need to be collected from various sources to build a good WDS model. The information comes from a number of sources, e.g. paper maps, GIS, SCADA, work management systems, etc., often stored in a number of formats. It is important that all sources are thoroughly checked for accuracy and consistency before the data are used in model building. For example, if GIS records are used to build the basic structure of the model, distribution network records drawings could be used to highlight any apparent

Figure 6.11 (a) A minimum-cost solution versus (b) a highly reliable solution
Modified with from permission Farmani *et al.* (2006). © IWA Publishing

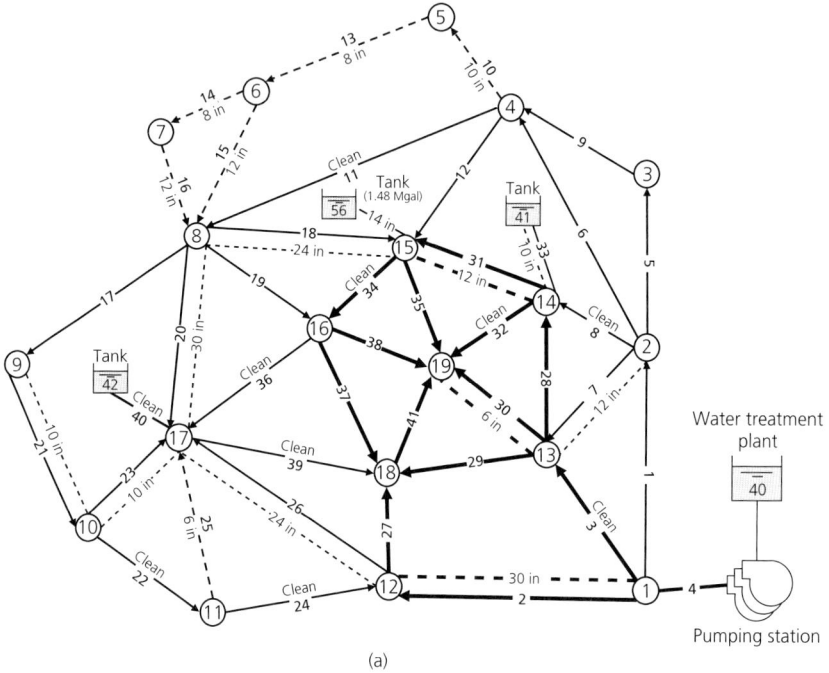

(a)

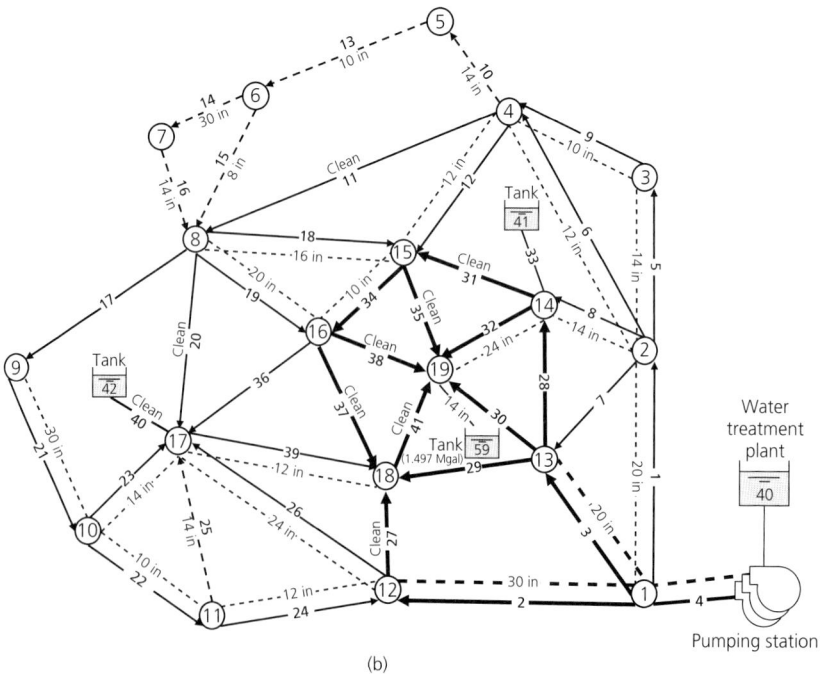

(b)

Figure 6.12 (a) Water GIS map and (b) the resulting model

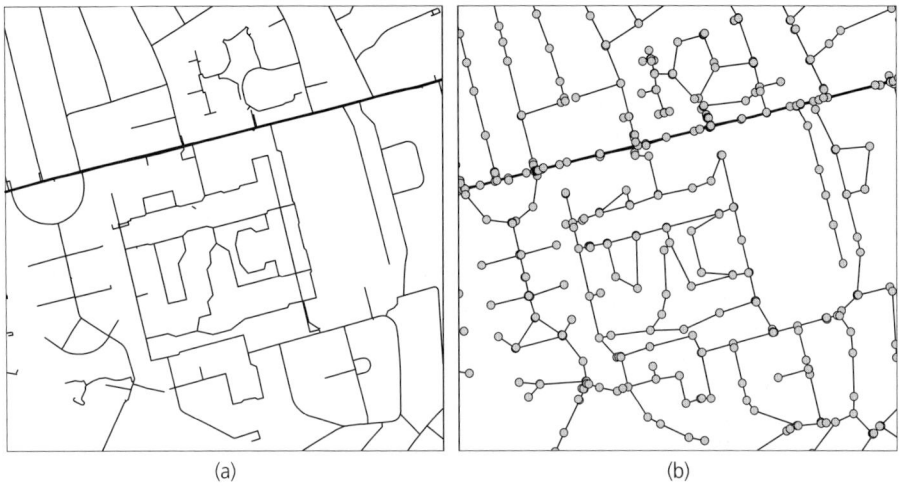

(a)	(b)

anomalies, errors or omissions prior to commencing model building. An example of a GIS map imported into modelling software is given in Figure 6.12.

As can be seen from Figure 6.12b, even if automated GIS import into the network model is possible, the model created from the GIS data has a much greater level of detail than needed. For example, a number of adjacent nodes on short pipes could be removed (to create 'logical pipes'), as they are not normally needed for most modelling applications. However, such alterations to the hydraulic model are not normally implemented in the GIS data, which poses a problem of model maintenance once the situation on the ground, and consequently in the GIS, changes.

Additionally, local knowledge of the structure of the system and its operations available from the personnel on the ground may provide valuable insight during model building. It is also interesting to note that the latest trend is to use readily available data, such as from SCADA systems, and link it to models, although, as with GISs, we are still some time away from the situation where the data are automatically loaded into the water model.

As each network is represented as a set of nodes and links, data associated with each of these network elements need to be collected and entered into the analysis software. This, for example, means having the correct pipe diameters, in the proper locations and with appropriate connectivity to neighbouring pipes. Further information needed includes service reservoirs with their appropriate shapes and dimensions, pump and valve information including settings and controls, elevation data for junctions and good data for demands (Speight et al., 2009).

Nodal points, or simply nodes, indentify pipe junctions, locations of changes in pipe characteristics (e.g. diameter, material), connections to reservoirs, pumps and valves,

Figure 6.13 DEM with a WDS

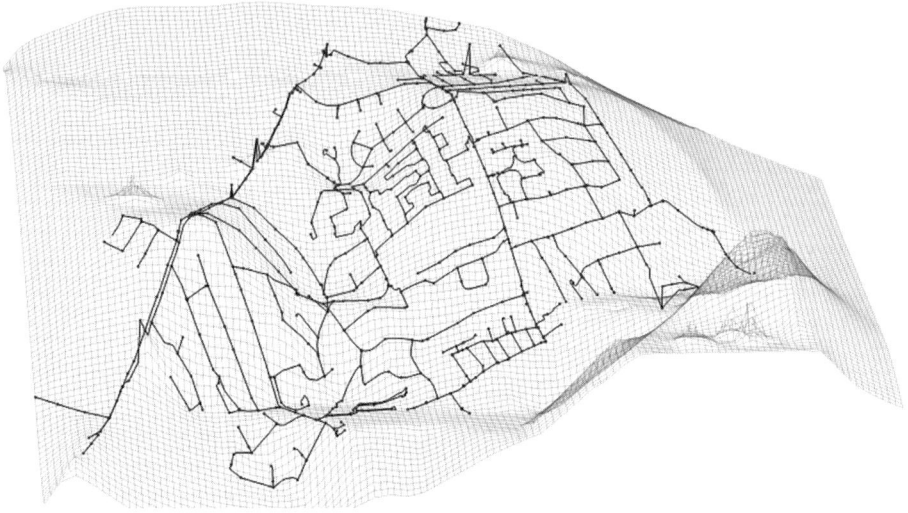

centres of demand, SCADA points and field measurement points. Topographic maps or digital elevation models (DEMs) are commonly used for establishing ground elevations for each of the nodes (Figure 6.13). These can be verified by global positioning system (GPS) measurements for important nodes, e.g. field test points, reservoirs and pump stations.

All water demands are allocated at network nodes. This is an approximation of the real situation where service connections are distributed along the pipe (not normally at model nodes), but is commonly utilised in network modelling software. The general rule for demand allocation is that nodal demands of the same type are lumped together in a demand area for a number of customers near a particular node. The demand along a pipe is split between the adjacent nodes, so that the boundary of the two demand areas passes equidistantly between the nodes. Different methods of allocation could be employed, ranging from developing demand areas based on postcode information and different types of consumers to the assignment of geocoded customer meters to the nearest nodes. The pattern of demand chosen for each customer type (e.g. commercial/industrial, domestic, etc.) should be determined either based on field tests (e.g. for large metered users) or based on the local knowledge of the likely pattern of usage. Typically, a 30 minute or 1 hour time step is used for the demand pattern. However, it should be checked that true peak conditions are not lost due to coarse time steps. It should be pointed out that service reservoirs and valves are normally modelled as nodes. Detailed information on these elements is needed for building a realistic model of the network system.

Link data, such as pipe diameters and roughness values, should also be entered into the modelling software. Pipe diameters are normally taken from the network drawings, GIS or asset databases, although care must be taken to use internal dimensions (i.e. the distance from one inner wall to the opposite) and not nominal diameters. Internal diameter may also change due to tuberculation, i.e. the build-up of deposit (e.g. scale or rust) in a pipe. Depending on the software conventions, pumps are sometimes modelled as links or nodes. Pump tests are a preferred way of obtaining the most important information about the pumps – efficiency curves – necessary for inclusion in the modelling software.

Nowadays, most modelling tools allow the use of both head loss equations, i.e. Hazen–Williams and Darcy–Weisbach (Colebrook–White). Regardless of which equation is used, roughness values cannot be measured and need to be estimated through calibration. First estimates of pipe roughness values can be obtained from standard tables using available information on pipe material, size, age and internal conditions. These values are then refined during the calibration process. Creating a well-calibrated model is more of an art than a science. Calibration is often limited to an operating range and a degree of accuracy. Hence, calibrating certain node pressures against field-recorded pressures to within a specified precision under specific operating conditions does not mean that predictions of within 1 m are guaranteed if the model is used to examine significantly different operating regimes. This is particularly true where high flows are involved, such as with re-zoning, significant pumping regime changes or exceptional demand conditions.

It is also worth remembering that measurement locations can only be used to calibrate the network upstream of the logging point; downstream, the network remains uncalibrated. Furthermore, calibrating pressures to within a pre-specified tolerance relative to the overall pressure does not guarantee a similar accuracy for flow at that point. Finally, attempting to calibrate a network at low-demand periods (i.e. during the winter) or a network with a low overall hydraulic gradient is less likely to give accurate predictions if the network is better utilised with higher flows in the future. For example, shut valves at low flows are more likely to go undetected during the calibration process.

A high degree of interest in automated calibration has been shown in the past by researchers, but it has been considerably less covered by practitioners, who mostly rely on trial-and-error procedures (Savić *et al.*, 2009). However, the recent development of commercial modelling packages which include an automated (optimisation) calibration module in the available software (Wu and Clark, 2009) has the potential to change the attitudes of practitioners.

6.6   Model calibration
6.6.1   Introduction
A WDS model must be compared with field measurements on the real system to demonstrate that it adequately predicts the behaviour of the system under a range of conditions for an extended period of time, or, in other words, that it behaves similarly to the real system. This applies to both models simulating water quantity and quality, which are widely used by planners, water utility personnel, consultants and many others involved in the analysis, design, operation or maintenance of WDSs. The iterative process of

modifying the model parameters until the output from the model matches an observed set of data is called *calibration*. The goal of calibration is 'to reduce uncertainty in model parameters to a level such that the accuracy of the model is commensurate with the type of decisions that will be made based on model predictions' (EPA, 2005). Pipe friction characteristics and nodal demands are the most commonly encountered calibration parameters.

Calibration methods are as old as network modelling; however, there are no general guidelines on how to calibrate a model, nor there are standards for defining what a well-calibrated model is. Furthermore, the level of calibration effort required will depend on the intended use of the model, as, for example, water quality models would require a greater degree of calibration than models used for strategic analysis, e.g. master planning.

In addition to the distribution system hydraulic models, water quality models are being used more often in everyday practice, and, thus, have to be calibrated. Such a water quality model relies on the flow and velocity information calculated by the hydraulic model, and, consequently, suffers if the hydraulic model is not properly calibrated (i.e. if inaccurate flow and velocity estimates from the hydraulic model are used). Tracer tests and online water quality data are starting to be used for water quality modelling applications. However, the cost associated with this more-intensive field work is often a barrier to its implementation (Speight and Khanal, 2009).

6.6.2   Field measurements

Historically, most water utilities relied primarily on field tests for calibration of their hydraulic models, but with the advent of SCADA, this type of information is being used more and more to augment the field tests. Well-calibrated measurement equipment is a must for a credible model calibration exercise. This is achieved by proper main-tenance and calibration of measuring devices, which should be performed at regular intervals. The most important field measurements are simultaneous flow and pressure readings at selected nodes in the network. Additionally, monitoring of control valve operations, reservoir levels and major users' consumption meters should also be done, preferably at the same time. Finally, pump flows/heads should be monitored and actual pump characteristics developed, as calibration should rely not only on pump manufacturer's curves but also on actually performed pump characteristics tests. The field test should be carried out for a period of time long enough to account for different demand/flow patterns. This will typically mean a full week, but could be extended if necessary.

Generally speaking, the more data that are used during calibration, the more confidence one can have in model predictions. However, there is a limit to the amount of field data that are available, often because of the costs involved in taking field measurements. Sampling design concerns the question of where to collect data during the field test. In other words, where should measurement equipment be placed in the network in order to collect data that, when used for the calibration of the model, will yield the best results. When considering the number and locations of measurement devices, the modeller should attempt to achieve the best trade-off between the field test cost and model prediction accuracy. This should avoid unnecessary expense in collecting redundant data, i.e. data whose information contribution is already contained within another measurement.

6.6.3 Calibration approaches

Numerous calibration procedures for water network models have been developed since the 1970s. Generally, practical calibration methods may be grouped into two categories (Savić et al., 2009):

1. iterative (trial-and-error) procedure models
2. implicit models (or optimisation models).

Iterative calibration models are based on specifically developed, trial-and-error procedures (Rahal et al., 1980; Walski, 1983; Bhave, 1988) where unknown parameters are updated at each trial using pressures and flows obtained from the simulation model. Once good agreement between the predicted results and field data is obtained, the process stops. Implicit calibration refers to a situation where an automated optimisation technique is used with a simulation model (Savić et al., 2009). The optimisation initially sets all input parameters and then, in each further iteration, updates them in order to minimise discrepancies between the model predictions and the corresponding field data. New parameters are then passed onto the simulation model, which in turn passes back obtained model predicted heads, flows, reservoir levels, etc. The optimisation tool employs an objective function to minimise the differences between measured and model-predicted variables. The recent development of commercial modelling packages, which include an automated (optimisation) calibration module in the available software (Wu and Clark, 2009), has the potential to improve the way that practitioners perform calibration and to reduce the cost of calibration.

Recently, Bayesian-based approaches to the calibration of WDS models have been developed (e.g. Kapelan et al., 2007). These new models improve upon the traditional parameter-fitting approaches by being able to obtain, in a single model run, not only the calibration parameter values but also the information on errors associated with these estimates and the corresponding model predictions (e.g. pressures and flows). In this way, a modeller knows which model predictions to trust more (or less). The Bayesian approach also provides a natural framework for maintenance, i.e. for the periodic recalibration of the simulation models.

6.6.4 Some calibration issues

Calibration of a WDS model is often an ill-posed problem. In practice, ill-posedness is typically manifested as non-uniqueness of the problem solution, and it is usually a consequence of the inadequate quantity and/or quality of measurement data, but it may also be a consequence of an inadequate model parameterisation scheme being used. A classic example of ill-posedness is the case of calibrating a WDS hydraulic model for unknown pipe roughnesses by collecting pressure measurement data at low-demand periods (i.e. during the winter) or in a network with a low overall hydraulic gradient. This is unlikely to give accurate calibration results, because nodal pressures are simply not sensitive enough to the changes in pipe roughness values due to low head losses in the network (a consequence of low demands, i.e. low flows/velocities in pipes). The model calibrated this way is less likely to have good predictions if it is utilised with higher flows. Several improvements can be made to condition an ill-posed problem, e.g. re-parameterisation of the WDS hydraulic model, provision of additional measurements under different conditions and/or incorporation of independent, prior information in calibration parameters (Kapelan et al., 2004).

In existing calibration practice, the calibration problem is considered solved once a relatively good model fit is obtained. However, this fit is usually evaluated by the visual plots of model predictions versus measurements and/or, in the best case scenario, measured by some objective function value obtained. This, however, may lead to unreliable results and, in some cases, to incorrect conclusions (Kapelan, 2010). A more thorough assessment of the results obtained should be performed. In addition to the determination of parameter values which is conventionally done, the following could be evaluated (Kapelan, 2010): (1) check the well-posedness of the calibration problem (to determine whether the obtained solution is identifiable/unique and, generally, if it is meaningful at all); (2) check the model fit using various statistics and analysis other than the objective function value and graphs (to thoroughly evaluate the model fit); (3) check the parameter and model prediction uncertainties, i.e. errors including residual analysis (to determine how reliable calibration results are).

The issue of sampling design is one which is closely related to the problem of calibration. The objective of sampling design for calibration is to, in the general case, determine what (e.g. pressure and/or flow), where (which pipe/node), how frequently (every 15 minutes, every minute, etc.), for how long (days, weeks, permanently) and under what conditions (regular demand, fire fighting conditions, etc.) to measure so that the data collected can be used to calibrate the relevant prediction model as accurately as possible. A number of different approaches have been developed in the past to achieve this (Yu and Powell, 1994; Bush and Uber, 1998; Kapelan et al., 2003).

6.6.5   Model maintenance

Over time, the predictions of a calibrated model become less certain, as changes to the real network occur. The accuracy of the model and its effective lifespan is normally determined by the amount of change in the network, making it difficult to set a universal number. These changes can usually be categorised in three groups:

1. network reconfiguration – changes to valve status (open/closed) or re-zoning
2. asset changes – new mains, pumping stations, pressure-reducing valves, etc., plus mains renewal
3. demand changes and development growth – changes in large consumers, new housing estates and mains.

Model maintenance is a means of maintaining the accuracy and extending the lifespan of a model to avoid rebuilding the model for every new project or application. New assets (e.g. pumping stations and mains) can be added to the existing model, and a limited calibration carried out on the new assets. Likewise, minor re-zoning or new housing can be added to the existing model.

Unrecorded valve changes can be accommodated by a recalibration of the model, and changes in commercial consumption by updating metered demands. However, after a period of time, when these become significant, it is likely to be more economic to rebuild and recalibrate the model. The costs associated with maintaining a distribution system model may be more easily justified if it is used for a variety of applications by a water utility.

REFERENCES

Bergeron, L. (1961) *Waterhammer in Hydraulics and Wave Surges in Electricity*. Chichester: John Wiley.

Bhave, P. R. (1988) Calibrating water distribution network models. *Journal of Environmental Engineering, ASCE* 114(1): 120–136.

Bush, C. A. and Uber, J. G. (1998) Sampling design methods for water distribution model calibration. *Journal of Water Resources Planning and Management, ASCE* 124(6): 334–344.

Brdys, M. A. and Chen, K. (1993) Joint state and parameter estimation of dynamic water supply systems with unknown but bounded uncertainty. In: Coulbeck, B. (ed.), *Integrated Computer Applications in Water Supply*, Vol. 1, pp. 335–355. Baldock: Research Studies Press.

EPA (2005) *Water Distribution System Analysis: Field Studies, Modeling and Management. A Reference Guide for Utilities*. Cincinnati, OH: Environmental Protection Agency.

Farmani, R., Walters, G. A. and Savić, D. A. (2006) Evolutionary multi-objective optimisation of the design and operation of water distribution network: total cost vs. reliability vs. water quality. *Journal of Hydroinformatics* 8(3): 165–179.

Filion, Y. R. and Karney, B. W. (2002) Sources of error in network modeling. *Journal of the American Water Works Association* 95(2): 119–130.

Kapelan, Z. (2010) *Calibration of Water Distribution System Hydraulic Models*. Saarbrücken: Lambert Academic Publishing.

Kapelan, Z., Savić, D. A. and Walters, G. A. (2003) Multi-objective sampling design for water distribution model calibration. *Journal of Water Resources Planning and Management, ASCE* 129(6): 466–479.

Kapelan, Z., Savić, D. A. and Walters, G. A. (2004) Incorporation of prior information on parameters in inverse transient analysis for leak detection and roughness calibration. *Urban Water* 1(2): 129–143.

Kapelan, Z., Savić, D. A. and Walters, G. A. (2007) Calibration of WDS hydraulic models using the Bayesian recursive procedure. *Journal of Hydraulic Engineering, ASCE* 133(8): 927–936.

Mays, L. W. (2000) *Water Distribution Systems Handbook*. New York: McGraw-Hill.

Obradovic, D. and Lonsdale, P. (1998) *Public Water Supply*. London: Spon.

Press, W. H., Flannery, B. P., Teukolsky, S. A. and Vetterling, W. T. (1990) *Numerical Recipes: The Art of Scientific Computing*. Cambridge: Cambridge University Press.

Rahal, C. M., Sterling, M. J. H. and Coulbeck, B. (1980) Parameter tuning for simulation models of water distribution networks. *Proceedings of the Institution of Civil Engineers* 69: 751–762.

Rossman, L. A. (2000) *EPANET 2: Users Manual.* Cincinnati, OH: Environmental Protection Agency.

Savić, D. A., Kapelan, Z. and Jonkergouw, P. (2009) *Quo vadis* water distribution model calibration? *Urban Water Journal* 6(1): 3–22.

Speight, V. and Khanal, N. (2009) Model calibration and current usage in practice. *Urban Water Journal* 6(1): 23–28.

Speight, V., Khanal, N., Savić, D. A., Kapelan, Z. and Jonkergouw, P. (2009) *Guidelines for Developing, Calibrating and Using Hydraulic Models.* Denver, CO: Water Research Foundation.

Walski, T. (1983) A technique for calibrating network models. *Journal of Water Resources Planning and Management, ASCE* 109(4): 360–372.

Walski, T. M., Brill, E. D., Gessler, J., Goulter, I. C., Jeppson, R. M., Lansey, K., Lee, H. L., Liebman, J. C., Mays, L., Morgan, D. R. and Ormsbee, L. (1987) Battle of the network models: epilogue. *Journal of Water Resources Planning and Management, ASCE* 113(2): 191–203.

Walski, T. M., Chase, D. V., Savić, D. A., Grayman, W. M., Beckwith, S. and Coelle, E. (2003) *Advanced Water Distribution Modeling and Management.* Waterbury, CT: Haestad Methods Press.

Wu, Z. Y. and Clark, C. (2009) Evolving effective hydraulic model for municipal water systems. *Water Resources Management* 23: 117–136.

Wylie, E. B. and Streeter, V. L. (1978) *Fluid Transients.* New York: McGraw-Hill.

Xu, C. and Goulter, I. C. (1998) Probabilistic model for water distribution reliability. *Journal of Water Resources Planning and Management, ASCE* 124(4): 218–228.

Yu, G. and Powell, R. S. (1994) Optimal design of meter placement in water distribution systems. *International Journal of Systems Science* 25(12): 2155–2166.

Zadeh, L. (1965) Fuzzy sets. *Information and Control* 8(3): 338–353.

Water Distribution Systems
ISBN: 978-0-7277-4112-7

ICE Publishing: All rights reserved
doi: 10.1680/wds.41127.171

ice
Institution of Civil Engineers

publishing

Chapter 7
Design of water distribution systems

Kalanithy Vairavamoorthy University of Birmingham, UK
Seneshaw Tsegaye University of Birmingham, UK
Harrison Mutikanga National Water and Sewerage Corporation, Kampala, Uganda
Frank Grimshaw Severn Trent PLC, Birmingham, UK

7.1 Introduction

Water distribution systems (WDSs) vary from simple to complex. The main objective of all water systems is to supply safe water for the cheapest cost. These systems are designed based on least-cost and enhanced reliability considerations, and design principles should satisfy both hydraulic and engineering requirements. The hydraulic requirements include pressure, velocity, sufficient flow, minimum operational cost, etc., and the engineering requirements involve selection of durable materials, system component configuration, ease of access to components, etc.

The chapter begins by considering the design objectives of WDSs. It describes performance indicators that can be used to drive WDS design. These include risk of low pressures and unplanned supply interruptions, and water quality standards, etc. The chapter then introduces optimisation techniques and how they can be applied in the planning and design of WDSs. A description will be provided on the role of multi-objective decision-making when optimising two or more conflicting objectives, and how this can be used to investigate the trade-offs involved between the conflicting objectives. The chapter concludes by providing a discussion on the need to develop WDSs that can cope with future uncertainties and change requirements.

7.2 WDS requirements
7.2.1 Design objectives

The primary purpose of a WDS is to provide good potable water to the public in sufficient quantities and pressure at all times. A WDS must therefore be properly designed to sustainably meet the objective at the least cost possible. A well-designed WDS should minimise operational costs, supply water for fire protection and provide a sufficient level of redundancy to support the minimum level of service during emergency conditions (i.e. power loss or water main failure). Achieving these objectives requires acquisition of basic information about the users, including historical water usage, population trends, planned growth, topography, and availability of water resources, to name just a few. This information can then be used to plan and design the WDS to provide sufficient water of good quality and adequate pressure.

7.2.2 Performance indicators and levels of service

WDSs often represent a large proportion of capital investment in water supply. This investment is buried in the ground and deteriorates with time. Because of the key role played by the WDS, it is critical that performance indicators (PIs) are established to evaluate its performance towards delivering the required levels of service.

Three major performance criteria are used to evaluate a WDS: *adequacy, serviceability* and *efficiency*. Adequacy refers to the delivery of an acceptable quantity and quality of water to the customer. Serviceability (reliability or dependability) measures the ability of the distribution system to consistently deliver an acceptable quantity and quality of water, while efficiency measures how well resources such as water and energy are utilised to produce the service. The task of measuring and evaluating performance is accomplished by performance assessment systems through well-defined PIs. One of the key requirements of a good PI is that it must be clearly defined, with a concise meaning and easy to understand, even by non-specialists – particularly by consumers. The level of service provided should be looked at from three perspectives:

- the regulator's perspective
- the customer's perspective
- the water supplier's perspective.

The levels of service PIs provided by the WDS are classified as structural, hydraulic, water quality and customer satisfaction. The PIs corresponding to each performance criterion are as follows (Deb *et al.*, 1995; Coelho and Alegre, 1998; Ofwat, 2009):

- *adequacy* – pressure, flow, water quality, customer satisfaction/complaints, response time to customer complaints
- *serviceability* – service interruptions, inoperable valves and hydrants, main breaks, water quality violations
- *efficiency* – leakage, metering functionality, pumping efficiency.

The indictors used for measuring the appropriate levels of service vary according to the local conditions. In England and Wales, the Water Act 1989 and the Water Industry Act 1991 established a regulatory framework which is based on the definition of quantifiable levels of service. In the UK the service level indicators are defined and regulated by the economic regulator, the Office of Water Services (Ofwat). For instance, the level-of-service indicator for pressure is assessed in terms of the number of connected properties that have received, and are likely to continue to receive, pressure below the reference level when demand for water is at a normal level. The reference level of service is defined as 10 m head of pressure at a boundary stop tap with a flow of 9 l/min (Ofwat, 2009).

7.2.2.1 Adequacy

The adequacy of a WDS is measured in terms of how well the customers of the system are served. Hydraulic performance measures relate to the delivery of adequate supplies of water, and are measured in terms of pressure and flow. Pressure in a WDS is measured in terms of the number of customers at risk of being supplied at pressure levels lower than the reference level. Water quality performance measures are assessed by the number of water quality violations that the system experiences. Customer satisfaction can be measured by surveys of customers, or in terms of complaints and the response rate to complaints or system resiliency (how quickly the system recovers from failure to meet the customer's needs).

7.2.2.2 Serviceability
Serviceability is a measure of the consistency of service that customers of the system experience, and is related to how well WDS assets are managed now and in the future. The structural performance measures include service interruptions, mains bursts, and hydrant and valve functionality. In terms of dependability, the water quality performance measure of interest is water quality violations of extended duration. Supply interruptions are measured in terms of the number of customers affected by interruptions to the supply lasting longer than the reference duration, without adequate warning and appropriate justification by the water service providers.

7.2.2.3 Efficiency
The efficiency of a WDS is measured in terms of three structural performance measures. The leakage–energy nexus measures how well water resources are utilised. Leakage, a significant component of water losses, represents a major waste of a precious resource, but the benefits of reducing leakage need to be balanced against the costs of finding and fixing leaks or replacing mains. The excessive use of energy for pumping can be a symptom of distribution system problems (undersized mains, tuberculation, main breaks, leakage, worn-out impellers, etc.). Customer metering under-registration due to inaccurate meters does not promote efficient use of water, and results in substantial revenue loss to the water service providers. All these may result in increased capital and operating costs, which may affect customers as tariff increases.

7.2.3 Basic design principles
In order to design a WDS, key basic principles must be adhered to, and these include the following (Mays, 2000):

- Water demand assessment and projections (design period and peak factors).
- Storage and water-balancing requirements.
- Analysis of the WDS using hydraulic modelling techniques:
  – laws of hydraulics (energy and continuity equations)
  – application of pipe head loss formulae (Colebrook–White or Hazen–Williams).
- Optimisation of the design.

7.3 Optimal design of WDSs
Designing a WDS is accomplished by trial and error methods of different alternative scenarios using a network solver. Because of the complex interactions between components, identifying changes to improve a design can be difficult even for small systems. This approach may not guarantee an optimal design of the WDS. This section highlights research efforts being made to obtain optimal designs for WDSs. The following section highlights the basic concepts in the optimal design of WDSs.

7.3.1   Problem formulation

Optimisation of a WDS is often viewed as the selection of the least net present value (NPV) combination of component sizes and settings such that the criteria of demands and other design constraints are satisfied (Zecchin *et al.*, 2005). However, many real-world decision-making problems need to achieve several objectives such as minimise risks, maximise reliability, minimise deviations from desired levels, minimise cost, etc. (Savić, 2002), and it is not always easy to attribute monetary values to these objectives. The monetary values usually include the costs for construction, operation and maintenance. According to Mays (2000), the main constraints are supplying the desired demands with an adequate pressure head being maintained at withdrawal locations. Also, the flow of water in a distribution network and the nodal pressure heads must satisfy the governing laws of conservation of energy and mass. Problem formulation is normally the most important part of the optimisation process. It involves the selection of design variables, constraints and objective functions.

7.3.1.1   Design variable

A design variable for the optimisation problem is any quantity or choice directly under the control of the designer. It involves many forms, as WDSs comprise many components and performance criteria. For example, the selection of diameters for all the pipes, pump types and locations, the sizing and locating of tanks, valve pressure settings and valve locations.

7.3.1.2   Constraints

A constraint is a condition that must be satisfied in order for the design to be feasible, and constraints can reflect resource limitations, user requirements, or bounds on the validity of the analysis models. The constraints in the WDS optimisation problem could be specified to include minimum and maximum allowable pressures at each demand point, minimum and maximum velocity constraints for each of the pipes, and water quality requirements. Further constraints may be added for materials and for different rehabilitation alternatives (cleaning, relining or both) (Walski *et al.*, 2003).

7.3.1.3   Objective functions

An objective function is a numerical value that is to be minimised or maximised. The optimal design of a water distribution network is often viewed as the least-cost optimisation problem. However, there are other possible objectives, such as network reliability, redundancy and water quality that can be included in the optimisation process. Conventionally, water engineers have treated these problems as single-objective optimisation problems instead of multi-objective ones. This type of optimisation is useful to provide decision-makers with insights into the nature of the problem, but usually it cannot provide a wide range of alternatives that trade different objectives against each other (Savić, 2002).

Once the design variables, constraints, objectives and the relationships between them have been chosen, the optimisation problem can be expressed mathematically. For example, the mathematical formulation of the WDS optimisation model for finding the least-cost combination of pipe sizes can be stated as

$$\min f(D_1, \ldots, D_{\text{npipe}} = \sum_{k \in \text{npipe}} C(D_k, L_k) \tag{7.1}$$

subject to

$$\sum Q_{\text{in}} - \sum Q_{\text{out}} = Q_e \tag{7.2}$$

$$\sum_{k \in \text{Loop}l} \Delta H_k = 0, \quad \forall l \in \text{NL} \tag{7.3}$$

$$\Delta H_k = H_{\text{u/snode},k} - H_{\text{d/snode},k} = \omega \frac{L_k}{C_k^\beta D_k^y} Q|Q_k|^{\beta - 1}, \quad \forall k \in \text{npipe} \tag{7.4}$$

$$H_{iN} \geqslant H_{\min iN}, \quad \forall iN \in \text{NN} \tag{7.5}$$

$$D_k \in [D], \quad \forall k \in \text{npipe} \tag{7.6}$$

where $D_1, \ldots, D_{\text{npipe}}$ are npipe discrete pipe diameter decisions selected from the set of commercial pipe sizes (D) and $C_k(D_k, L_k)$ is the cost of pipe k with diameter D_k and length L_k. Equation (7.2) represents conservation of mass for each node, where Q_{in} and Q_{out} are the flow into and out of the node, respectively, and Q_e is the external inflow (negative) or demand (positive) at each node. Equation (7.3) represents the conservation of energy around a loop, where ΔH_k denotes the head loss in pipe k and NL is the total number of loops in the system. The head loss in each pipe is the head difference between connected nodes (equation (7.4)), where C is the Hazen–Williams roughness coefficient (the Darcy–Weisbach equation can also be used). Equation (7.5) requires the total node pressure H for any node iN (where the total number of nodes is NN) is equal to or greater than a pre-specified minimum pressure $H_{\min}$.

7.3.2 Applications of multiple-objective optimisation to WDSs
The design of a WDS is often viewed as a single-objective, least-cost optimisation problem, with pipe diameters acting as the primary decision variables. The problem of choosing the best possible set of network improvements to make with a limited budget is a large optimisation problem to which conventional optimisation techniques are poorly suited. The exponential growth of the problem size with the increase in the number of discrete decisions persuades designers to use multi-objective optimisation approaches in WDS design. Three major improvements of the multi-objective optimisation approach have been identified by Cohon (1978):

1. A wider range of alternatives is usually identified when a multi-objective methodology is employed.
2. Consideration of multiple objectives promotes more appropriate roles for the participants in the planning and decision-making processes, i.e. *analyst* or *modeller* (who generates alternative solutions) and *decision-maker* (who uses the solutions generated by the analyst to make informed decisions).
3. Models of a problem will be more realistic if many objectives are considered.

Figure 7.1 Example of a dual-objective trade-off curve obtained in a single run of an optimisation routine

Data from Kapelan *et al.* (2005)

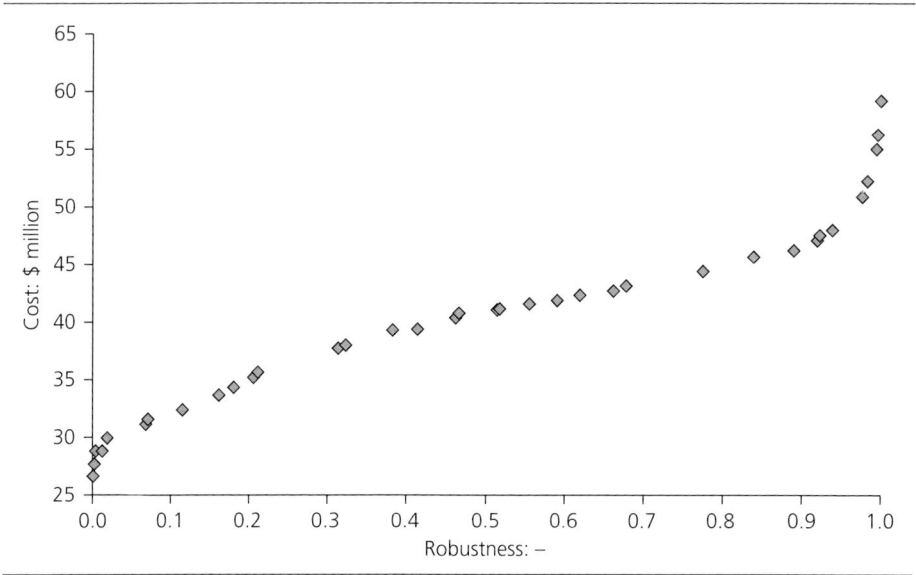

Optimisation techniques are tools to develop useful information for the decision-makers. However, single-objective models require that all design objectives must be measurable in terms of a single fitness function which depends on prior ordering of objectives (i.e. a weighting scheme). Thus, single-objective approaches place the burden of decision-making on the shoulders of the analyst. However, multi-objective optimisation allows decision-makers to assign relative values based on trade-off curves between different objectives (Savić, 2002).

An example of a dual-objective trade-off curve obtained in a single run of an optimisation routine (Kapelan *et al.*, 2005) is shown in Figure 7.1. This curve defines an efficient frontier, that is, the minimum network design cost against the level of system robustness achieved or, equivalently, the minimum robustness that can be attained for various levels of investment. Several points can be noted here; for example, that due to the discrete nature of the network design problem, the solutions on the trade-off curve are represented as discrete points, and to reach a robustness level of, say, about 80%, the minimum investment cannot be less than $45 million. Furthermore, the decision-maker could also benefit from the knowledge that increasing the robustness from 80% to 99% would require an increase in investment of almost $15 million.

In recent decades, the focus of optimisation for WDSs has also shifted from the use of traditional optimisation methods, such as linear programming, to decomposition methods, non-linear programming and the use of heuristics derived from nature

(HDNs), such as genetic algorithms (GAs), simulated annealing (SA) and, more recently, ant colony optimisation (ACO). These optimisation techniques encourage the implementation of multi-objective formulation in the planning and design of WDSs.

Multi-objective optimisation techniques have been successfully applied to the planning, design and management of WDSs. For example, Savić et al. (1997) have used a multi-objective optimisation technique for pump scheduling. The pump scheduling problem is formulated as two objectives: the minimisation of energy costs and the minimisation of pump switches. These multi-criterion methods provide a choice of trade-off solutions from which a decision-maker can select a suitable one to implement. Hence, this approach is used to find a spread of good trade-off solutions with respect to all objectives, and it achieves remarkable reductions in operation costs by optimising the pump-scheduling problem.

Improvement in WDS performance can be achieved through replacing, rehabilitating, duplicating or repairing some of the pipes or other components (pumps, tanks, etc.). Halhal et al. (1997) implemented a multi-objective approach (capital cost and benefit as dual objectives) to enhance the performance of an existing water distribution system. Pipes are considered in improving the performance by increasing the hydraulic capacity of the network, by cleaning, relining, duplicating or replacing existing pipes, increasing the physical integrity of the network by replacing structurally weak pipes; by increasing system flexibility, by adding additional pipe links; and by improving water quality, by removing or relining old (cast iron) pipes.

A study by Zecchin et al. (2005) developed parametric guidelines for the application of ACO algorithms to WDSs, and found this approach to outperform other HDNs for two well-known case studies. Zecchin et al. also highlight the high requirement for calibration in using HDNs because of the drawbacks associated with their searching behaviour, and, hence, performance is governed by a set of non-deterministic. Babayan et al. (2006) also use two objectives (the cost of the network design/rehabilitation and the probability of network failure) to formulate the problem associated with stochastic WDS design. The problem is solved using a multi-objective GA after converting it to an equivalent, simplified, deterministic optimisation problem. The design method was tested and successfully applied to a case study.

7.3.3 Example applications of optimisation for WDS performance, least cost and reliability design

7.3.3.1 Application of optimisations to WDS design

Multi-objective optimisation has been applied by the Regional Municipality of York (Canada) to develop the water supply for the area that arose due to an expected rapid development and doubling of population over a 35 year period taken from 1996 as the base. This enabled the optimal selection of feasible solutions from the options of different infrastructure configurations (Morley et al., 2001). Given the complexity of the problem, the GA approach was applied to a calibrated hydraulic model, to carry out extensive optimisation of the regional infrastructure requirements that were needed to meet the demands in 2031. The optimisation applied the common processes basically used for similar problems elsewhere. These included:

- a GA with an objective function defined on a set of decision variables, e.g. pipe diameters, reservoir storage requirements, etc.
- a calibrated model of the system to simulate its hydraulic behaviour and to ensure that continuity and head loss equations are satisfied at all times (hard constraints)
- a penalty term to penalise an insufficient level of service (soft constraints), e.g. pressures at nodes, imbalance of reservoir flows, etc.

The GA was programmed to ensure that the overall solution reflected cost-effectiveness on a whole-life cost basis. It took into account expected operating and maintenance costs as well as capital costs. Failure to meet a set of performance criteria (e.g. acceptable pressure ranges and storage requirements) was penalised during the optimisation process.

The optimal result proposed the construction of 85 new transmission mains added to the existing 750 mains in the system. Six new pumping stations were proposed and three old pumping stations were identified for expansion, with a total of 42 new pumps. The optimised plan also recommended three pumping stations to be decommissioned. Seven new ground reservoirs and elevated tanks were proposed (total volume of 78 million litres), with two existing elevated tanks identified for decommissioning.

This enabled the authority to realise considerable savings; for instance, the phased optimal solution for 2011 would cost $102 million instead of the previous manual solution cost of $156 million – a saving of $54 million or 35% (1997 Canadian dollars). The estimated cost for the 2031 planning horizon was found to be $194.42 million. The size of the problem solved by the GA, which is estimated to be 10^{357}, against that of the age of the universe, estimated at 14×10^9 years, demonstrates its superiority in dealing with complex, non-linear and discrete optimisation problems. It is not surprising that an impressive improvement over the manual solution has been achieved by the optimisation approach.

7.3.3.2 Determining the optimal level of service

Customer expectations of the reliability of water supply are likely to rise over time, leading to a need to improve the distribution system to reduce interruptions and ensure adequate pressure. For example, in the UK, there have been increased concerns about ensuring reliable supplies, following rising concerns about security and the prolonged loss of supplies to 138 000 households in Gloucestershire in 2007, as a result of flooding. This requires additions to the system to reduce the number of communities which are dependent on a single mains link from the network. However, the scale and pace of improvements need to be balanced against other potential service improvements, such as improving the taste and odour of water supplies, and the need to ensure customers' bills are affordable.

To review water prices in UK for 2009, water companies balanced service improvements and changes in bills by establishing customer willingness to pay for improvements, established through surveys using choice experiments. This involved presenting customers

Table 7.1 Level of performance and customers willing to pay

Service measure	Current performance	Potential change	Household willingness to pay	Non-household willingness to pay
Interruptions – number per year	11 500	7 500	£7.26	£1.76
	11 500	3 500	£14.53	£3.51
Low pressure – number of customers at risk	10 000	15 000	−£8.17	−£1.38
	10 000	5 000	£8.17	£1.38
	10 000	2 000	£8.17	£2.21

Data from Severn Trent Water (2009)

with alternative packages of services and bills, covering all the main potential areas of improvement to water and sewerage. Customers' choices of the best package allowed estimation, through statistical analysis, of the willingness to pay for improvements in each of the service measures. An example of the Severn Trent Water company's results for service measures affected by network performance is shown Table 7.1.

These willingness to pay results are used in optimisation modelling. For example, for a project to improve the resilience of the distribution system, the benefits of the project were assessed in terms of:

- the probability of water supply interruption before and after the scheme
- the numbers of customers affected.

This gives an average number of customers affected by interruptions per year, which is multiplied by the willingness to pay, to give a total benefit from the scheme. The NPV of the benefits was then compared with the NPV of costs (including the cost of carbon) to determine whether there were net benefits from the project and therefore whether it should be included in company plans.

7.3.3.3   Optimisation for the rehabilitation of WDSs

Improving the performance of a WDS involves a high cost while available financial resources are limited. Thus, there is a need for the development of optimal rehabilitation plans. Many objectives can be incorporated in rehabilitation decision models. For example, work by Cheung et al. (2003) demonstrates a multi-objective optimisation for the rehabilitation of WDS, recognising the multi-objective nature of the problem. The problem formulation is based on a three-objective network rehabilitation function such as the minimisation of cost and pressure deficit and the maximisation of hydraulic benefit, considering various combinations of rehabilitation choices. The individual objectives are expressed as follows.

Minimise cost: $F_1 = \sum_{l=\chi} c_l L_l + \sum_{k=\pi} c_k L_k$ (7.7)

where $l=$ the index of the pipes to be rehabilitated (cleaned or left unaltered)
$k=$ the index of the new pipes (replaced or duplicated)
$\chi=$ the set of alternatives related to the pipes requiring rehabilitation
$\pi=$ the set of alternatives for new pipes
$L=$ the length of the pipe
$c_l=$ the rehabilitation unit costs
$c_k=$ the unit costs of new pipes

The decision problem corresponds to the identification of pipes to be added in parallel or as a new pipe.

The pressure deficit is the sum of the maximum nodal deficits on the network for each demand pattern.

$$\text{Minimise pressure deficit: } F_2 = \sum_{i=1}^{LC} \max(H_j - H_{j\min})_i \quad j = 1, 2, \ldots, \text{nn} \tag{7.8}$$

where $j=$ the index of nodes
nn $=$ the total number of nodes in the system
$H_j=$ the energy at node j
$H_{j\min}=$ the required minimum energy at node j
LC $=$ the number of demand patterns considered (three demand patterns are investigated: peak, average and minimum demands)

Finally, a modified hydraulic benefit formulation is used as the third objective function. It is quantified as the difference between the pressure deficiencies in the network before improvement (DEFO) and after improvement (DEFP) represented by each solution found, which is calculated by

$$\text{DEFO/DEFP} = \gamma \sum_{j \in \chi} |H_{\min} - H|_j Q_j \tag{7.9}$$

$$\text{Maximise hydraulic benefit: } F_3 = \text{DEFO} - \text{DEFP} \tag{7.10}$$

where $\gamma=$ the specific weight of water
$\chi=$ the set of nodes related to the energy below the minimum required energy at node j
$H=$ the energy at node j
$H_{\min}=$ the required minimum energy at node j
$Q=$ the demand at node j

The study employed an elitist multi-objective evolutionary algorithm, called the strength Pareto evolutionary algorithm (SPEA), to generate a series of non-dominated solutions. The method uses these elite solutions to participate in the genetic operations along with the current population, in the hope of influencing the population to steer towards good regions in the search space. The rehabilitation analyses were conducted on a simple hypothetical network. The paper demonstrated the advantage of SPEA methods in visualising trade-offs (costs, pressure deficit and hydraulic benefit) and choosing a satisfactory solution, and showed the benefit of multi-objective optimisation techniques in decision-making for water distribution systems problems. Moreover, the importance of choosing appropriated recombination and mutation operators for reading a stable Pareto front was suggested by the authors to improve results using multi-objective evolutionary algorithms for WDSs.

7.4   Planning of WDSs under uncertainty

Projections of future global change pressures are plagued with uncertainties, and one of the major challenges that water service providers face is designing WDSs with incomplete information concerning the future global change pressures. Based on the possibility of returns above or below the expected, uncertainty has a good and a bad side (Bernanke, 1983).

Traditional planning of WDS development and management has been based on deterministic assumptions when future conditions are not known with certainty.

The level of investment for these systems is often very high (measured in billions of dollars), while decision-making has major and long-lasting consequences and is fraught with risk and uncertainty (Savić, 2005). Planners have been dealing with uncertainty and risk associated with global change pressures in recent decades, and have contributed to a significant shift away from the traditional design principles towards the design of WDSs under uncertainty (Babayan *et al.*, 2005, 2006; Giustolisi *et al.*, 2009).

7.4.1   Global change pressures affecting the future design of WDSs

WDSs are facing a range of dynamic global change pressures such as climate changes and variability, population growth and urbanisation, changes in public behaviour and socio-economic conditions, ageing and deterioration of buried infrastructure, technological development, governance, privatisation, etc. The major pressures include climate change, population growth and urbanisation, and ageing and deteriorating of water distribution infrastructure (Khatri and Vairavamoorthy, 2007).

In order to develop sustainable urban water systems, one must recognise the global change pressures. Hence, there is a need for us to pay attention to these changes in the context of how these systems will be designed and operated in an ever-changing environment. However, investment decisions are still one of the major challenges for WDSs which are performing in an inevitably changing environment.

7.4.1.1   Climate change

Climate change will affect the availability of water by interrupting the water cycle process. Although the regional distribution is uncertain, precipitation is expected to increase in higher latitudes, particularly in winter. Potential evapotranspiration (ET) rises with air temperature. Consequently, even in areas with increased precipitation, higher ET rates may lead to reduced run-off, implying a possible reduction in renewable water supplies. The frequency and severity of droughts could also increase in some areas as a result of a decrease in the total rainfall, more frequent dry spells, and higher ET.

Figure 7.2 Demand variation during the year – example for the UK (Worcestershire weekly distribution input)

Data from Severn Trent Water (2009)

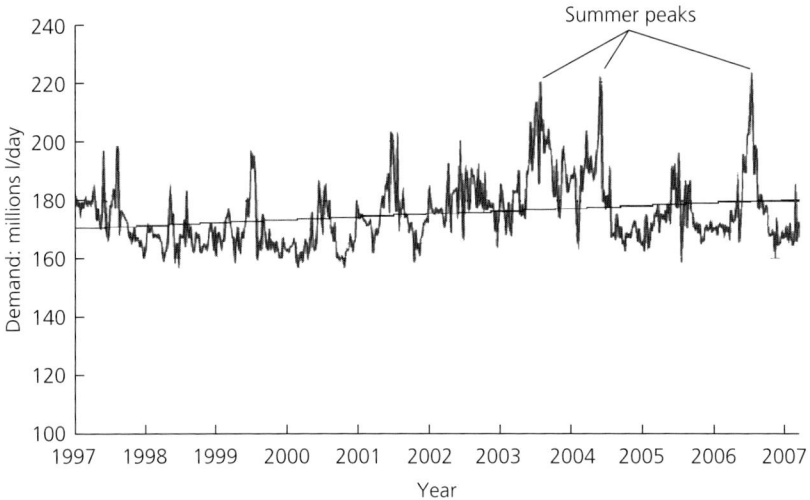

The periods of highest demand on a distribution system in the UK and in many other countries are during hot, dry spells, due to demand for uses such as garden watering. Figure 7.2 shows an example from the UK of how demand varies during the year.

With hotter, drier summers, the peaks can be expected to increase. This increases the likelihood that customers will experience low pressure, and, where service reservoir capacity is inadequate, the supply of water may completely fail. This is likely to require investment in increased capacity.

7.4.1.2   Population growth and urbanisation

Population growth and urbanisation is one of the world's major challenges in developing water distribution infrastructure. The numbers and size of the cities in the world are increasing due to the higher rate of urbanisation. In 1950, New York City and Tokyo were the only two cities with a population of over 10 million inhabitants. By 2015, it is expected that there will be 23 cities with a population over 10 million, and 19 of these will be in developing countries. In 2000, there were 22 cities with a population of between 5 and 10 million. Cities in developing countries are already facing enormous backlogs in infrastructure and services and confronted with insufficient water supply, deteriorating sanitation and environmental pollution. Population growth and rapid urbanisation will create a greater demand for water while simultaneously decreasing the ability of ecosystems to provide more-regular and cleaner supplies. Moreover, a rapid increase in built-up areas disturbs the local hydrological cycle and environment by reducing the natural infiltration opportunity. In addition, uncertainty in the prediction of population growth and urbanisation may cause variation in the water demand, which will affect system operation and create many severe problems. For example, high risk will be associated with a network which conveys insufficient water to meet people's living needs and industrial consumption. Under these uncertainties related to population growth and urbanisation, we need to think about increasing a system's ability to deal with these unpredicted global drives.

7.4.1.3 Ageing and deterioration of infrastructure systems

In some cities worldwide, the concept of asset management in water distributions has been neglected for many years. A significant proportion of WDSs are old and malfunctioning due to deterioration. This is also associated with a lack of records and data about the location and condition of the infrastructure and a lack of efficient decision support tools or management of the infrastructure (Misiunas, 2005). Deterioration of WDSs will result in high rates of water losses and higher chances of infiltration and exfiltration of water. This will create a higher probability of drinking water contamination and outbreak of water-borne diseases (Vairavamoorthy et al., 2008).

The escalating deterioration of WDSs also threatens our ability to provide safe and sufficient drinking water services for current and future generations. Deterioration is an exogenous factor which is coupled with many uncertainties. Therefore, this highlights the need for effective decisions against these drivers and the uncertainties associated with them.

7.4.2   Design of WDSs under uncertainty

On the one hand, it is foreseeable that uncertain future drivers will change the basic conditions of a WDS itself during its long operational life span. A WDS is affected by several uncertain future global drivers, discussed above. As a consequence, during its life span, fundamental alterations to the basic conditions are expected. On the other hand, long-lasting decisions for the planning of urban water systems have to be made, even if it is expected that the bases for these decisions will change. Hence, it is necessary to design a WDS that can adapt to the expected future global changes. Moreover, the possible future developments caused by global change are associated with several uncertainties. Additionally, there are uncertainties in the input data of the predicting models presently used, and additional uncertainties occur during the downscaling of the models at the local level.

For example, after construction had begun of a large project to expand water supplies in the Skane region in Sweden, water consumption unexpectedly stopped increasing. There has been nearly no growth in consumption for 10 years in the region, and there are multiple causes for this: a decreased birth rate, environmental regulations to reduce industrial water use, higher prices for water and restrictive government development policies. This history of water development for the Skane region highlights two major factors associated with the failure of the project: prediction of future uncertainty drives and long-term planning. This illustrates the hazard of conventional planning based on a deterministic growth forecast when a long period is considered (Erlenkotter et al., 1989). In most cases, such an uncertain shift in future water consumption may lead to an unnecessary investment and/or cause system performance problems. Therefore, the design of WDSs that operate in an uncertain world requires decisions to be made against a background of various sources of uncertainty.

The recognition of future uncertainty in both design requirements and the operating environment is the most important issue in WDS planning and management, and is a significant shift away from traditional practice (Hassan and de Neufville, 2006). Babayan *et al.* (2005) considered the uncertainty associated with water demand when predicting the behaviour of a system. The research focused on the design a WDS with minimum cost while meeting the pressure requirements in terms of a given robustness level under uncertain demand. A stochastic WDS design methodology was used to obtain robust and economic solutions for water distribution network design, where the robustness of the network is defined as its ability to provide an adequate supply to customers despite fluctuations in some or all of the design parameters.

Babayan *et al.* (2006) developed a multi-objective optimisation approach to formulate the problem associated with a stochastic (i.e. robust) WDS design under uncertain (future water consumption and pipe roughness) variables. The problem formulation is based on two parameters, such as minimisation of cost of the network design/ rehabilitation and probability of network failure. The most uncertain parameters, future water consumption and pipe roughness, are considered as an independent variable with a pre-specified probability density function (PDF). The problem is solved using GAs after converting it to an equivalent, simplified deterministic optimisation problem. The methodology was tested and compared using the well-known problem in the literature of New York tunnel reinforcement, and showed that neglecting uncertainty in the design process may lead to serious under-design of water distribution networks.

Giustolisi *et al.* (2009) proposed a procedure for robust design through a multi-objective (minimisation of design cost, maximisation of WDS robustness) approach, considering nodal demands and pipe roughness as uncertain variables. The research follows a two-step design procedure for computational efficiency: a deterministic design (i.e. con-strained least-cost design procedure) as the first step and, then, the deterministic solutions as an initial population to solve the robust design problem multi-objectively, implementing the minimisation of design costs and the maximisation of WDS robustness as objective functions. This research is a significant achievement in the design of WDSs under uncertainties.

Both infrastructure (i.e. WDSs) and living organisms struggle to survive in an ever-changing environment. So, it could be beneficial for WDS designers to replicate how living things function and interact with their environment. According to Darwin's theory of evolution, individuals or species that are better equipped to adapt to changing environments tend to live longer. Similarly, systems that are better equipped to adapt to changing environments will outlast more-rigid systems (Saleh *et al.*, 2003). Thus, if a system is to be designed for an extended design life and value delivery, the ability to cope with uncertainty and change has to be embedded in the system. In general, the design of and investment decisions for WDSs should consider the inevitable future alterations and look for ways of adding value from them. One fundamental way of doing this is by designing for flexibility.

7.5 Introduction to flexible designs for WDSs

One of the challenges that water service providers face is to design WDSs with incomplete information concerning future global change pressures. Relevant strategies are needed for coping with both negative (worse than expected) as well as positive (better than expected) outcomes associated with uncertainty (Dean, 1951). Three basic strategies have been identified by de Neufville (2004), including reducing the uncertainty in the system, increasing system robustness and increasing system flexibility.

Flexibility is proposed as a key property to be embedded (Schulz et al., 2000) in high-value assets, particularly if they are to be designed for an increased or longer design life. It has also been cited by many scholars as a key goal for dealing with uncertainty in the design of complex systems (Saleh et al., 2001; de Neufville, 2004). As postulated by Silver and de Weck (2007) and Zhao and Tseng (2003), increasing the flexibility of a system provides a potential solution to deal with uncertainties acting on systems which are required to adapt/evolve to future stages. Scholtes (2007) also recognises flexibility as way to transform a risk associated with uncertainty into an opportunity.

Uncertainty is not always a negative to be mitigated: it can also be a positive to be exploited (de Neufville, 2004). Based on the possibility of returns below or above expected, uncertainty has a good and a bad side (Bernanke, 1983). Flexibility offers an opportunity to exploit the benefits of uncertainty and to enhance the ability to act or to respond to future change requirements in a cost-effective manner. However, system flexibility has not yet received sufficient attention in the design of WDSs.

7.5.1 Definition of flexibility in WDSs

In recent years, flexibility has become a key concept in many fields, such as manu-facturing, software engineering, architecture, finance, etc. The theoretical background and definition of flexibility have been discussed by many researchers. However, very little effort has been made to define it formally and clearly for WDSs. Allen et al. (2001) define flexibility as the ease of changing the requirements of a system with a relatively small increase in complexity (and rework). Saleh et al. (2001) define the flexibility of a design as 'the property of a system that allows it to respond to changes in its initial objectives and requirements – both in terms of capabilities and attributes – occurring after the system has been fielded, i.e., is in operation, in a timely and cost-effective way'. Shah et al. (2008) characterise flexibility as 'the ability of a system to respond to potential internal or external changes affecting its value delivery, in a timely and cost-effective manner'. Fricke and Schulz (2005) also define flexibility as a 'system ability to be changed easily. External changes have to be implemented to cope with changing environments'. Lack of a clear definition of flexibility is the major problem in addressing its distinct features. Most of the confusion about the concept of flexibility comes from the subtle distinction between systems features. In general, three major problems in the existing flexibility theory can be identified:

1. Incompatibility of the existing definition of flexibility between one system and another.
2. The lack of a description for a measure of flexibility, or the ability to rank different designs according to their flexibility.
3. The overlap of the concept of flexibility with other properties describing change: changeability (adaptability, robustness, etc.). Adaptability characterises the ability of a system to adapt itself towards changing environments (Fricke and Schulz, 2005). Adaptation is an internally initiated change, while flexibility is externally initiated (Shah *et al.*, 2008). Robustness is the property of a system which allows it to satisfy a fixed set of requirements, despite changes occurring in the environment or within the system itself (Saleh *et al.*, 2001).

According to Upton (1994), constructing a definition of flexibility is not straightforward, since definitions are often coloured by a particular situation or problem. There is, therefore, a need to recast the existing definition of flexibility to suit urban water systems (i.e. WDSs). Based on its general definitions, flexibility can be defined for the field of urban water management as follows: the ability of urban water systems to use their active capacity to act or to respond to relevant alterations in a performance-efficient, timely and cost-effective way. This definition covers most of the important characteristics of flexibility, such as to overcome alterations, the capacity for change and the characteristics of the change process, and the metrics of flexibility, such as the cost of change, the duration of change and system performance.

7.5.2   Designing for flexibility

In order to design flexible WDSs that have the capability to cope with future alterations and to enhance the ability of a system to utilise the positive side of uncertainty, the following basic features should be addressed:

- Drivers: flexibility to *what* and *when*?
- Option: *what* flexibility and *where* to embed?
- Level: *how* much flexibility?

These basic features can be addressed through four major stages: uncertainty modelling; option identification and/or system alternatives; flexibility generation and evaluation; and decision-making. Similar approaches have been followed in the design of flexible systems in different disciplines (Ramirez, 2002; Cardin and de Neufville, 2008; Shah *et al.*, 2008). For example, Shah *et al.* (2008) developed a three 'D' (dice, design and decision, and discounting) concept in response to the common problem of uncertainty facing designs. 'Dice' represents the uncertain future within which the engineering solution will be delivering benefit. 'Designs and decisions' represents designers' control over current design choices and, as the design allows, over choices in the future, in response to the resolution of uncertainty. 'Discounting' is used to represent comparison between future benefits and costs associated with subsequent contingent decisions.

With respect to addressing the objectives of designing flexible WDSs, the four steps listed above can be followed. Uncertainty modelling addresses the issue with respect to *Flexibility to what and when?* It identifies and quantifies the key sources of uncertainties in WDSs. The distribution of uncertain parameters should also be limited to the number of possible future states. Option identification addresses the issue of *What flexibility and where to embed?* in the system. It attempts to identify the 'best' sets of options in a WDS that most likely offer better lifetime flexibility in the uncertain environment. The flexibility generation and evaluation stages analyse different system alternatives for the described future states. Flexibility-based decision-making should look at the 'best' set of alternatives (systems) that deliver good lifetime benefit. This is achieved through comparing with a non-flexible system. The flexibility generation and decision-making stages should attempt to address the issue of *How much flexibility?* These four major steps in designing flexible WDSs are discussed briefly below.

7.5.2.1 Uncertainty description and modelling

Future conditions will certainly differ from past sequences. A statistical analysis of recorded sequences and the stochastic generation of various possible future sequences are done to account for this. Since the statistical characteristics are themselves uncertain, there is no assurance that generated sequences are representative of the range of sequences that might occur in the future (Beard, 1982). However, the description and characterisation of these unknown conditions is the most important factor in the design of flexible WDSs. The capacity of uncertainty to be resolved in future is usually understood as the characteristic that allows it to generate value (Ramirez, 2002). Uncertainty is identified as a key element of flexibility. It creates both risks and opportunities in a system, and it is with the existence of uncertainty that flexibility becomes valuable (Nilchiani, 2005).

Uncertainties can be modelled by a number of different methods. These methods are classified into analytical and sampling approaches. The analytical approaches include first-order second-moment, second-order second-moment, etc., and the sampling methods include Monte Carlo simulation. The sampling-type methods are much more computationally demanding. The choice of the particular method depends on the information available, and none of the methods give a precise result (Nilchiani, 2005).

7.5.2.2 Option identification

In finance, an option is defined as the 'right but not the obligation' to perform an action. The action could be expanding, contracting, deferring or abandoning. The key feature of an option is that cost of exercising the option of using your right to perform an action. It is in this respect that an option has value (de Neufville, 2001). Real options are options that relate to physical assets rather than to financial instruments. Real options can be categorised as those that are either 'in' or 'on' projects. In exploring flexibility in engineering systems, options are also identified as flexibility 'in' and 'on' a system. Flexibility 'in' a system is a technical aspect of the design that enables the system to adapt to its environment, while 'on' a system relates to management decisions without altering technical components (de Neufville, 2002). For example, the flexibility to defer WDS expansion for a specific phase is non-technical, and is flexibility 'on' a system. Most of the sources for flexibility 'on' a WDS are well known. These sources include investment deferral, multistage deployment, expansion, etc.

The design of flexible systems which have the ability to contend with an ever-changing environment often needs an identification of flexible options in the system. The identification of potential flexible options has been discussed by several researchers. Cardin and Neufville (2008) define flexible design options as a physical component enabling flexibility 'in' a system. Several techniques have been used to identify the flexible options in a system, such as change propagation analysis (Eckert *et al.*, 2004), the sensitivity design structure matrix (Kalligeros, 2006), the interview method (Shah *et al.*, 2008), etc. However, the appropriateness of the method depends on the type of system and the source of uncertainty, and needs to be explored.

7.5.2.3 Generation and valuation of flexibility

Generating flexibility in a system design is an investment problem: a premium has to be paid so that an option can be exercised later. The investment decision depends on the trade off between the cost of capturing the options and the expected benefit that may arise from future uncertainties. The estimation of the value of flexibility has three major elements (de Neufville, 2002):

1. estimation of the loss associated with the system without flexibility
2. calculation of the value of the flexible options
3. identification of the strategies for exploiting the options, to permit the best use of the flexibility built into the system.

The generation and valuation of flexibility should consider the changes that result in a change in value delivery. Changes and responses to changes can be measured in the form of costs and monetary or non-monetary benefits. The criteria for generating and measuring flexibility can be deduced from the definition of flexibility (Eckart *et al.*, 2010):

- *Capability for change*. The capability for change indicates for which uncertain future states a change of the system is possible.
- *Performance of the system*. Flexibility should guarantee that future alterations have minor impacts on the system performance.
- *Costs of change*. Due to the long operational life-span of WDSs, the costs for several changes in the whole life-span of systems should be considered. The life-cycle costs should consider the cost of planning, development and management of a system, the cost of embedding flexibility 'in' or 'on' a system and damage costs associated with missing/delayed action on altering conditions.
- *Duration of change*. The duration of change is the period which is required to adapt the system to new requirements. This matrix may not be a significant factor, as the time is usually sufficient to react to alterations in a WDS.

Different techniques can be used to value flexibility and identify a strategy that effectively hedges risks and ensures optimality in terms of the criteria listed above, such as discounted cash flow (DCF), decision analysis (DA) and real options analysis (ROA). DCF-based approaches, such as NPV, the internal rate of return and the payback period, require the analyst to estimate the net cash flows during the design life. These methods presume one line of action from time zero, and evaluate the project when fixed. These techniques do not consider the dynamics of the project in the valuation (Arboleda and Abraham,

2006). ROA and DA are better suited to deal with uncertainty. Both of these techniques view projects as processes that management can continually modify in the light of changes in the environment. They promote a change from a deterministic type of valuation to a dynamic planning process, which encourages flexible designs that effectively deal with the uncertainties in the environment (Ramirez, 2002). According to Nilchiani and Hastings (2007), the choice of the valuation method depends on how well the methodology can capture a specific type of uncertainty in a system. The valuation of flexibility should also be coupled with a decision-making process.

7.5.2.4   Decision-making under uncertainty

Investment decisions are one of the major challenges for urban water infrastructure (i.e. WDSs) operating in an environment that is, inevitably, changing. Shah *et al.* (2008) discuss attempts at and challenges of decision-making under uncertainty. The goal of these designers is to develop engineering solutions that meet their needs both now and in the future, while that future is uncertain. They therefore suggest a solution that will deliver high value in a variety of different possible futures. They also attempt to create designs that allow them to make changes and adjustments to the engineering solution so that they can maximise value once the future is known. Since they must make design choices now on the promise of future benefits, their decisions are based on the perception of the value of the future benefits as seen at the time of decision.

Flexibility-based decision-making should follow a general quantitative approach to making decisions which are suitable for a wide range of future uncertain states. Decisions are based on a set of possible future conditions that are likely to occur, a list of considerations, a known result for each alternative under each condition (state of nature), an estimated probability for each future condition, and a set of decision criteria under which decisions are made to select the best alternative. Usually, decision criteria under uncertainty are chosen by selecting the alternative which has the best pay-off based on one of the following decision criteria:

- maximin – the 'best worst' pay-off establishes the minimum outcome
- maximax – the 'best best' pay-off establishes the best possible outcome
- Laplace – the 'best average' pay-off establishes the average pay-off, assuming each future condition is equally likely
- minimax regret – the 'best worst' regret minimises the difference between the realised pay-off and the best pay-off for each future condition

For a large design space, the decision-making process may demand an optimisation approach aimed at optimising the value of decision criteria. In addition, it requires a specific system to be chosen as a baseline (usually a non-flexible system), to determine the whole-life economic gain as well as the associated regret.

REFERENCES

Allen, T., Moses, J., Hastings, D., Lloyd, S., Little, J., McGowan, D., Magee, C., Moavenza-deh, F., Nightingale, D., Roos, D. and Whitney, D. (2001) *ESD Terms and Definitions.* Cambridge, MA: Massachusetts Institute of Technology, Engineering Systems Division.

Arboleda, C. A. and Abraham, D. M. (2006) Evaluation of flexibility in capital investments of infrastructure systems. *Engineering Construction and Architectural Management* 13(3): 254.

Babayan, A., Kapelan, Z., Savić, D. and Walters, G. (2005) Least-cost design of water distribution networks under demand uncertainty. *Journal of Water Resources Planning and Management* 131(5): 375–382.

Babayan, A. V., Savić, D. A. and Walters, G. A. (2006) Multi-objective optimisation of water distribution system design under uncertain demand and pipe roughness. In: Castelletti, A. and Soncini-Sessa, R. (eds), *Topics on System Analysis and Integrated Water Resource Management*, pp. 161–172. Amsterdam: Elsevier.

Beard, L. R. (1982) Flexibility – a key to the management of risk and uncertainty in water supply. In: *Optimal Allocation of Water Resources*, No. 135, pp. 177–183. Paris: International Association of Hydrological Sciences.

Bernanke, B. S. (1983) Irreversibility, uncertainty, and cyclical investment. *Quarterly Journal of Economics* 98(1): 85–106.

Cardin, M.-A. and de Neufville, R. (2008) *A Survey of State-of-the-art Methodologies and a Framework for Identifying and Valuing Flexible Design Opportunities in Engineering Systems*. Cambridge, MA: Massachusetts Institute of Technology.

Cheung, P. B., Reis, L. F. R. and Carrijo, I. B. (2003) Multi-objective optimisation to the rehabilitation of a water distribution network. network. In: Maksimovic, C., Butler D. and Memon, F. A. (eds), *Advances in Water Supply Management*, pp. 315–325. Rotterdam: Balkema.

Coelho, S. T. and Alegre, H. (1998) Performance indicators for water supply and wastewater drainage systems. In: *ICTH Report, LNEC*. Lisbon [in Portuguese].

Cohon, J. L. (1978) *Multiobjective Programming and Planning*. New York: Academic Press.

de Neufville, R. (2001) Real options: dealing with uncertainty in systems planning and design. In: *5th International Conference on Technology Policy and Innovation*. Delft.

de Neufville, R. (2002) Architecting/designing engineering systems using real options. In: *Engineering Systems Division Internal Symposium, Massachusetts Institute of Technology*. Cambridge, MA.

de Neufville, R. (2004) *Uncertainty Management for Engineering Systems Planning and Design*. Cambridge, MA: Massachusetts Institute of Technology.

Dean, J. (1951) *Capital Budgeting: Top-management Policy on Plant, Equipment, and Product Development*. New York: Columbia University Press.

Deb, A. K., Hasit, Y. J. and Grablutz, F. M. (1995) *Distribution System Performance Evaluation*. Denver, CO: AWWA.

Eckart, J., Sieker, H. and Vairavamoorthy, K. (2010) *Flexible Urban Drainage Systems*. Water Convention Singapore International Water Week.

Eckert, C., Clarkson, P. J. and Zanker, W. (2004) Change and customisation in complex engineering domains. *Research Engineering Design* 15(1): 1–21.

Erlenkotter, D., Sethi, S. and Okada, N. (1989) Planning for surprise: water resources development under demand and supply uncertainty I. The general model. *Management Science* 35(2): 149–163.

Fricke, E. and Schulz, A. P. (2005) Design for changeability (DfC): principles to enable changes in systems throughout their entire lifecycle. *Systems Engineering, New York* 8(4): 342.

Giustolisi, O., Laucelli, D. and Colombo, A. F. (2009) Deterministic versus stochastic design of water distribution networks. *Journal of Water Resources Planning and Management, ACSE* 135(2): 117–127.

Halhal, D., Walters, G. A., Ouazar, D. and Savić, D. A. (1997) Water network rehabilitation with structured messy genetic algorithm. *Journal of Water Resources Planning and Management, ASCE* 123(3): 137–146.

Hassan, R. and de Neufville, R. (2006) Design of engineering systems under uncertainty via real options and heuristic optimisation. [Unpublished paper.] Cambridge, MA: Massachusetts Institute of Technology.

Kalligeros, K. (2006) *Platforms and Real Options in Large-scale Engineering Systems.* Cambridge, MA: Massachusetts Institute of Technology.

Kapelan, Z. S., Savić, D. A. and Walters, G. A. (2005) Multiobjective design of water distribution systems under uncertainty. *Water Resources Research* 41: W11407–W11421.

Khatri, K. B. and Vairavamoorthy, K. (2007) Challenges for urban water supply and sanitation in the developing countries. In: *Symposium on Water for a Changing World – Enhancing Local Knowledge and Capacity.* Delft.

Mays, L. W. (2000) *Water Distribution Systems Handbook.* New York: McGraw-Hill.

Misiunas, D. (2005) *Failure Monitoring and Asset Condition Assessment in Water Supply Systems.* Lund: Department of Industrial Electrical Engineering and Automation, Lund University.

Morley, M. S., Atkinson, R. M., Savi, D. A. and Walters, G. A. (2001) GAnet: genetic algorithm platform for pipe network optimisation. *Advances in Engineering Software* 32(6): 467–475.

Nilchiani, R. (2005) *Measuring Space Systems Flexibility: A Comprehensive Six-element Framework.* Cambridge, MA: Massachusetts Institute of Technology.

Nilchiani, R. and Hastings, D. E. (2007) Measuring the value of flexibility in space systems: a six-element framework. *Systems Engineering* 10(1): 26–44.

Ofwat (2009) *Service and Delivery – Performance of the Water Companies in England and Wales 2008–09. Supporting Information.* Birmingham: Ofwat.

Ramirez, N. (2002) *Valuing Flexibility in Infrastructure Developments: The Bogota Water Supply Expansion Plan.* Cambridge, MA: Massachusetts Institute of Technology.

Saleh, J. H., Hastings, D. E. and Newman, D. J. (2001) *Extracting the Essence of Flexibility in System Design.* Cambridge, MA: Massachusetts Institute of Technology.

Saleh, J. H., Hastings, D. E. and Newman, D. J. (2003) Flexibility in system design and implications for aerospace systems. *Acta Astronautica* 53(12): 927–944.

Savić, D. (2002) Single-objective vs. multiobjective optimisation for integrated decision support. In: Rizzoli, A. E. and Jakeman, A. J. (eds), *Integrated Assessment and Decision Support. Proceedings of the First Biennial Meeting of International Environmental Modelling and software Society*, pp. 7–12. Lugano.

Savić, D. A. (2005) *Coping with Risk and Uncertainty in Urban Water Infrastructure Rehabilitation Planning.* Exeter: School of Engineering, Computer Science and Mathematics, University of Exeter.

Savić, D., Walters, G. and Schwab, M. (1997) Multiobjective genetic algorithms for pump scheduling in water supply. In: *AISB International Workshop on Evolutionary Computing. Lecture Notes in Computer Science*, No. 1305, pp. 227–236. Berlin: Springer-Verlag.

Scholtes, S. (2007) *Flexibility: The Secret to Transforming Risks into Opportunities.* www.eng.cam.ac.uk/~ss248/publications/BusinessDigest.pdf [accessed 01.09.10].

Schulz, A. P., Fricke, E. and Igenbergs, E. (2000) Enabling changes in systems throughout the entire life-cycle – key to success? In: *10th annual INCOSE Conference.* Minneapolis, MN.

Severn Trent Water (2009) *Severn Trent Water Final Business Plan April 2009.* Chapter C1/B5. Coventry: Severn Trent Water.

Shah, N. B., Viscito, L., Wilds, J., Ross, A. M. and Hastings, D. (2008) Quantifying flexibility for architecting changeable systems. In: *MIT System Design and Management Thesis Seminar.* Cambridge, MA: Massachusetts Institute of Technology.

Silver, M. R. and de Weck, O. L. (2007) Time-expanded decision networks: a framework for designing evolvable complex systems. *System Engineering, New York* 10(2): 167.

Upton, D. M. (1994) The management of manufacturing flexibility. *California Management Review* 36: 72–72.

Vairavamoorthy, K., Gorantiwar, S. D. and Mohan, S. (2008) Intermittent water supply under water scarcity situations. *Water International* 32(1): 121–132.

Walski, T. M., Chase, D. V., Savić, D. A., Grayman, W., Beckwith, S. and Koelle, E. (2003) *Advanced Water Distribution Modeling and Management.* Waterbury, CT: Haestad Press.

Zecchin, A. C., Simpson, A. R., Maier, H. R. and Nixon, J. B. (2005) Parametric study for an ant algorithm applied to water distribution system optimisation. *IEEE Transactions on Evolutionary Computation* 9(2): 175–191.

Zhao, T. and Tseng, C. L. (2003) Valuing flexibility in infrastructure expansion. *Journal of Infrastructure Systems* 9(3): 89–97.

REFERENCED LEGISLATION

The Water Act 1989, Chapter 15. London: HMSO.
The Water Industry Act 1991, Chapter 56. London: HMSO.

Water Distribution Systems
ISBN: 978-0-7277-4112-7

ICE Publishing: All rights reserved
doi: 10.1680/wds.41127.193

Chapter 8

Operation, maintenance and performance

Joby Boxall University of Sheffield, UK
John Machell University of Sheffield, UK
Neil Dewis Yorkshire Water Services, Bradford, UK
Ken Gedman Consultant, UK
Adrian Saul University of Sheffield, UK

8.1   Introduction

Since their inception by the Romans, water supply systems have developed in piecemeal fashion around available sources to meet customer demands and public health requirements. Driven by different needs over the years, each system has been built from a variety of materials, using different design and construction techniques and a whole host of piecemeal layouts. In addition, little focus was given to the individual source-water quality characteristics of the water they conveyed or of the treatment process, which were often poor. As a consequence, such poor-quality water may have caused the system materials to deteriorate at different rates via a number of physical, chemical and biological mechanisms. Piecemeal development, system complexity, aging asset stock with low renewal rates, and asset deterioration are the primary causes of many of the current operational, maintenance and performance challenges that face the water industry today.

Recent regulatory and customer-driven service level agreements, especially for continuity of supply, leakage and water quality, have challenged the industry to be more efficient and provide good value for money and return for stakeholder investment in an environment of escalating cost. In order to meet these challenges, the industry is changing from a reactive to a proactive management style, and beginning to employ new technologies and methods to reduce water losses and to optimise energy use.

This chapter commences by describing the piecemeal development of water distribution pipe networks to provide a clean supply of wholesome drinking water to ensure public health through to the more recent operational drivers associated with water quantity (e.g. reduced leakage and minimal energy use) and water quality (e.g. a reduced number of customer contacts). It is this piecemeal development, rather than idealised design, as presented in the last chapter, that is often the primary cause of many of the operational, maintenance and performance challenges that face the water industry.

A focus has also been given to an overview of the regulations and standards, which range from a need to meet stringent regulatory standards to the softer measures of customer expectations and customer-oriented care. The main body of the chapter is structured around the operation and maintenance cycle under the 'MAIDE' concept of five core elements:

- monitoring
- analysis
- interventions
- decision
- evaluation.

Under these headings, the relevant issues, techniques and approaches are discussed with reference to both the quantity of water, including loss of supply and leakage, and the quality of water. The chapter is developed first by reference to historical tried-and-tested methodologies, then through the latest developments and approaches to optimise system performance, operation and maintenance.

8.2   Historic development of networks and regulation

The principle function of a distribution network is to supply wholesome water to the point of use, when required, as efficiently as possible. The early development of water distribution dates back to the first century AD, when the 1.2 million inhabitants of Rome enjoyed a supply of 900 l/day. In the UK, relatively simple systems were expanded at the time of the Industrial Revolution to meet the needs of populations as they grew, such that by the 1960s over 65% of the distribution network in the UK had been constructed. These systems were constructed using basic engineering principles that transferred water from a source to the point of delivery, with the inclusion of storage to meet peak demands and some resilience incorporated into the system to ensure the security of supply to strategic locations. Since this time, significant further piecemeal development of the networks has taken place (approximately 30% of the system), driven by changing demand due to population growth and the need to improve the networks to meet regulatory standards. In this latter respect, three notable acts have contributed significantly to the progressive design and development of the distribution networks:

- The Water Act 1945
- The Water Act 1973
- The Water Act 1989.

The Water Act 1945 focused largely on providing a blueprint for the development of water distribution systems, with guidance and clarity on how to modify and expand the existing network. However, the Acts of 1973 and 1989 had a greater impact on the strategies for operation and maintenance of the distribution network, and resulted in a change in the structure of the industry. The Water Act 1973 brought together the Regional Water Authorities, which allowed greater connectivity of the disparate piecemeal systems. This was followed by the Water Act 1989, which heralded the last significant organisational change to the water industry in England and Wales that was the privatisation of the water companies, bringing financial resources previously denied under the PSBR (public sector borrowing requirement) regime but accompanied by far tighter financial discipline from the capital markets. It had a huge influence.

It is clear, therefore, that the current operation and maintenance of water distribution systems is complicated and has been influenced by the many different design criteria, regulatory standards, and construction methods and materials. However, it may be argued that possibly the most fundamental issue associated with the operation and maintenance of the systems is that they are underground and that, if they work well, the concept of 'out of sight out of mind' has often been applied.

8.3 Monitoring

The operation and management of water distribution systems produces an inherent conflict in that there is a need to provide a continuous supply of water to meet demand while at the same time delivering wholesome and high-quality water to the customer's tap. Despite the initial purpose of the water supply system to ensure public health, conservative design approaches have often resulted in over-engineered solutions; for example, the occasional need to ensure the delivery of high demands for fire-fighting have led to systems that may be considered to be oversized. Such networks contain significantly more water than is required to meet the usual daily demands. Many systems contain significant redundancy, helping to ensure the continuity of supply under most event scenarios. For example, such systems are heavily meshed with numerous loops and valves, which facilitates maintenance and repair whereby, at the time of a malfunction event (e.g. bursts, discolouration, etc.), small numbers and lengths of pipe, and hence customers, may be isolated to undertake a repair. Such over-engineered systems also have an advantage in that they are adaptable to meet the changing requirements of customers, to accommodate population growth and the capacity to supply new developments. In contrast, it is known that the quality of water deteriorates with time, and hence to maintain water quality it would be preferable to have smaller-diameter pipes that supply only the peak demands, plus minimal headroom for events and network developments, as these smaller pipes produce higher flow velocities, thereby reducing the residence time and age of the of water within the system. This reduces the potential of the water to deteriorate. However, it should be noted that it is feasible to optimise the operation and the management of over-engineered systems with redundancy, such that areas of low velocity and high residence times are minimised (Prasad and Walters, 2006).

To balance the conflict between the quantity and quality issues, the water industry in England and Wales is regulated by two independent bodies:

- The Office of Water Services (Ofwat) is responsible for economic regulation, flow and pressure standards, and policing for flow and pressure.
- The Drinking Water Inspectorate (DWI) is responsible for all aspects of water quality.

These two bodies are effectively responsible for the regulation of the operation and maintenance activities that water companies undertake. Additionally, there are three other bodies that influence the performance, targets, and operational and maintenance techniques that are used by water companies:

- The Consumer Council for Water in the UK was set up in 2005 to provide a strong voice for water and sewerage consumers in England and Wales; with a mission to ensure that the consumers' voice was heard and to seek to ensure that customers were represented at the heart of the industry's decision making.
- Water UK is the association that represents all UK water and wastewater companies and suppliers at a national and European level. It provides a framework for the water industry to engage with government, regulators, stakeholder organisations and the public, and actively seeks to develop policy and improve understanding in areas that involve the industry, its customers and stakeholders
- The Water Regulations Advisory Scheme (WRAS) is primarily concerned with the materials and products that are used in the water supply industry. It has a significant influence on the products and tools available for the operation and maintenance of water supply systems.

8.3.1 Quantity

Ofwat regulates the 21 regional water companies in England and Wales. Of these, 10 companies provide both potable water and sewerage services while 11 companies supply only potable water. Ofwat has many functions, these include:

- ensuring that the companies provide customers with a good-quality and efficient service at a fair price
- monitoring the companies' performance and taking action, including enforcement, to protect consumers' interests
- setting the companies challenging efficiency targets.

The Guaranteed Standards Scheme (GSS) sets out the requirements for water companies to maintain a minimum static pressure head of 7 m in the communication pipe serving the premises supplied with water. The Water Industry Act 1991, Part III, Chapter II, 'Means of supply', section 65, requires that there is a pressure of not less than 10 m head (1 bar) at the external stop tap of a property *and* that the flow should not be less than 9 l/min within the property. Ofwat requires companies to report the number of properties that receive pressure lower than this standard annually, and companies compile a list of the number of properties at risk of low pressure, known as the Director General Measure 2 (DG2) register. This level of service does not, however, override the statutory duty to deliver a continuous supply of water at a pressure that is able to reach the upper floors of properties.

In addition, companies have to provide a continuous supply of water, without interruption. Here, Ofwat requires companies to maintain a register, known as the Director General Measure 3 (DG3), which records the number of properties that experience unplanned interruptions to their supply of greater than 6 hours duration. Incidents of supply interruptions are excluded if they are caused by a third party, or are a result of planned maintenance work and customers have been given reasonable advance warning.

In reality, the measurement of flow and pressure delivered to every customer is not feasible, particularly as the measurement of flow is generally complex and requires a permanent installation of equipment. Pressure is more readily measured, and most operational staff carry suitable devices to manually obtain pressure data from any customer tap, fire hydrant or other accessible tapping point in a distribution system. Such manually obtained spot sample pressure data are usually not recorded formally, as they are used primarily to inform current local operational decisions. For regulatory and more strategic management of the system, both flow and pressure are continuously measured at strategic points. In practice, therefore, the current status of the system at each individual customer's tap is unknown, and hence water companies usually only become aware of the quality of service at an individual premises through customer contacts and complaints. Consequently, failure to meet the regulatory standards or the delivery of a poor quality of service is only reported to the company once it has occurred. Most companies invest significant effort in policies and procedures for handling customer contacts as well as categorising, including both quantity and quality issues, and recording them for future analysis. The importance and weighting placed on customer contacts reflects the UK attitude of water supply as a service industry.

Water supply systems consist of treatment, storage and distribution assets joined and controlled by pumps, valves and other operational components (see Chapter 5). For management and reporting purposes, these systems are broken down into hierarchical areas comprising production management zones, distribution management areas (DMA) and sub-DMAs, created for example for pressure control (see Chapter 5). Flow and pressure data are collected from all these areas in order to provide operational and regulatory data. The layout of a typical network and of the typical locations at which flow data are recorded are shown in Figure 8.1.

Flow is usually monitored in trunk mains, at the inlets and outlets of service reservoirs, pumping stations, some large-diameter distribution mains, DMA inlets and exports, at some sub-DMA inlets and exports, and major consumer premises. Pressure is also measured at these locations and at critical pressure monitoring points inside DMAs, specifically selected as locations where the first indications of pressure problems may be anticipated to occur. Such current industry practice for pressure monitoring does not necessarily identify the most sensitive location for the detection of a leak or a burst event, and, as a consequence, methodologies are being developed to identify 'optimal' spatial resolution for sensor location in order to capture event data irrespective of where in a DMA the event occurs (Farley et al., 2008). Analysis of such data can help to determine where the event actually took place. Mounce et al. (2010) present the online application and verification of an automated event detection system, based on artificial intelligence self-learning algorithms. An example event detection using this technique is shown in Figure 8.2, measured at the DMA level.

For hydraulic design and monitoring purposes, and more recently the need to accurately determine leakage levels, manufacturers have developed flow and pressure instruments to a very high level of sophistication. Off-the-shelf instruments are capable of measuring flow and pressure to an accuracy of 0.1%, and even 0.01% if the instruments are rated correctly for purpose and manufacturers' installation procedures are strictly followed. The installation, calibration and maintenance requirements of flow instruments are far

Figure 8.1 Division of a distribution system into DMAs, including main flow meter locations
Reproduced with permission from Morrison (2004). © IWA Publishing

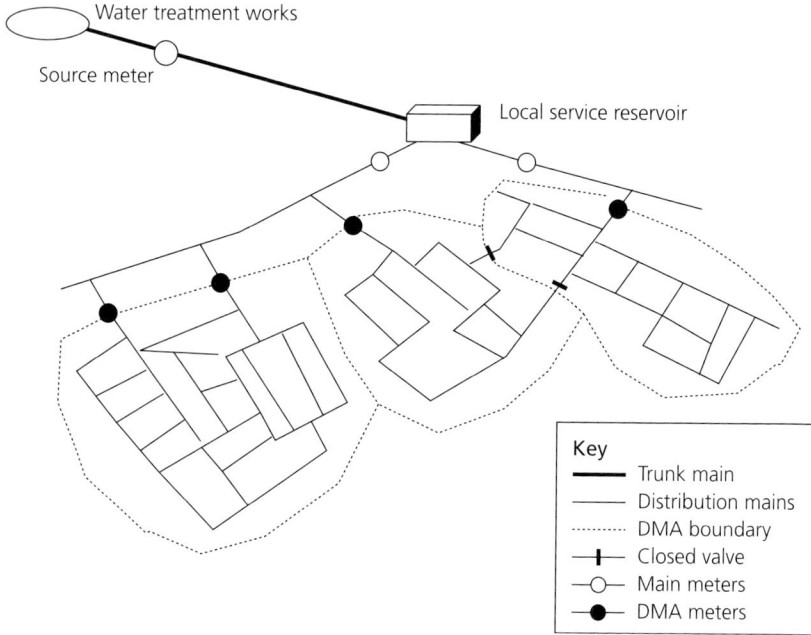

more onerous when compared with those for pressure equipment, so it is simpler and more cost-effective to collect pressure data than flow data. However, irrespective of the type of instrumentation used, it is recommended that regular testing and calibration of the instrumentation is undertaken to ensure that optimum accuracy of measurement is achieved.

Continuous sampling from fixed instruments produces a time series of data that represents all the changes in the measured parameters at the point in the system at which the measurement is being made, over the duration of the measurement period. The unofficial industry standard for the collection interval of both flow and pressure data, at most locations, is 15 minutes. The basis and justification for this is somewhat obscure, but it does provide a reasonable balance between the volume of data and the definition of daily patterns. However, ideally, the frequency of data collection should be different for different applications. A good representation of the overall pressure dynamics within a network can be observed at 15 minute intervals, but, clearly, this is insufficient for rapidly changing events; for example, the shape and amplitude of pressure transients cannot be resolved with data points less than a 10th of a second apart. Flow dynamics can also be well represented by 15 minute data points, but higher frequencies allow a much better understanding of the system hydraulics; for example, it is feasible to assess the different contributions to the overall flow from different types of demand such as domestic and industrial.

198

Figure 8.2 Example flow time series plot and event detection using an artificial intelligence (AI) technique

Redrawn with permission from Mounce *et al.* (2010). © ASCE

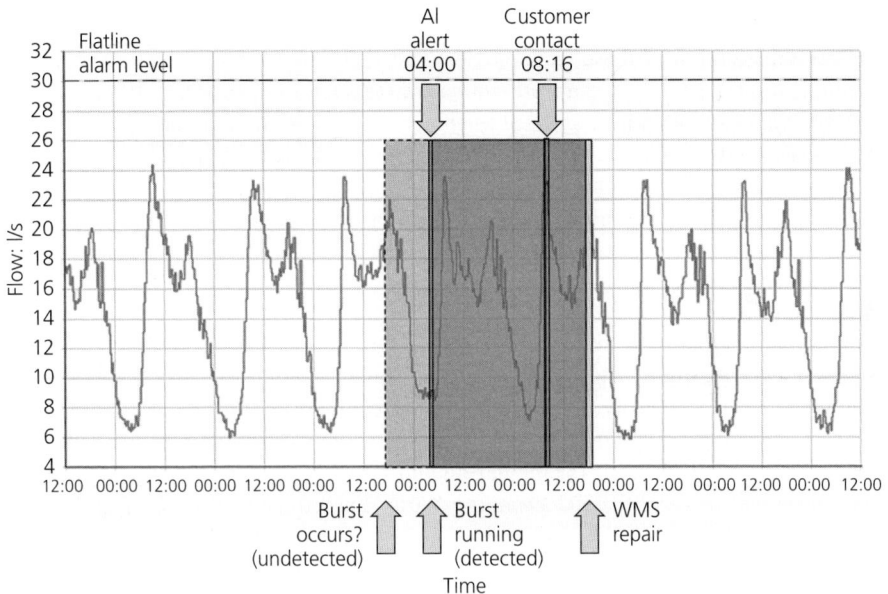

The output of most flow measurement instrumentation is an electronic pulse per unit of flow. The number of these pulses is generally counted over, and then divided by, the data resolution period. Hence, most flow data is in the form of time-averaged values. This 'smooths' the data, but, in doing so, disguises potentially valuable information. Pressure instrumentation, however, generally provides an instantaneous value, which may be preferable, but such measurements can give rise to analysis problems as dynamic data may be subject to noise. In such cases, the instantaneous reading could be at a level not representative of the general pressure trend.

Data recorded at permanently installed flow and pressure monitors is subsequently required for analysis. To facilitate this, most flow and pressure instrumentation is, or can be, equipped with a memory, such that it can store data. Which can then be obtained either manually at the site or remotely by using communication technology. Manual data collection is widely practised, particularly from DMA flow meters and critical pressure-monitoring points. Such data are commonly collected at monthly intervals, when it is also possible to check the performance of the instrumentation and the batteries. Pressure and flow instrumentation at large installations, such as pumping stations and service reservoirs, where power is available, have commonly been connected to centralised online data storage via supervisory control and data acquisition (SCADA) systems and public service telephone networks. Such connectivity facilitates regular data transfer; historically at 4-hourly intervals. More recently, the diminishing costs of automated data transfer, such as by GSM and GPRS systems, is allowing all types of recorded data to be transferred from a large number of monitors, many at remote locations in the distribution system, and is now a viable alternative to manual collection. GSM-based technology has typically been implemented to expedite daily data transfer, while the latest GPRS technology has been set up to promote 30 minute data transfer, although higher frequencies are feasible. Power and cost-effective communications are key to the successful application of these technologies, and there are now various low-power and communication technologies currently under trial (Stoianov *et al.*, 2006). It is anticipated that these systems will address some of the power and connectivity issues associated with GSM/GPRS that should further improve the potential for, and ease of, data transfer in the future. There are also current initiatives underway to standardise the format and protocols for such data transfer that will also improve connectivity and compatibility issues.

8.3.2   Quality

Drinking water quality is an issue of concern for human health in both developing and developed countries. The risks arise from infectious agents, toxic chemicals and radiological hazards. The World Health Organisation (WHO) produces, and regularly reviews, international standards on water quality and human health in the form of guidelines that are used as the basis for regulation and standard setting. These guidelines have been interpreted to form the EU Drinking Water Directive (98/83/EC). This Directive has formed the basis of strict standards for the quality of the public supply and laid down in national regulations by the DWI – which closely checks and regulates water quality in England and Wales.

The drinking water quality criteria are set out in the Water Supply (Water Quality) Regulations 2000, which stipulate a maximum concentration or acceptable level of a large number of substances. In addition to these quality standards, the regulations stipulate the minimum frequency at which water samples should be collected and analysed, and describe the methodology to create appropriate sampling 'zones'. The regulations also provide advice on analysis, reporting, approval of materials for use in contact with drinking water, and the necessary actions if the standards are breached. Since the regulations came into force, they have been regularly subjected to amendment to take account of a number of EU Directives. The most important of these are the Water Supply (Water Quality) (Amendment) Regulations 1989 and 1999, which take account of the need to regulate *Cryptosporidium*, nitrates and pesticides.

The standards defining the wholesomeness of the water supplied are also prescribed by the DWI. The object of these standards is to stimulate improvement in drinking water quality and to encourage countries of advanced economic and technological capability in Europe to attain higher standards than the minimum ones specified in international standards for drinking water. The latter standards are considered to be necessary and attainable by every country. At the same time, the industrial development and intensive agriculture of some European countries create hazards to water supplies not always encountered in other regions. Hence, stricter standards are demanded and justified. More recently, in the Water Act 2003, additional emphasis has been placed on consumer acceptability. Consideration needs to be given to any effect on the taste, odour and appearance of the water supplied to consumers. It is desirable, for example, that consumers continue to receive water of a hardness and mineral content that they are accustomed to receiving (DWI, 2004). The guidance also states that consideration must be given to maintaining any existing quality agreements that the water company has with non-domestic customers. Some manufacturing processes require water with a fairly constant chemical composition. It might be possible for them to handle changes in composition, providing there is adequate consultation and advance warning of the changes. However, frequent or unplanned changes need to be avoided (DWI, 2004).

UK water suppliers place the highest priority on assuring the quality of water provided to their customers. Drinking water quality compliance in England and Wales to meet the strict UK and European standards is 99.96%. This impressive figure highlights the importance, as well as the effectiveness, of the procedures used to ensure water quality. These procedures are typically based on risk assessment and risk management, and are being further developed to work in a more holistic way, focusing on the whole of the water supply chain, from source to tap (Water Science and Technology Board *et al.*, 2005). This concept has given rise to the term 'water safety plans' (WSPs). Such plans are endorsed by the WHO. The WSP approach also requires that other stakeholders play their part in assuring water quality. The source-to-tap concept is a complex issue, as all aspects of water quality, from the management and the protection of water sources through to the delivery of acceptable water quality in buildings, including households, involves many stakeholders, some of whom are unaware of their roles and responsibilities.

The concentrations and levels of regulated substances are set out in tables of grouped parameters including aesthetic (e.g. colour, turbidity, taste and odour), microbiological (e.g. indicator organisms, pathogenic organisms and viruses), chemical (e.g. iron, manganese and lead) and disinfection residual (e.g. chlorine). These tables are subdivided into directive and national levels and concentrations. Different standards are applied for the quality of water at different points in the network, primarily at the exit from the water treatment works, service reservoirs and at the point of use, usually the customer's tap. At treatment works and service reservoirs, a combination of online and discrete sampling is used to obtain water quality data, while at the customer's tap, data is derived from a random programme of discrete sampling. Hence, samples are collected from customers' properties, selected at random from within the water supply zones defined by the water companies. A water supply zone is a region containing not more than 100 000 population (see Chapter 5). Sampling rates for different parameters vary, and are generally defined with respect to the total population of the zone; for example, coliform bacteria and disinfection residual require 12 annual samples per 5000 population. Full details are set out in the Water Supply (Water Quality) Regulations 2000. These discrete manually collected samples produce a small amount of 'spot' data that is representative of the quality of the water at a single time and duration at a given location. Such snapshots of system conditions provide no understanding of the system status prior to or after the data was recorded. The analysis of the discrete samples has to be completed by accredited laboratories that are recognised by the UK Accreditation Service (UKAS).

This agency is the sole body recognised by the UK government to assess, against internationally agreed standards, organisations that provide certification, testing, inspection and calibration services. Once the samples have been analysed by a UKAS-accredited laboratory, the analytical results are returned to the water companies. Failure to meet the standards is judged by reference to the defined standards. Failure must be reported to the DWI. When a failure is recognised, or there are a large number of customer contacts, it is usual for companies to stimulate a programme of investigative sampling, over and above the routine samples. This is intended to capture the cause, extent and severity of the failure and to inform incident reporting to the DWI, covered in Section 8.4.

Instruments continuously measuring water quality parameters generally require significant maintenance and the use of triple-validation systems to ensure confidence and quality assurance. Application is generally limited to locations at treatment works and service reservoirs, and measurement parameters include temperature, turbidity, colour, iron, manganese, free and total chlorine, pH, redox potential, dissolved oxygen and conductivity. Technologies for the continuous measurement of water quality parameters are improving, with recent advances take advantage of refinements in optical, thin- and thick-film printing on ceramic substrates, and microfluidics, to provide continuous reagent-free probes. Some instruments even combine multiple parameters onto a single probe that can be inserted directly into mains or installed at customer boundary fittings. Such instruments currently remain unproven on a cost–benefit basis, and application has not been widely taken up. It is envisioned and widely anticipated that, one day, these new instruments will be embedded into the pipes of water distribution systems at the time they are manufactured. These so-called 'smart' pipes will enhance and move forward operational practice so that pipe replacement programmes and other maintenance activities will be explicitly based on a network of sensors continuously relaying operational and regulatory data to the point of use.

8.4   Analysis

Much of the data analysis that is formally performed by water companies in England and Wales is used for regulatory reporting to Ofwat and the DWI.

Ofwat regulates to make sure that water companies provide customers with a good-quality, efficient service at a fair price; it monitors the companies' performance and takes action, including enforcement, to protect consumers' interests, and it sets challenging efficiency targets for the companies. Each year, the water companies are placed in 'bands' from A to E, with A being the most efficient. However, for price-setting purposes, each band is subdivided into an upper and lower part. This information is then used to set efficiency targets for each company. The results are published as league tables of relative operating efficiency bands and ranks. A major component in deriving these tables is the overall performance assessment (OPA), developed by Ofwat. The OPA was designed to capture all aspects of service and to incentivise good overall service through comparative competition and transparent performance scores. A major aim of the data analysis undertaken by water companies in England and Wales is therefore to inform these rankings. The main water distribution system measures, requiring data analysis, used in quantifying OPA scores are the DG2 and DG3 registers (defined in Section 8.3) and leakage indicators.

In a similar way, the metric used by the DWI is the 'drinking water compliance indicator', which reports the number of regulatory water quality samples and parameters that failed to meet the standards. As described later, such failures invoke investigational sampling and analysis to determine the cause of failure. Ofwat includes the drinking water compliance indicator as a factor in the OPA evaluations.

Although not formally considered in OPA scores, the water industry in England and Wales also undertakes significant data analysis to inform pre- and post-rehabilitation assessment (PPRA). This was introduced in 1996 as the process to comply with the requirements of the Section 19 undertakings, which were introduced to facilitate investment following previous under-investment in infrastructure. PPRA is required to show that the water quality before renovation was unsatisfactory, thereby identifying that remedial work was required and that water quality after renovation was satisfactory. Hence, there is a need to compare water quality before and after renovation to demonstrate that the renovation has been effective.

Beyond Section 19 undertakings, water companies in England and Wales are formally required to submit and gain approval for distribution operation and maintenance strategies (DOMS), as first set out in the *DWI Information Letter 15/2002* (DWI, 2002). These plans are reviewed regularly and are considered as part of the water company 5-yearly asset management plan (AMP) submissions to Ofwat. The purpose of a water company DOMS is 'to ensure customers are supplied with water of consistent or improving quality, through appropriate operation and maintenance of the distribution system'. They cover strategies for the proactive management of drinking water distribution systems, so that water companies meet and continue to meet drinking water quality standards. They typically include but are not limited to the following:

■ arrangements for proactive, periodic, medium-term, system-by-system investigations
■ monitoring of water quality at a local level, leading to timely responsive maintenance
■ control of operational activities related to risks to water quality
■ regular inspection and maintenance for certain components of the distribution system related to risks to water quality.

8.4.1 Quantity

The most common instrumentation used to gather quantity data from water distribution networks are flow meters and pressure transducers. Such data is usually captured as a continuous time series that is analysed to identify performance shortfalls. Pressure data tend to be used primarily for monitoring pressure-critical areas of DMAs and inlets and outlets of pumping stations and other critical assets. Pressure data indicates when low-pressure events are, or are not, occurring and hence inform when pressure standards of service (DG2) are met and that pumping stations are operating correctly or not. Analysis of pressure data is typically performed by examination of threshold levels, set with respect to regulatory standards and site-specific operational experience.

Pressure is also an important data source used to inform network modelling (see Chapter 6). Flow data is used primarily for leakage management, but also in resource, statistical, hydraulic and water quality models.

In general, UK water companies are proficient at ensuring that their customers receive a continuous supply of water. However, during transit through the distribution network, a significant proportion of the water entering the system is lost through leaks. In the UK, as much as 30% of distributed water is lost through leaks, with a national average of around 22%. Although leakage levels in England and Wales have dropped from a maximum of 5112 million l/day in 1994–1995 to 3608 million l/day in 2004–2005 (Ofwat, 2005), they are still high; for example, the existing levels of leakage could still supply about 24 million domestic users. Data from the Environment Agency for 2004–2005 provide another perspective on leakage in England and Wales, in that industrial usage and leakage account for broadly similar volumes of water from the public water supply. Across Europe, reported urban leakage rates vary greatly, between 3% and 50% (European Environment Agency, 2003). Hence, with these volumes of water being lost, there is considerable potential for significant efficiency gains through leakage management and reduction.

Current UK leakage management practice is geared to achieving an 'economic level of leakage' (ELL). ELL is defined as the point where the cost of identifying and repairing leaks is such that to go beyond this level of expenditure would incur unacceptably high costs per leak repaired. The best practice principles in the ELL calculation are set out in the Tripartite Study report (Ofwat, 2002). It is worth noting that as the cost of energy changes, so the ELL will vary, and that in the future this measure may become more inclusive of social and environmental measures, a sustainability-based metric (House of Lords, 2006). Before leakage can be precisely estimated, accurate measurements of the volume of water entering the network and the volume used by all domestic and industrial users are required. However, leakage figures are built upon estimates of domestic and industry use components which themselves are subject to potentially large estimation errors. For example, accurate measurement of the domestic demand is not possible because only ~26% of household supplies in England and Wales is metered and although major industrial demand is monitored; all other components are estimated.

Water balance is a commonly used leakage detection method. This involves the detailed accounting of water flow into, and out of, the distribution network or some part(s) of it. At the level of the whole system, this consists of a total water supply balance, i.e. the summation of all water consumed (metered and un-metered) and not consumed (leakage, theft, exports, etc.), which is compared with the total system input. Instead of monitoring the entire system, individual zones can be monitored, such as in district flow metering. These balances help to identify areas of the network that exhibit excessive leakage. However, they do not provide information about the location of leaks. Local location techniques are commonly used for this (WRc, 1994).

The UK water industry initiated a significant research programme in the early 1990s to provide guidance on leakage management. The results of this work were published in

Managing Leakage (WRc, 1994). One of the most important outputs of this research work was the concept of understanding the components of leakage and how these could be estimated. This provided practitioners with the capability of modelling leakage and the factors that affect it. The resulting Bursts And Background Estimates (BABE) methodology is now one of the commonly used methods for leakage estimation, and is incorporated within the International Water Association (IWA) leakage methodology to determine unavoidable annual real losses and the infrastructure leakage index (Lambert, 2003; IWA Water Loss Task Force, 2005). It is clear from observations of the application of this method that background leakage represents a substantial proportion of the overall leakage estimate in many networks. Leakage that is not an identifiable burst is labelled as background leakage.

Flow data can also be used to directly evaluate leakage. A widely used operational approach is an analysis of measured minimum night flows, i.e. the lowest flow supplied to a hydraulically isolated DMA. It is generally evaluated during the night hours, between midnight and 5:00 am, and is a measure of leakage as well as certain minimum night consumption. Estimating distribution losses from minimum night flow relies on the accurate estimation of the additional components that contribute to night flows. By monitoring night flows continuously, unusual changes in water volumes can be detected.

Flow and pressure data analysis for leakage detection and reporting tends to be a manual or semi-manual process with inherent inefficiencies that are prone to human error. A result of this is that leaks which are not obvious can potentially run undetected for extended periods (sometimes several months). Advances in instrumentation, telemetry systems and communications are leading to more data of a higher quality being gathered by water companies. Hence, automated data analysis techniques are being developed to deal with these larger data volumes and to add consistency to, and remove human error from, the analysis. In addition, the constant flow of high-quality data is enabling water companies to develop online data analysis techniques using artificial intelligence, or machine learning. Currently, these are being applied to analyse DMA data, in near real time, in order to automatically detect bursts and leakage (Mounce *et al.*, 2007, 2008, 2010).

As introduced in Chapter 6, hydraulic models of water distribution systems can be produced to provide a mathematical representation of the non-linear dynamics of the hydraulic performance of a water distribution network. Such software is predominately used in an 'off-line' form for activities such as strategic supply analysis, design of control strategies, network extensions, and maintenance planning. However, for use in day-to-day operational management the models need to be populated with the most up-to-date data possible. A network model driven by the latest data can provide invaluable information on system performance; for example, by extrapolating data from a relatively few measurement points the hydraulic performance at every point in the model may be estimated. Preliminary efforts made to link simulation models with real-time data from telemetry systems have been reasonably successful (Orr *et al.*, 1999; Skipworth *et al.*, 1999). Improvements in data collection transfer and warehousing technologies have now

205

enabled a new generation of online modelling techniques to be developed. A sufficiently accurate model can now be regularly updated with flow and pressure data sent directly from the field, and can be used to detect many features of system performance, including hydraulic events such as low pressure or illegal hydrant or valve operations as well as automatically monitoring standards and levels of service (Machell *et al.*, 2010). It is anticipated that such models will gain popularity and become widespread in their use, enabling almost real-time optimisation of network operations in the future.

8.4.2 Quality

The DWI has the power to prosecute companies which fail to meet the required water quality standards, and it is feasible that a water company could lose its operating license for a serious breach of the water quality standards of service. With drinking water compliance in England and Wales being around 99.96%, prosecutions are rare. However, should such a prosecution be taken out against a water company for an offence under section 70 of the Water Industry Act 1991 – supplying water unfit for human consumption – a defence must be provided by the water company to show that it took all reasonable steps and exercised 'all due diligence' for securing that the water was fit for human consumption on leaving its pipes or that it was not intended for human consumption.

The major task in the use of the water quality data is to report whether the quality meets the regulatory requirements, primarily through the comparison of analytical results of discrete sampling with the defined standards. Results of this are then reported to the DWI for the purpose of demonstrating compliance. The information provided is compiled into drinking water quality data tables that indicate the extent to which the companies have, or have not, met each of the drinking water standards in force at the time of submission, and are used for inter-company performance ranking.

Water quality data is also available to the public. There is a statutory requirement that customers of water companies may enquire about their water supply through the Drinking Water Register, providing access to drinking water quality information for every water supply zone in a company. Data analysis for this purpose requires that the records include the name of the zone, the population of the zone, the water treatment works supplying water to the zone, the details of any undertakings, and the results of analysis of water samples.

Water quality sample results are also fed back to operational teams, to inform of deteriorating quality in order that remedial actions can be taken before standards of service are impacted. It is at this level that data from continuous water quality instruments is most often used, to supplement discrete sample data, since absolute values are no longer the primary concern but, rather, it is changes in the trend and pattern of the data with time that become important. Such data use might result in a quick remedial action or the bringing forward of planned maintenance. Water quality data are also used to generate warnings about impending problems such as discolouration, and can initiate a programme of planned maintenance such as mains flushing or service reservoir cleaning.

When water quality data analysis indicates an event, water companies have a responsibility to inform the DWI as soon as possible. This is covered under the Water Undertakers (Information) Direction 2004. An event is defined as 'having affected, or likely to affect drinking water quality, or sufficiency of supplies and, where as a result, there may be a risk to consumers' health'. When notified of an event, DWI inspectors assess the information provided to establish if the event is an 'incident'. If deemed to be an incident, a detailed report is usually required from the company, including all possible data analysis. At this stage, significant added value can be obtained from the analysis of continuously monitored data to supplement the regulatory-driven discrete sampling. The data analysis contained in the report should be such that the inspector can determine:

- the cause
- whether the incident was avoidable
- the company's response
- how the incident was handled
- lessons that can be learned for the future
- if there were any breaches of enforceable regulations
- whether the company supplied water that was unfit for human consumption.

The outcome of such assessment can range from a simple letter, with or without recommendations, through to prosecution proceedings or a caution for a criminal offence.

In respect of the reporting of incidents to the DWI, the response to incident management and contingency planning is covered under 'drinking water safety plans'. This is a source-to-tap risk management approach that is key to the way in which water companies ensure a continuous supply of safe drinking water, both now and for the future. Understanding risk enables effective controls to be implemented to safeguard water quality. Managing risk means that the companies can anticipate problems to protect public health. When assessing risk to drinking water quality, the likelihood and consequence of hazardous events are ranked using a matrix to derive a risk score. This process is then repeated to take into account any control measures, to give an estimate of the residual risk. Summary reports are then submitted to the DWI, as required by the amended Water Supply (Water Quality) Regulations 2000. Information from the risk assessments is generally incorporated into water company business plans, which are submitted to Ofwat, and define the level of investment for maintaining and enhancing water supplies during the AMP planning periods. Identified risks are continually reviewed, and the effectiveness of controls is monitored to ensure that the drinking water safety plan remains effective.

As with quantity, network modelling can also be a valuable tool in the analysis and interpretation of water quality data. The requirement is again for an up-to-date, preferably online, solution. Applications can range from source tracking, to identify sections of networks supplied from different sources and areas receiving mixed water, through the simulation of water age (hypothesised to be essential for the accurate simulation of all the various kinetic reactions and interactions that occur within the complex distribution system that acts as a highly variable high-surface-area reactor with high residence time;

American Water Works Association, 2002; Machell *et al.*, 2009), through to modelling of specific reactions. Modelling of disinfection residuals and the formation of disinfection by-products has received specific research attention (Rossman *et al.*, 1994; Rossman and Boulos, 1996; Kirmeyer, 2000; Mutoti *et al.*, 2007). Water quality modelling relies on the use of an accurate and calibrated hydraulic model that utilises the solution of hydraulic transport and tracking algorithms as the basis for simulating the quality changes to parcels of water as they travel through water distribution systems. Water quality models are gradually being adopted as investigative tools to better understand how water quality changes in distribution systems and to investigate the reasons for failed water quality samples, and as predictive tools to determine the age of water, chlorine, trihalomethanes and other chemical concentrations that are created or increase within the bulk water flow during transit through the pipe network. Predictive modelling tools such as PODDS (Boxall and Dewis, 2003; Boxall and Saul, 2005) are used to assess the potential discolouration response to changes in hydraulic conditions. Because the mechanisms that determine the quality of the water emerging at a customer's tap are many and complex, current modelling approaches are known to have shortfalls. However, much research work is currently being undertaken, with longer-term deliverables, to fully understand these mechanisms and to develop models that accurately reflect the complex processes and quality relationships that occur in water distribution systems. These models will ultimately be written into the industry-standard software for use by practising engineers and scientists.

8.5 Interventions

In this section, the major pipeline intervention options that are commonly applied to water distribution systems are outlined.

8.5.1 Quantity

8.5.1.1 Mains renewal and replacement

Several techniques have been developed for mains renewal and replacement to minimise impacts to customers, society and the environment. This section outlines different options for mains replacement and the circumstances in which these may be applied. Relining techniques are covered under the quality heading, as this is the major driver for such techniques, although they can also offer structural performance improvement.

OPEN CUT

The most obvious approach to mains replacement is the excavation of a trench to lay the new water main (Figure 8.3). This approach may be selected as the most effective option, depending on the location and the method of reinstatement. Common factors to consider are the route, the location and size of the multi-utility buried infrastructure, the impacts to traffic, shops and businesses and pedestrian access, and the size (diameter) of the main to be replaced, i.e. the new main is likely to be of a larger diameter than that of the existing pipe.

DIRECTIONAL DRILLING

In certain situations it may be more appropriate to use directional drilling rather than replacing the pipe in a trench (Figure 8.4). This involves drilling a pilot tunnel with

Figure 8.3 Open cut
Reproduced with permission from Thames Water Utilities Ltd

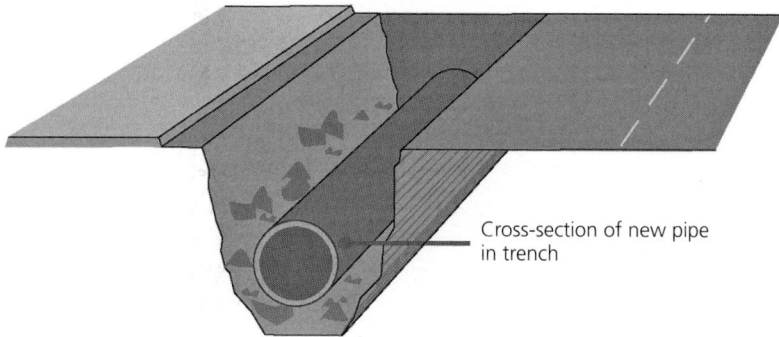

Cross-section of new pipe
in trench

a precision-guided drilling rig, then pulling the new pipe back through the tunnel. This option minimises the impact on the highway, and is often used in situations where mains need to traverse a strategic crossing such as a watercourse or other such significant obstacle.

PIPE BURSTING
Pipe bursting can be applied in most circumstances where the new main is of a similar diameter to the existing main. Work is done underground, with only two small pits dug at each end of the section of pipe that is being replaced (Figure 8.5). In simple terms, a steel rod is pushed into the existing old main, which is then used to pull a cutting tool back through the pipe. This breaks up the old pipe as the rod

Figure 8.4 Directional drilling
Reproduced with permission from Thames Water Utilities Ltd

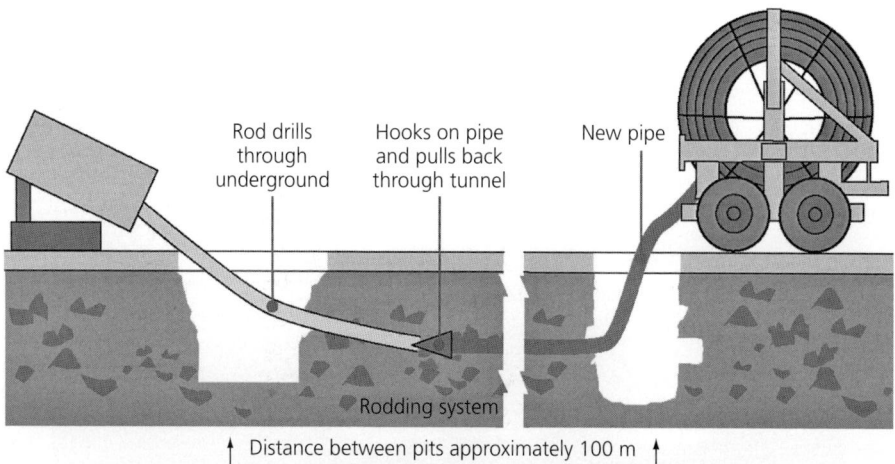

Rod drills through underground

Hooks on pipe and pulls back through tunnel

New pipe

Rodding system

Distance between pits approximately 100 m

Figure 8.5 Pipe bursting
Reproduced with permission from Thames Water Utilities Ltd

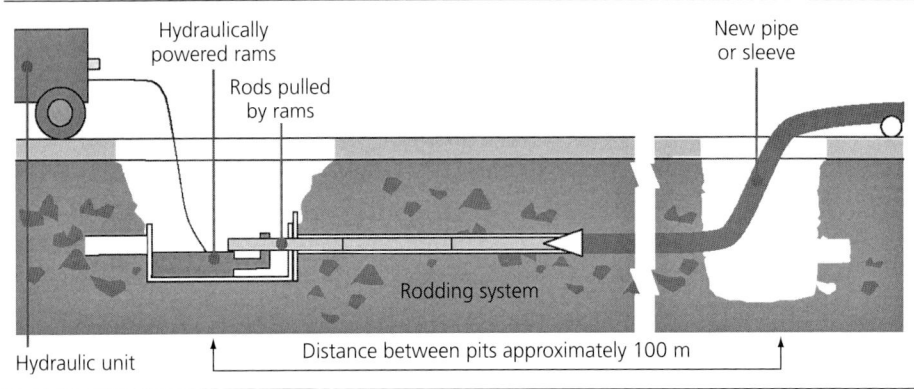

returns, and at the same time the new pipe is pulled into the space left by the old pipe. This approach offers the advantage of minimising the above-ground impact, access to the highway.

INSERTING A NEW PIPE

The further option for mains renewal is to insert or 'sleeve' – a new plastic pipe inside that of the current main (Figure 8.6). The obvious constraint on this approach is whether the reduction in diameter will compromise the hydraulics, thus preventing delivery of an adequate level of service. Network analysis should be used to model this impact and test the applicability of the approach. However, in most cases, although the pipe

Figure 8.6 Inserting a new pipe
Reproduced with permission from Thames Water Utilities Ltd

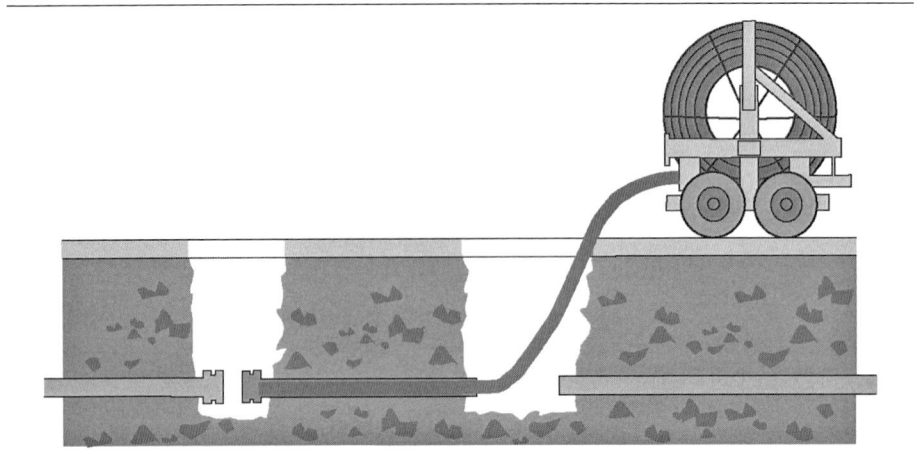

would be of a reduced diameter, the new smoother pipe often results in little or no impact or reduction in the level of service.

8.5.1.2 Pump scheduling, valve regulation and system optimisation

The rising cost of energy is one of the greatest challenges facing the water industry, and investment in the assets which offer the opportunity for increased operational efficiency is one way to reduce such costs. However, it should be recognised that energy savings may also be realised through optimising the operation of existing assets and infrastructure, and, hence, it is important to consider both approaches.

Energy cost reduction can be realised in several ways by:

- moving energy consumption to cheaper tariff bands
- reducing peak demand charges
- running pumps more efficiently
- choosing the shortest path from the source to the destination
- choosing the lowest-cost source of water (raw and treated).

Operators need to choose the lowest-cost solution without sacrificing reliability or levels of service. The production of treated water needs to be directed to the cheapest sources of water, while pumping should be moved to the lowest-cost tariff periods. Where possible, the storage capacity of service reservoirs should be optimised, using greater volumes of turnover, to satisfy demand during expensive electricity periods. Such an approach will realise benefits in terms of:

- energy cost savings
- efficiency gains and, therefore, carbon emissions reductions
- improved water quality through lower water age and improved turnover in reservoirs
- a planned rather than reactive operating culture
- a structured and measured approach, which enables risk assessment of operating decisions.

MOVING ENERGY CONSUMPTION IN TIME

Energy companies typically charge more for energy consumption when demand is high. Scheduling pumps and flows into lower-cost energy periods, while still satisfying storage and pressure requirements, will reduce energy expenditure aligned to these tariff structures.

REDUCING PEAK DEMAND CHARGES

A peak demand charge is a penalty fee imposed by energy companies for the highest electrical load recorded in a month. A proactive and measured approach to pump scheduling helps companies avoid peak demand charges by choosing the minimum number of pumps required to run concurrently.

IMPROVING PUMPING EFFICIENCY

Pumps are designed to run most efficiently at a particular pressure and flow rate. Understanding how to operate individual pumps and combinations of pumps at the

211

most efficient point on their efficiency curves will lead to optimised scheduling, resulting in more-efficient use requiring less electricity to pump the same volume of water.

SELECTING THE LOWEST-COST SOURCE OR PATH

As water transmission and distribution systems have evolved over many years, there are often a number of routes by which water can be delivered to a customer. Often, more than one treatment plant can service the same area, providing back-up for planned and unplanned operational events. Network models can be used to assess the use of different sources, to reliably meet demand, and can be used to optimise and select the lowest-cost source to supply customers within a water supply zone. Similarly, a combination of modelling and near-real-time information taken from the network can be used to optimise variations in the demand pattern during the day to ensure the lowest-cost path to customers at all times of the day; for example, by optimising the use of storage tanks.

There are a number of technical challenges to overcome when attempts are made to optimise transmission using valve and pump operations. These include:

- use of multiple sources of water, storage or demand points
- reacting to unanticipated changes of production capacities
- maximising the use of the lowest-cost source water within a number of constraints e.g. abstraction volume limits or raw water quality
- operation of fixed or variable-speed pumps and parallel pump sets
- predicting daily patterns of water consumption in near real time
- continually re-optimising pumping schedules
- interfacing to SCADA and telemetry systems to collect near-real-time data
- implementing changes remotely whilst scheduling the availability of pumps and storage for planned operations
- maximising the use of off-peak electricity periods within any constraints.

A number of tools and solutions exist to enable this mode of operation, which include all or various combinations of:

- Network models that offer the opportunity to run offline, hypothetical scenarios for strategic planning. These are particularly useful for planning maintenance and refurbishment work on both infrastructure and non-infrastructure assets, and assessing the risks of undertaking the work.
- Technology to retrieve real-time, or near-real-time, data directly from the network, to better understand current boundary conditions and dynamic element operation within the network.
- A network model, or models, linked with an optimiser, to continually generate optimised operational scenarios that take account of changes to pump and valve combination and operations, and other dynamic element characteristics. They can predict changes towards an appropriate time horizon of, for example, 24–48 hours ahead.

- Technology to control assets remotely through interfaces driven by optimised pump and valve schedules, with back-up functionality for manual or closed-loop operation; i.e. safe mode, interventions.

8.5.1.3 Pressure and demand management

Pressure management or the use of pressure reduction as a method of reducing leakage is an accepted approach that is widely used across the industry. Water losses are reduced as a consequence of:

- leakage being pressure-dependent – as defined by the standard orifice equation, discharge through an orifice is a function of the size of the opening and the square root of the pressure head, and thus reducing the pressure will reduce the volume of leakage
- maintaining a stable network – by reducing variations in pressure, there is a lower propensity for mains, fittings and connections to rupture, thus minimising the main sources of leakage.

A practical guide to pressure management for leak reduction is provided by Thornton (2003), including extension of the standard orifice equation to consider fixed- and variable-area leaks and the impact of pressure variations as well as absolute values.

Other benefits of pressure management are:

- Reduced un-regulated demand or 'open-tap' use. The lower the network pressure, the less water escapes through open taps, hosepipes and similar.
- Reduced source water demand. The required distribution input flow will be reduced as a result of leakage reduction. This will also provide subsequent reductions in the power and chemical costs associated water production.
- Environmental benefits. Reduced water input will also reduce water resource requirement and hence lower the environmental impact on impounding reservoirs and aquifers. This could have the additional benefit of reducing the need for new resource development.

Pressure management or reduction is generally implemented through the installation of a pressure-reducing valve (PRV). Typically installed on a bypass of lower diameter than the host main, water is constricted within the valve body and directed through an inner chamber controlled by an adjustable spring-loaded diaphragm and disc. As the inlet pressure fluctuates, the PRV ensures a constant flow of water at a functional pressure, as long as the supply pressure does not drop below the pre-set outlet pressure of the valve. There are a number of types of PRV that may be used for pressure management, and hence in any pressure management scheme the performance of the valve should match the specific operational requirements and conditions of the network.

When implementing pressure management schemes, it is important to consider the following:

■ Customer perception. Where pressures are reduced, some customers may perceive a degradation of service, even though adequate levels of service are maintained. When the PRV scheme is commissioned, it is recommended that outlet pressure is reduced stepswise. The number of steps, and magnitude of each step, should be governed by the overall magnitude of planned pressure reduction.

■ Sensitive or vulnerable customers that may be significantly affected by pressure reduction. These may include commercial users with fire suppression systems, or dialysis patients requiring a specific minimum pressure for systems to operate. Consideration should particularly be given to ensuring the provision of adequate fire-fighting flows in the network following the implementation of a PRV scheme.

■ Maintenance. The PRVs should be included in regular planned inspection and maintenance schedules throughout their asset life. The cost of maintaining a PRV may be equal to and often exceed the cost of a new valve. The design of the installation should facilitate ease of maintenance and eventual replacement. Not maintaining the PRV will lead to an unstable network and a reduction in overall leakage benefits.

The scale of a pressure management scheme is an important consideration and account should be taken of both the overall magnitude of the pressure reduction and the number of properties affected by the scheme. Generally, the larger the number of properties, the greater the resulting benefits and the cost–benefit of the scheme. Ideally, the number of pressure management schemes in any network should be minimised through rationalisation thereby reducing the total number of PRVs to be monitored and maintained. However, there is often a reluctance to treat pressure management in a more holistic way due to levels of historical investment in pressure management and the quantified risk of level of service failures to large numbers of customers.

Other opportunities for demand management are realised from:

■ Promotion of water efficiency, through structured and targeted campaigns aimed at increasing user awareness of water use and how this can be minimised as a consequence of changes in behaviour, the use of grey or untreated sources for activities such as irrigation, and the application of various forms of water-saving devices fitted to points of consumption, such as taps, toilet cisterns and shower heads. Water companies have now been tasked with the need to drive water efficiency with their customers to achieve regulatory targets.

■ Water metering, with customers being charged for the volume of water used and disposed of via the sewerage network. In general, awareness of charging for the actual water used drives customer behaviour to reduce consumption.

■ Rationalised and optimised pumping, using reduced pumping delivery heads to reduce pressures across the water distribution network. This can be applied where customers' available pressure can be reduced, similar to the application of a PRV for pressure management, or where greater turnovers of storage volumes at services reservoirs can be achieved, by, for example, reducing pumping durations. The reduction in pressure results in reduced volumes of water lost through leakage.

8.5.2   Quality

This section focuses on the approaches and techniques associated with undertaking mains rehabilitation, usually driven primarily by water quality considerations.

8.5.2.1   Mains relining, swabbing and scouring

The approaches listed and discussed below are seen as aggressive techniques to return the internal condition of mains to varying degrees approaching their original 'as-installed' condition:

- pigging or swabbing
- air-scouring
- relining.

PIGGING OR SWABBING

Pigging and swabbing are probably the most effective and proven approaches to cleaning pipelines to improve serviceability and to reduce or defer capital investment programmes, and are capable of removing bio-films, sediments, tuberculation and other scales. A bullet-shaped pig or swab is inserted into the main, and is driven by the hydraulic force of mains pressure. As the pig or swab is pushed through the main, its abrasive force on the pipe wall cleans the pipe, with the residual debris pushed along the pipe in front of the device.

A main may be pigged or swabbed several times (progressive pigging/swabbing), where the 'aggressiveness' of the pig or swab may be gradually increased, based on the number of repeats and the type or size of the water main being cleaned. Careful selection is required, as a too-aggressive clean can exacerbate water quality problems by removing tuberculation from unlined iron pipes and causing accelerated corrosion and bleeding of material into the supply.

Deploying pigs or swabs requires launch and receiving pits. For distribution-size mains (i.e. 4–6 in. (100–150 mm)), hydrants often provide ready-made access. For mains of larger diameter, it is often necessary to install 'Y' or 'T' entry facilities, and these have to be specially installed as part of preparatory enabling works.

A newly developed and recently patented approach to pigging/swabbing is the use of a crushed ice slurry to replace the pig or swab (University of Bristol, 2010). A slug of ice is inserted into the main, and is driven by mains pressure and then flushed out with the resulting debris. The consistency or density of the slug is varied, depending on the required abrasiveness. Currently, this approach is limited to pigging mains of less than 12 in. (300 mm) in diameter, but it is hoped that the technique may be applied to larger diameters in future.

AIR-SCOURING

Air-scouring is another effective technique for mains cleaning. By isolating a section of water main between an entry and exit point, typically between two hydrants, the section is purged of standing water with a high volume of low-pressure air at a high velocity. By opening the valve upstream of the entry point, slugs of water are then produced, allowing the passage of controlled volumes of water into the section. The compressed air, mixing with the water slugs, creates a highly turbulent disturbance that travels through the main, removing sediment and bio-film material. Air-scouring is not as abrasive as pigging or swabbing, and has the distinct advantage of, typically, being a third of the cost of pigging or swabbing.

RELINING

Following the abrasive cleaning of any iron main, it is recommended that the cleaned main is relined, to provide an inert surface to reduce any further potential for internal corrosion (Figure 8.7).

Historically, mains were lined with a cement mortar (typically 4 mm thickness), sprayed on the cleaned pipe wall. Water quality and water chemistry can have an impact on this form of liner, leading in some cases to the lime in the mortar leaching away, destroying the liner and leaving a sand residue. The curing time for this form of lining means that it can take a day or more before the main can be successfully re-commissioned.

More recently, epoxy or polyurethane (PU) spray lining systems have been developed. A thin lining of resin (typically 1 mm thickness) is sprayed onto the cleaned surface of the main. Semi-structural liners with a greater thickness (typically 2–3.5 mm) are being developed to reinforce the structural capabilities of the host main. A particular benefit of epoxy liners is the much-reduced curing time before the main can be re-commissioned: typically, epoxy- or PU-lined mains can be re-commissioned within 2–8 hours of relining.

Attention needs to be paid to maintaining service connections when applying liners. Commonly, these are reinstated by the deployment of special drilling pigs; however, some of the latest epoxy and PU sprays employ a vaporising technique such that service connections are not covered by the lining.

8.5.2.2 Mains flushing

Historically, water companies have flushed water mains in response to customer complaints or following maintenance work. More recently, companies have been implementing regular, structured flushing programmes as part of their strategy to maintain water quality in the distribution system. Flushing can be used to remedy aged water in the network and, where implemented aggressively, scour and clean internal pipe surfaces to reduce the risk of discolouration events occurring.

Current research suggests that suspended materials accumulate over the entire circumference of distribution mains in cohesive layers. The daily demand patterns and the resultant shear stress are used to describe how the material accumulates in layers. These layers are remobilised within the main by increasing the shear stress above that found in the daily pattern (Boxall and Saul, 2005; Vreeburg and Boxall, 2007; Husband et al., 2008). Discolouration events often occur due to an increase in shear stress such as caused by a burst or valve operation that may be some distance from where the contact(s) occurs.

216

Figure 8.7 Relining
Reproduced with the permission. © WRc

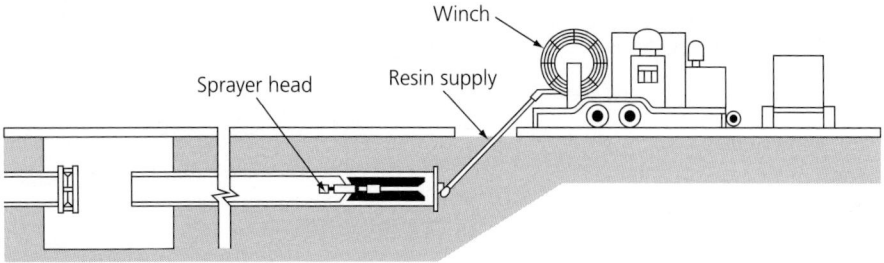

Epoxy resin system

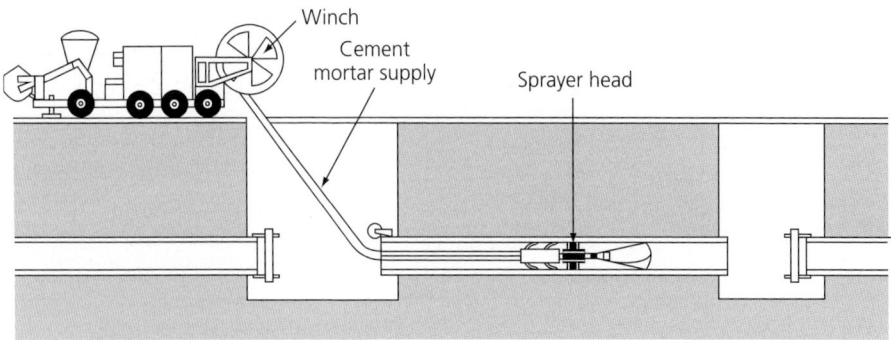

Cement mortar system

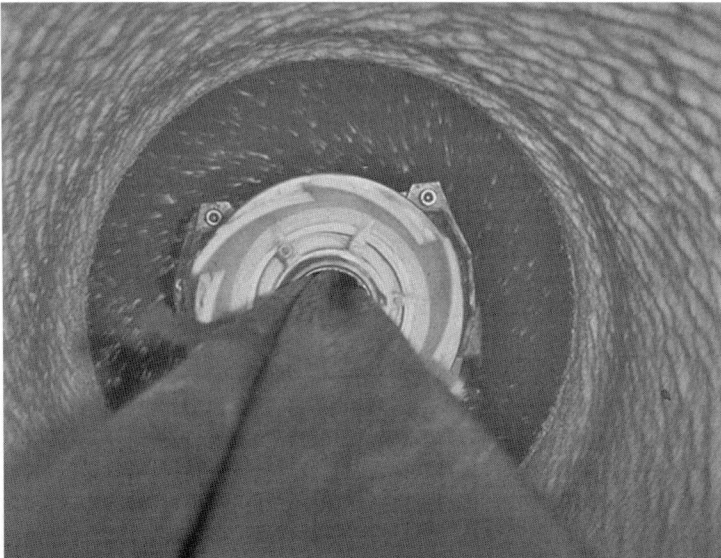

Figure 8.8 Monitored flushing operation being undertaken by the University of Sheffield

Mains are flushed over specific lengths, typically by the opening of a fire hydrant with a stand pipe attached. A flow meter may also be attached, to monitor and regulate the aggressiveness of the flushing operation. Water quality samples may be taken during the course of the flush for further analysis. Flushing operations are typically conducted during the night, to minimise disruption to customers. Figure 8.8 shows a monitored, night-time, flushing operation being conducted by the University of Sheffield in association with a UK water company.

Modelling tools are used to assess those pipes which have the greatest increase in shear stress, and therefore discolouration potential due to increases in demand in other pipes. Such tools can be used to improve the flushing process, assess the required flushing velocities and focus on specific lengths of main to improve the effectiveness of the flushing operations.

Further information on mains rehabilitation techniques can be found in WRc (1989).

8.6   Decisions

The decision to operate and maintain the distribution network is made in accordance with the needs of the consumers it supplies, and is governed by the regulatory standards set out to protect defined levels of service. In the first instance, the design of the network should be appropriate to provide a wholesome supply of water at a sufficient quantity on demand. The decision-making process is instigated either proactively or reactively.

Proactive operation and maintenance is triggered where the operator takes action to meet expected changes; for example, a change in demand due to seasonal tourism. The operator will take action in accordance with either a locally or globally prescribed operating regime, defined within an operation and maintenance schedule. Reactive operation and maintenance is the more common approach. The network is operated and maintained in response to it failing to meet the defined operational criteria or when deviation from performance standards is observed through monitoring. However, the consumer is often the first to observe that the level of service is not being achieved. Such customer observation may be clearly identified where the level of service has been completely lost, such as in the case of a burst main, or it may be a more subjective assessment of the acceptability of the product, e.g. its aesthetic appearance.

In practice, a typical water company will react both proactively and reactively to operate and maintain the network to achieve the required level of service. The key levels of service around which a typical operation and maintenance schedule is designed are, in England and Wales, collectively known as the performance standards, and are measured through the OPA (Section 8.4).

How a water company takes decisions to maintain or improve system performance against standards will vary from one organisation to another. However, because quantity and quality performance, and hence problems, are closely related, the decision process does not strictly differentiate between quantity and quality aspects, and investigations to determine the cause and to identify a solution therefore tend to include both parameters. Broadly speaking, the following functions within an organisation will be involved in the decision-making process.

8.6.1 Customer contacts

The majority of decisions will be initiated following contact from the customer. Most companies provide an operational contact centre to deal with questions and observations from customers and to resolve them. In some cases, problem resolution may be possible on first contact, where known historic performance issues are held on corporate databases, but, for the majority of contacts, the customer observations, such as discolouration, visible leaks, etc., are recorded and forwarded to appropriate company sections to be dealt with.

8.6.2 Field management

Where the performance issue is identified before the customer becomes aware of it, the problem will be passed directly to a 'field management function' for further investigation. A field manager will often have direct access to resources and business processes which can be used to identify the issue through desktop evaluation or, if further information is required, for example where there is a loss of supply, by deployment of a field-based technician. Advances in technology and business processes now mean that this procedure is mostly automated, using work management and scheduling tools that select the most appropriate resource to respond. The field-based technicians attend the site to try to understand further the nature of the issue faced by the consumer. Where this issue can be easily resolved, for example by reconfiguring the network or undertaking minor maintenance such as replacement of a stop tap, the query is closed immediately, and feedback is provided directly to the consumer or relayed to them through the operational contact centre.

8.6.3 Network asset management

Where resolution of the problem is more complicated, for example where the consumer is experiencing intermittent pressure variations, the issue is usually referred to a network or asset management function. Most water companies have a performance or asset management function that is subdivided into sections which focus on specific issues such as leakage, water quality or structural integrity. These functions provide a more proactive approach to network operation and maintenance. Within this function, the distribution engineer will often have an overarching strategic plan such as the AMP, and defined within this plan will be the specific performance standards and levels of service that the water company intends to maintain or exceed. The AMP is supported by more detailed operation and maintenance schedules such as DOMSs. These strategies will set out the timeframe over which the network will be monitored, evaluated and maintained. For simple network configurations, a distribution engineer may utilise corporate performance figures such as flow and pressure data and asset records to determine the nature of the issue. Where the issue is more complicated and it is necessary to undertake additional flow and pressure monitoring in conjunction with hydraulic analysis to identify the problem, support may be required from specialist analysts. Once the cause has been identified, the distribution engineer will promote a solution to the problem. The solution, and those responsible for its implementation, will depend on the complexity of the solution procedure. For minor maintenance such as re-zoning, the solution can be implemented by the field-based technicians or partner organisations that undertake repair and maintenance. However, where more complex solutions are required, involving capital maintenance on the network, the problem will often be passed to investment functions that will prioritise the work according to cost–benefit analysis, and implement the solution through an investment and resource management programme.

8.6.4 Additional monitoring and analysis

Where the distribution engineer identifies a divergence from the defined performance standards or levels of service, additional analysis will be required. In order to make the monitoring and evaluation of the network more manageable, it will typically be subdivided into water supply systems. These are notional hydraulically defined areas with one or more source waters and which may contain one or more water quality zones. In turn, these are made up of DMAs. The DMA concept, in its current form, was introduced during the 1990s to improve the control, management and reporting of water losses in the network but, today, it is often the foundation of all proactive monitoring and operation.

The key pieces of information used in the decision-making process are flow, pressure, consumer contacts, water quality sample analysis and network data, such as asset records, consumption records, connections, burst records and pipe samples. All of this information will be at the disposal of the distribution engineer when investigating variance from performance standards. The quality of this information is variable from one water company to the next. Where good-quality data are available, accessible and well-maintained, the ability of the distribution engineer to monitor, evaluate and make good decisions with regard to the operation and maintenance of the network is greatly enhanced. The distribution engineer will access and evaluate data according to the needs of the required operational and maintenance activity and the performance standards which are being managed.

In the case of leakage, this is an ongoing process, with flow, pressure and demand information constantly being collected and processed to identify differences between the observed and target leakage level. The most commonly utilised approach is to monitor the nightline (see Section 8.4.1). Much of the analysis and leakage detection is still undertaken manually, but, as technology advances, there is potential for a greater degree of automated data analysis and decision-making.

For water quality, the processes associated with the operation and maintenance of the system are applied over a longer timeframe, often assessing water quality data and consumer contacts over a period of months and years, to identify trends in network deterioration. Similarly, pressure variation and structural integrity will be assessed over a longer timeframe to identify trends. The distribution engineer will then decide whether to take small corrective interventions or develop longer-term investment programmes to address a range of performance issues, thereby making the investment much more cost-effective.

The challenges facing distribution engineers, operations staff and managers over the next 10 years relate to how they can more effectively make decisions across large and complex water networks considered as a whole, without the need to subdivide networks to the extent done today. In addition, the expectations of the consumer are increasing all the time, and changes to the regulatory framework, such as the introduction of the service incentive mechanism in England and Wales, will undoubtedly lead to a need for new tools and techniques. The ability, in the future, to locate, diagnose and repair faults in the network without interruption to the supply and disruption to the public will become increasingly necessary.

8.7   Evaluation

The water industry in the UK, including regulators and water companies, has clearly positioned itself as a service industry, where the customer experience is a primary consideration. A key factor in achieving and improving this service provision is the need to move from a reactive to a proactive strategy for the effective operation and maintenance of distribution systems. At the present time, it is usually the customer who is the first to know of a service problem, and customer contacts alert the water companies, which subsequently react to the event. This approach has recognised shortfalls, and this has moved the industry to invest in research that is directed at the introduction of a better understanding of system performance in near real time and to stimulate a proactive operation and maintenance strategy. There is, therefore, a move towards near-real-time monitoring and modelling, where systems are appropriately monitored and data analysed to provide the best-available knowledge of the system status such that failures are anticipated and decisions made to intervene prior to service interruption or failure. Such an advance requires a paradigm shift in approach and a move away from data collection and analysis driven by regulatory requirements, to a system where near-real-time data collection and analysis is used to best inform water companies of the condition and performance of their, significant, infrastructure asset base.

Reacting to water supply system events, e.g. low water pressure or poor water quality, as and when they occur, with little to no anticipation of the events, describes the current operational and maintenance strategy in most UK water companies. The usual outcome of such a reactive strategy is to undertake a repair, initiate maintenance or change operational practice in order to return the supply system to again meet the regulatory requirements. It is recognised that there are major problems with this approach, in that, by waiting for an event to occur, the company is not fully in control of the performance of the system, and standards of service are often compromised. The customer often experiences poor or failing service for long periods before the company is aware or able to react. Once aware, it is then necessary to carry out an investigation and subsequently propose, plan and implement a solution. All this takes time, during which the effects of the impact of the event may become widespread and the solution procedure more complex and costly. Once the solution has been implemented, further observations and analysis are completed to ensure that the system status again meets compliance standards. As a consequence, the reactive management approach is one where it is usual to quickly get the system resources back into production, whether it be supply, infrastructure or people. What is also clear is that, over the years, water companies have developed significant expertise in reactive management, and now have the ability to:

- rapidly identify the root cause of events
- devise different and creative possible solutions that are proven and innovative
- select the best remedial option and quickly implement the chosen solution, calmly and in control in the midst of a crisis, to resolve the problem.

Some argue 'Why change?', but current thinking and advances in technology and communications highlight that there is considerable potential for the industry to move to more proactive strategies for the maintenance and management of system performance and operation.

Proactive planning and management involve anticipating system events such as operational problems, water quality failures and customer demands, prior to their occurrence. This requires forward planning to agree the desired future operational state of the system and subsequently to implement a strategy and schemes that will deliver and create the preferred future state. This approach allows a company to actively control the outcome of planning and investment and to actively shape the future. To move to such a proactive approach also provides the opportunity to reduce the potential of future threats by managing risk, to capitalise on a potential future opportunity and subsequently to optimise efficiency gain and cost–benefit. Proactive planning and management involves:

- planning for the short and long term, to meet company, regulatory, operational and customer service requirements
- taking calculated risks
- encouraging innovation
- working closely with technical and customer service staff and regulators
- undertaking customer satisfaction surveys.

Clearly, the proactive approach is the one that provides the most opportunity for operational stability and control, and tools and methods are constantly being improved, and some are now being implemented (Savić *et al.*, 2008; Bicik *et al.*, 2009).

In order to manage proactively, it is essential to understand all aspects of the supply system, including the current system status and the way in which operation and maintenance strategies influence this performance. The various regulatory requirements provide a base framework for the performance requirements of the system, and developments in instrumentation, communications technologies, computing power and data-processing methods are enabling companies to better understand the hydraulic and water quality characteristics of systems operation in near real time. Hence, these systems provide companies with a timely feed of data that are transformed into information on which future operation and maintenance decisions may be based, and such that the proactive management of water supply systems may become a reality. In time, these systems will be the normal mode of system operation, and will vastly improve over the coming years.

The vision for the future is real-time operational control of all aspects of water supply in order to continually optimise storage, flow, pressure and water quality; minimise energy use and leakage; and reduce carbon emissions. This will be achieved by introducing small, cheap, robust instrumentation at key locations that will have intelligent data analysis geared to record and report only data that are outside normal operational bounds. Continued technological development will drive the introduction of smart assets that have local measuring capabilities built in at the manufacturing stage and the embedding of data-processing functionality with self-diagnostic algorithms. Transmission of data may be through the fabric of the assets or even the body of the water within the system itself. Data from the field will be passed directly to hydraulic, water quality and performance models, which, in turn, will feed into operational models and decision support systems. The models and systems will be capable of accepting operational feedback and will automatically control key assets to continually optimise system operation.

The future is bright, but will require significant investment to maintain the current provision of safe and wholesome water throughout the UK water distribution systems.

REFERENCES

American Water Works Association with assistance from Economic Engineering Services (2002) *Effects of Water Age on Distribution System Water Quality*. Washington, DC: Office of Ground Water and Drinking Water, Environmental Protection Agency. www.epa.gov/safewater/disinfection/tcr/pdfs/whitepaper_tcr_waterdistribution.pdf [accessed 02.04.07].

Bicik, J., Kapelan, Z. and Savić, D. A. (2009) Operational perspective of the impact of failures in water distribution systems. In: *Proceedings of the World Environmental and Water Resources Congress 2009*. Kansas City, MI.

Boxall, J. B. and Dewis, N. (2003) Identification of discolouration risk through simplified modelling. In: *ASCE ERWI World Water and Environmental Water Resources*. Anchorage, AK.

Boxall, J. B. and Saul, A. J. (2005) Modelling discolouration in potable water distribution systems. *Journal Environmental Engineering, ASCE* 131(5): 716–725.

DWI (2002) *DWI Information Letter 15/2002*. London: Drinking Water Inspectorate.

DWI (2004) *Guidance on the Water Quality Aspects of Common Carriage*. London: Drinking Water Inspectorate.

European Environment Agency (2003) *Indicator Fact Sheet (WQ06). Water Use Efficiency (in Cities): Leakage*. Copenhagen: European Environment Agency.

Farley, B., Boxall, J. B. and Mounce, S. R. (2008) Optimal location of pressure meters for burst detection. In: *10th Annual International Symposium on Water Distribution Systems Analysis*. Kruger National Park.

House of Lords (2006) *Water Management*, Vol. I. HL Paper 191-I. London: TSO.

Husband, P. S., Boxall, J. B. and Saul, A. J. (2008) Laboratory studies investigating the processes leading to discolouration in water distribution networks. *Water Research* 42(16): 4309–4318.

IWA Water Loss Task Force (2005) *Best Practice Performance Indicators for Non-revenue Water and Water Loss Components: A Practical Approach*. London: International Water Association.

Kirmeyer, G. (2000) *Guidance Manual for Maintaining Distribution System Water Quality*. Denver, CO: American Water Works Association.

Lambert, A. (2003) Assessing non-revenue water and its components: a practical approach. *Water21*, 50–51.

Machell, J. M., Boxall, J. B., Saul, A. J. and Bramely, D. (2009) Improved representation of water age in distribution networks to inform water quality. *Journal of Water Resources Planning and Management, ASCE* 135(5): 381–382.

Machell, J., Mounce, S. R. and Boxall, J. B. (2010) Online modelling of water distribution systems: a UK case study. *Drinking Water Engineering and Science* 3: 21–27.

Morrison, J. (IWA Water Loss Task Force) (2004) Managing leakage by district metered areas: a practical approach. *Water21*, 44–47.

Mounce, S. R., Boxall, J. B. and Machell, J. (2007) An artificial neural network/fuzzy logic system for DMA flow meter data analysis providing burst identification and size estimation. In: Ulanicki, B., Vairavamoorthy, K., Butler, D., Bounds, P. L. M. and Memon, F. A. (eds), *Water Management Challenges in Global Change*, pp. 313–320. London: Taylor and Francis.

Mounce, S. R., Boxall, J. B. and Machell, J. (2008) Online application of ANN and fuzzy logic system for burst detection. In: *Proceedings of the 10th Water Distribution System Analysis Symposium*. Kruger National Park.

Mounce, S. R., Boxall, J. B. and Machell, J. (2010) Development and verification of an online artificial intelligence system for detection of bursts and other abnormal flows. *Journal of Water Resources Planning and Management, ASCE* 136(3): 309–318.

Mutoti, G., Dietz, J. D., Imran, S., Taylor, J. and Cooper, C. D. (2007) Development of a novel iron release flux model for distribution systems. *Journal of the American Water Works Association* 99(1): 102–111.

Ofwat (2002) *Best Practice Principles in the Economic Level of Leakage Calculation*. Birmingham: Tripartite Group, Ofwat.

Ofwat (2005) *Security of Supply, Leakage and the Efficient Use of Water: 2004–05 Report*. Birmingham: Ofwat.

Orr, C., Bouulos, P., Stern, C. and Liu, P. (1999) Developing real-time models of water distribution systems. In: Savić, D. and Walters, G. (eds), *Water Industry Systems: Modelling and Optimisation Applications*, Vol. 1. Baldock: Research Studies Press.

Prasad, T. D. and Walters, G. A. (2006) Minimizing residence times by rerouting flows to improve water quality in distribution networks. *Engineering Optimisation* 38(8): 923–939.

Rossman, L. A. and Boulos, P. F. (1996) Numerical methods for modeling water quality in distribution systems: a comparison. *Journal of Water Resources Planning and Management, ASCE* 122(2): 137–146.

Rossman, L. A., Clark, R. M. and Grayman, W. M. (1994) Modeling chlorine residuals in drinking water distribution systems. *Journal of Environmental Engineering, ASCE* 120(4): 803–820.

Savić, D. A., Boxall, J. B., Ulanicki, B., Kapelan, Z., Makropoulos, C., Fenner, R. *et al.* (2008) Project Neptune: improved operation of water distribution networks. In: *Proceedings of the 10th Annual Water Distribution Systems Analysis Conference*. Kruger National Park.

Skipworth, P. J., Saul, A. J. and Machell, J. (1999) Predicting water quality in distribution systems using artificial neural networks. *Proceedings of the Institution of Civil Engineers: Water Maritime and Energy* 136(1): 1–8.

Stoianov, I., Nachman, L., Whittle, A., Madden, S. and Kling, R. (2006) Sensor networks for monitoring water supply and sensor systems: lessons from Boston. In: *Proceedings of the 8th Annual Water Distribution System Analysis Symposium*. Cincinnati, OH.

Thornton, J. (IWA Water Loss Task Force) (2003) Managing leakage by managing pressure: a practical approach. *Water21*, 43–44.

University of Bristol (2010) *Ice Pigging*. www.bristol.ac.uk/red/techtransfer/scilicopps/icepig.html [accessed 20.05.10].

Vreeburg, J. and Boxall, J. B. (2007) Discoloration in potable water distribution systems: a review. *Water Research* 41(3): 519–529.

Water Science and Technology Board (Committee on Public Water Supply Distribution Systems: Assessing and Reducing Risks), National Research Council and National Academy of Sciences (2005) *Public Water Supply Distribution Systems: Assessing and Reducing Risks, First Report*. Washington, DC: National Academy Press.

WRc (1989) *Planning the Rehabilitation of Water Distribution Systems*. Swindon: Water Research Centre.

WRc (1994) *Managing Leakage*, Report A. Swindon: Water Research Centre.

REFERENCED LEGISLATION

The Drinking Water Directive. Council Directive 98/83/EC of 3 November 1998 on the quality of water intended for human consumption. *Official Journal of the European Communities* L330: 32–54.

The Water Act 2003. London: TSO.

The Water Industry Act 1999 (supersedes the Water Acts 1945, 1973, 1989, the Water Industry Act 1991). London: HMSO.

The Water Supply (Water Quality) (Amendment) Regulations 1989. London: HMSO.
The Water Supply (Water Quality) (Amendment) Regulations 1999. London: TSO.
The Water Supply (Water Quality) Regulations 2000. London: TSO.
The Water Undertakers (Information) Directive 2004. London: TSO.

Water Distribution Systems
ISBN: 978-0-7277-4112-7

ICE Publishing: All rights reserved
doi: 10.1680/wds.41127.227

ice

Institution of Civil Engineers

publishing

Chapter 9
Asset planning and management

Zoran Kapelan University of Exeter, UK
John K. Banyard Independent consultant, Warwick, UK
Mark Randall-Smith Water Engineer, Verwood, UK
Dragan A. Savić University of Exeter, UK

9.1   Introduction

The concept of asset management as a specific discipline is relatively modern in the (UK) water industry, perhaps going back to the mid-1980s. The privatisation of utilities (gas, electricity, telecommunications, etc., a group to which water companies are also normally considered to belong) brought conventional financial disciplines to previously 'nationalised' industries. Not only did the newly privatised utilities have to meet strict performance targets, they also had to do so within clearly defined financial constraints, monitored by their investors and other providers of capital. One of the disciplines introduced by a more-rigorous financial regime was the need to consider carefully the question of depreciation. All assets wear out, and conventional accounting practice requires this to be reflected in the value of the assets shown in the balance sheet. This is achieved by predicting the life of each asset and then reducing the value in the balance sheet to zero over the chosen life of the asset. It is essential that this book life is achieved, otherwise the whole residual value of the asset has to be written out of the accounts when the asset is taken out of service. Since financially this sum is subtracted from what would otherwise be profit, it can be extremely damaging to a company's performance if premature write off of assets is commonplace; however, that does not mean there can be unlimited spending to preserve assets, since that would also damage the profitability. Because the utilities are distinguished from other companies by having extremely large asset bases, this resulted in considerable attention being applied to the management of the asset base, and the discipline of asset management has emerged as a result of the importance of being able to control and forecast the serviceability of the asset base. Furthermore, it has been adopted by other organisations to assist in the management of large asset bases such as railways, and the concepts are widely adopted for utilities, whether they are state owned or private companies.

The objective of this chapter is to outline some of the main issues and potential solutions related to the planning and management of water supply and distribution system assets. This includes:

- relevant background information (see Section 9.2)
- regulatory aspects (see Section 9.3)
- main asset management drivers and issues (see Section 9.4)

- asset performance indicators (see Section 9.5)
- asset assessment techniques (see Section 9.6)
- interventions used to improve the condition/performance of assets (see Section 9.7)
- asset ageing and deterioration issues (see Section 9.8)
- asset failure consequence modelling (see Section 9.9)
- an integrated, whole-life-costing-based framework for making more-informed asset management decisions (see Section 9.10).

Finally, a summary is provided and future challenges identified in the last section of this chapter.

9.2 Background

The total estimated value of fixed tangible assets in the possession of all water (and sewerage) companies in England and Wales in 2007–2008 was approximately £251 billion (Ofwat, 2008a), with a turnover of £9.2 billion (£4.6 billion in the water service). According to the same source, the total annual capital investment was £4.9 billion (£2.46 billion in the water service), of which £2.5 billion was spent on maintaining the networks alone. The total annual operating expenditure was £3.39 billion (£1.87 billion in the water service). The quantity of assets that require careful planning and management is often vast, especially in the larger water utilities and companies. For example, the total length of water mains in the UK is approximately 350 000 km, which is roughly equal to the distance between the earth and the moon.

As understanding of the asset base has increased, so has the complexity of what is meant by asset management. The Institute of Asset Management (2009) describes it as follows:

> The management of physical assets (their selection, maintenance, inspection and renewal), plays a key role in determining the operational performance and profitability of industries that operate assets as part of their core business. Asset Management is the art and science of making the right decisions and optimising these processes. A common objective is to minimise the whole life cost of assets but there may be other critical factors such as risk or business continuity to be considered objectively in this decision making. This emerging professional discipline deals with the optimal management of physical asset systems and their life cycles. It represents a cross-disciplinary collaboration to achieve best net, sustained value for money in the selection, design/ acquisition operations maintenance and renewal/disposal of physical infrastructure and equipment.

An alternative and more concise definition is adopted by Montgomery Watson Harza's The Asset Group (MHW TAG) (see Box 9.1).

Box 9.1 Asset management definition by MWH TAG

'Asset management is a business discipline for managing the life cycle of assets to achieve a desired service level while mitigating risks and it encompasses management, financial, customer, engineering and other business processes.'

These definitions cover a number of important aspects of asset management:

■ First, asset management is a *business* discipline, and, as such, the success is measured using business indicators (e.g. profit).
■ Secondly, assets need to be managed through the whole *life cycle* of an asset, i.e. from planning and purchasing to installing, maintaining and, eventually, disposal.
■ Thirdly, one of the main asset management goals is to achieve the *target service level* at the lowest reasonable overall cost, while mitigating various *risks*.
■ Finally, it is a complex activity encompassing a large number of *diverse* processes, not all of which are of engineering nature (a fact often overlooked by engineers).

Because asset management is so complex, it is necessary to tease apart some of the different aspects to better understand them and how they interrelate with other elements. It is important, though, to always keep the above explanations in mind; it can be very misleading to base decisions on any single element.

For the purpose of asset management, all water supply and distribution system assets can be broadly classified into *infrastructure* assets (assets that are part of the water supply and distribution network) and *non-infrastructure* assets (e.g. headquarter buildings, vehicles, etc.). Accountants would normally differentiate between fixed assets and mobile plant, and so it can be immediately seen that reference to the normal accounting practices will not meet the need, and that a new system of classification will have to be established – one which uses data from the conventional accounting systems.

Another helpful classification is the one which separates the below- and above-ground assets. This important distinction is made because the location of assets relative to the ground level has a major effect on the way that asset condition and performance is monitored and also on the way that assets are maintained. The below-ground assets include primarily pipes, i.e. water trunk and distribution mains and service connection pipes. The above-ground assets include water sources (wells/boreholes, open reservoirs with dams, river captures, etc.), treatment works, various pumps/pumping stations, tanks/reservoirs, etc. – see Chapters 4 and 5 for detailed descriptions of these elements. The big difference is that 'above-ground assets' can be relatively easily physically inspected and accessed for maintenance. With buried pipes, the problems are very different: while it may be possible to use closed-circuit TV inspection for, in particular, the inside of sewers, the ground–pipe interface is always hidden. However, the pipe network is made up of what may be interpreted as a number of asset classes, each comprising an assumed set of homogeneous assets. Therefore, statistical techniques can be applied to help understand the overall serviceability of the network.

9.3   The regulatory framework

The issue of water supply and distribution has economic but also political, social and environmental dimensions. Also, the nature of the business is often referred to as a 'natural monopoly'; one cannot choose the water supplier, it is pre-determined on a geographic basis. As a consequence, the planning and management of water supply and distribution system assets are often regulated by the government.

The water industry in the England and Wales is regulated by the following main bodies:

- the Office of Water Services (Ofwat)
- the Environment Agency (EA)
- the Drinking Water Inspectorate (DWI).

All three regulators are established as statutory bodies, to empower them for practicable implementation of the legal requirements of Parliament and the policy aims of the government.

Ofwat is the economic regulator created under the Water Act 1989 as a non-ministerial government department directly responsible to Parliament. Ofwat was, therefore, created when the water industry in the UK was privatised, to protect the customers from the monopolistic nature of the business; i.e. effectively the role of the economic regulator is to make up for the lack of market competition. Ofwat currently regulates 21 water companies in England and Wales.

Ofwat's duties were laid down in section 2 of the Water Industry Act 1991 as updated by section 39 of the Water Act 2003. The main duties are as follows (Ofwat, 2009):

- to protect the interests of consumers, wherever appropriate by promoting effective competition
- to secure that the functions of all water companies are properly carried out and that they are able to finance their functions, in particular by securing a reasonable rate of return on their capital
- to secure that companies with water supply licences (i.e. those selling water to large business customers, known as licensees) properly carry out their functions.

Ofwat regulates companies' performance by a number of different instruments, including the following (Ofwat, 2009):

- Setting price limits that reflect what each company needs to charge to finance the provision of services to its customers. These are reviewed every 5 years, based on business plans submitted by the companies. The business plans are based *inter alia* on companies' asset planning and management plans, which specify in detail what each company will deliver during the analysed 5 year period in order to maintain/ improve the levels of different services.
- Monitoring the activities of the companies. Every year Ofwat asks the companies to provide information about the previous year in the so-called 'June return'. These statements provide Ofwat with details on a wide variety of activities, including levels of customer service, new additions to the network and leakage information. This information also allows Ofwat to compare performance levels between companies.
- Enforcing companies' licences. Companies operate under licences, granted by the Secretary of State for Environment, Food and Rural Affairs and by the Welsh Assembly Government, to provide water and sewerage services in England and Wales. The licences impose conditions on the companies which Ofwat enforces.

The EA is the environmental regulator. The EA's principal aims are to protect and improve the environment, and to promote sustainable development. The EA plays a central role in delivering the environmental priorities of central government and the Welsh Assembly Government through different functions and roles, including pollution control. The EA advises the government on environmental programmes and improvements required to meet the needs of national legislation but also the relevant EU Directives.

The DWI is the potable water quality regulator. The DWI is responsible for assessing the quality of drinking water in England and Wales, taking enforcement action if standards are not being met and appropriate action when water is unfit for human consumption. Legal standards are set out in the Water Supply (Water Quality) Regulations 2000, most of which are derived from the EU Drinking Water Directive.

Effectively, the DWI and the EA set the quality standards for both potable water and sewage discharges, against a background of statutory obligations, and are responsible for enforcing those standards. Ofwat is responsible for setting water charges that allow those standards to be achieved by an efficient water company that uses to the full the physical and financial facilities available to it. It is important to stress, yet again, that Ofwat takes an economic view, and is not obliged to support expensive solutions simply because they appeal to technologists. Equally, Ofwat will make assumptions about the sources of finance that would be available to a well-run company: it has no obligation to ensure the continuance of any company that has squandered its financial resources.

So far, the English and Welsh situation has been discussed with regard to Ofwat; the arrangements in both Scotland and Northern Ireland are similar but with their own independent regulatory bodies. This concept of independent regulators has become more common around the world, even though there are differences in structure and responsibility from country to country.

9.4 Asset management drivers

Asset management in the UK is subject to a number of different pressures coming from various sources:

- First, when acquiring assets, companies must realise that they will only be given the finance once through customer charges. If they select inappropriate assets, then the cost of replacement will fall on their owners (in England and Wales, the investors).
- Like all physical assets, water supply and distribution systems are ageing, and, as such, their condition and consequently performance is deteriorating with time, even though this can be slowed down or accelerated, depending on the maintenance regime.
- The performance of these systems may also deteriorate due to increased urbanisation (i.e. increased water demand, changing spatial and temporal patterns of demand, etc.).
- At the same time, customer expectations are increasing in terms of better service expected, e.g. fewer interruptions to supply, better water quality, reduced negative

impact on the environment, improved sustainability of various solutions, to name just a few.

The above pressures affect all water utilities (both public and private), but the effect can be increased by the regulators' own agenda; for example, an economic regulator may wish to emphasise customer service, or a quality regulator may wish to emphasise the need to control provision of discoloured water to customers. This introduces a problem that will be identified in more detail later in this chapter, but it is useful to introduce the idea now. The drivers highlighted by regulators do not necessarily identify all of the issues that need to be taken into account when seeking to optimise the management of the asset base. Regulators emphasise the issues that are currently of national importance, such as 'sustainability'; in doing so, they assume that the more-mundane issues such as maintaining borehole pumps are being properly addressed. If they are not, then the utility may well be disappointed when it seeks to increase charges in order to replace the pumps in a shorter timescale than its competitors. The consequence of this is that, for successful asset management, there is a need for more data than simply those which are required to respond to the regulators' demands.

To illustrate the complexity of this, a number of different issues can be considered that have been or are being addressed to improve the quality of the water service in the UK:

- water *supply/demand balance* management, including *security of water supply* and *demand efficiency* issues
- *leakage management*
- *water quality* management, including aesthetic, bacteriological and chemical aspects of the problem
- *hydraulic capacity*, including pressure-related issues
- *customer service*, including issues related to consumer bills, complaints, provision of information to consumers, etc.
- *sustainability*, including climate change, renewable energy, carbon footprint and other related issues.

Some of these are interrelated, such as water supply/demand balance, and leakage. Others, such as customer service, include issues that have no real relationship to asset management as defined here. For example, the speed of answering telephones in call centres may well depend on successful outsourcing of the service.

However, this does provide further illustration of the dangers of relying solely on the regulators' measures when managing an asset base. For call centre operatives to respond properly to operational problems, there needs to be good communication between the operational staff and the customer service staff. This will require IT systems such as geographic information systems (GISs). Regulators do not wish to micromanage utilities, so they do not specify the need for GISs, nor the type of GIS that should be adopted. However, without a good GIS it is most unlikely that customer service targets can be met; it is for management to make those decisions, not regulators.

9.5 Asset performance indicators (PIs)

The planning and management of water utility assets should be undertaken with the objective of achieving or maintaining a desired and defined level of service to the customers and the environment at the lowest reasonable cost. This, in turn, is achieved by actively controlling the performance of relevant assets and systems, which may involve improving the assets or simply operating them in the appropriate manner.

A wide range of PIs may be used to measure serviceability and performance of assets, including those found in water supply and distribution systems. These PIs are used to measure the current performance, but also to identify shortfalls against desired targets, which are then used, in turn, to drive future asset management plans.

PIs are also used for benchmarking, i.e. when comparing performances of different water companies, either within the UK or internationally. However, this can be misleading unless care is taken to ensure that all parties are measuring the same indicator in exactly the same way.

Different definitions of PIs exist, highlighting various aspects of their use. Some of the examples include:

- a standard and manageable way of measuring performance
- an indication as to how to progress towards a target
- a representation, numeric or otherwise, of the state of, or outcomes from, an organisation, or any of its parts or processes.

Ofwat has introduced high-level 'customer service' indicators to set target standards for the water companies, and are shown in Table 9.1 (original Director General (DG) indicators relevant for the clean water service) and Table 9.2 (additional indicators, introduced later on).

The method is called the 'overall performance assessment' (OPA) method, and is used to measure and incentivise performance across the broad range of services provided to consumers and the environment. It allows Ofwat to compare the quality of the overall service, and tells consumers and other interested parties about how their local water company has performed relative to other companies.

Ofwat measures the serviceability of the water mains networks (i.e. water infrastructure) by using the following indicators (Ofwat, 2008b,c):

- the extent of low-pressure problems (DG2)
- the scale of interruptions of supplies to consumers – unplanned interruptions to supplies greater than 12 hours (DG3)
- compliance in respect of the level of iron in water (mean zonal compliance)
- the number of burst water mains
- distribution losses (component of overall leakage).

Table 9.1 Ofwat's DG water and customer service indicators

Indicator	Title	Brief description
DG2	Risk of low pressure	The number of connected properties that have received, and are likely to continue to receive, pressure below the reference level when demand for water is at a normal level. The reference level of service is defined as 10 m head of pressure at a boundary stop tap with a flow of 9 l/min
DG3	Supply interruptions	This indicator shows the number of properties experiencing (unplanned) interruptions to their water supply for 3–6, 6–12, 12–24 and more than 24 hours
DG4	Restrictions on water use	This indicator shows the percentage of a company's population that has experienced water usage restrictions. Water usage restrictions can be divided into voluntary reductions (encouraged by a publicity campaign), hosepipe restrictions, Drought Orders restricting non-essential water use and Drought Orders imposing standpipes or rota cuts
DG6	Billing contacts	This indicator shows the total number of billing contacts that consumers made during the report year and the time each company took to respond to them. The time is measured in two bands: within 5 working days and in more than 10 working days
DG7	Written complaints	This indicator identifies the total number of written complaints received during the report year and the time taken to respond to them. The time is measured in two bands: within 10 working days and in more than 20 working days
DG8	Bills for metered customers	This indicator shows the percentage of metered consumers who receive at least one bill during the year based on a meter reading taken by either the water company (or its representative) or the consumer
DG9	Ease of telephone contact	The aim of this indicator is to identify the ease with which consumers can make telephone contact with their local water company and their satisfaction with the way the company handled their call

Ofwat measures the serviceability of water treatment works, service reservoirs and pumping stations (water non-infrastructure assets) by using the following indicators (Ofwat, 2008c):

■ the percentage of the total number of samples taken at water treatment works containing coliforms
■ the number of water treatment works where enforcement action was considered because of contravention of the coliforms standard

Table 9.2 Ofwat's other water service indicators

Drinking water quality: operational performance index (OPI)	DWI's OPI, based on the following six parameters: iron, manganese, aluminium, turbidity, faecal coliforms and trihalomethanes
Drinking water quality: percentage mean zonal compliance	The method that the DWI uses to assess and report on compliance with the drinking water standards is based on zones, and is known as percentage mean zonal compliance. This is calculated for eight parameters: *Escherichia coli*, odour, taste, nitrate, aluminium, iron, lead and pesticides
Security of supply index (SoSI)	Used to assess each company's compliance regarding its duty to ensure the security of its water supplies. At a company level, index scores reflect the size of any deficit against the company's estimate of target headroom in each of its resource zones and the proportion of consumers in each resource zone that are exposed to headroom deficits. For calculation details, see OFWAT (2008c). Once calculated, the SoSI values are banded as follows: A: no deficit against target headroom in any resource zone (SoSI = 100) B: marginal deficit against target headroom (SoSI = 90–99) C: significant deficit against target headroom (SoSI = 50–89) D: large deficit against target headroom (SoSI < 50)
Leakage performance	The amount of water leaked. A distinction is made between distribution, customer supply and service reservoir leakage. Reported in megalitres per day (Ml/day), litres per property per day (l/prop/day) and cubic metres per kilometre of main per day (m^3/km/day)
Water efficiency	Measured by customer supply pipe repairs and replacements, the number of cistern devices distributed to households, the number of household and non-household water audits and total savings/costs
Greenhouse gas (GHG) emissions	Measured by total operational GHG emissions, i.e. without GHG emissions released in the construction of assets or materials

- the number of service reservoirs with coliforms detected in more than 5% of samples, which is the current standard
- the number of water treatment works where turbidity (water clarity) exceeds a threshold value within the permitted range (these data can be affected by raw water quality at some types of treatment works, and thus do not provide a guide to serviceability in some cases)
- unplanned maintenance (company-specific).

When reporting the above indicator values, all companies are asked to provide the confidence grades for the reliability and accuracy of information submitted. The four Ofwat reliability bands are as follows:

- A: sound textual records, procedures, investigations or analysis properly documented and recognised as the best method of assessment.
- B: as A, but with minor shortcomings. Examples include old assessment, some missing documentation, some reliance on unconfirmed reports, or some use of extrapolation.
- C: extrapolation from limited samples for which grade A or B data are available.
- D: unconfirmed verbal reports, cursory inspections or analysis.

Ofwat's seven accuracy bands are as follows: $\pm 1\%$, $\pm 5\%$, $\pm 10\%$, $\pm 25\%$, $\pm 50\%$, $\pm 100\%$ and 'X' or very small numbers, where accuracy cannot be calculated or the error could be more than $\pm 100\%$.

The International Water Association (IWA) has developed an alternative PI system. The latest (second edition) of this PI system (Alegre et al., 2006) provides a library of more than 160 PIs covering both water and sewerage services, classified into the following main categories:

- water resources indicators
- personnel indicators
- physical assets indicators
- operational indicators
- quality of service indicators
- economic and financial indicators.

Each IWA PI has a number of its own properties (e.g. ID, description, units, etc.), the set of associated variables (used to calculate its value) and the contextual information (i.e. relevant to the PI analysed). Similarly to Ofwat, the IWA has confidence grades for the accuracy and reliability of PI values obtained. For technical use, four categories exist for data accuracy (0–5%, 5–20%, 20–50% and >50%), and three for data reliability (*, ** and ***). Accuracy and reliability are assessed at the variable level and then propagated to the relevant PI value. For the purpose of public communication, all PI values are simply classified as 'good', 'average' or 'bad'. The IWA system is, of course, an option for water utilities and their regulators around the world, or they may prefer to develop their own approaches.

If details of two of the Ofwat indicators are analysed, it can be seen clearly why companies cannot simply concentrate on these high-level indicators if they are to run their businesses successfully.

Let us start with the drinking water OPI in Table 9.2. This is taken from the DWI annual report, which actually contains performance against over 50 parameters defined by the EU, failure against which may result in regulatory action by the DWI against the company. In the case of *Cryptosporidium*, which does not even feature in the EU list, there is a strong likelihood of prosecution of the company if this limit is breached. So, simply managing the distribution system against the eight parameters is an almost certain route to failure.

A second example is slightly more complex but just as damaging to a company's long-term prospects. Let us consider the DG3 supply interruptions in Table 9.1. Imagine that there is a high-lift pumping station supplying a small town via a terminal reservoir. The station contains only two pumps: one duty, one standby. Initially, all works well: when the duty pump fails, the standby starts automatically, and there is no noticeable interruption to supply. The regulatory standard is being met, so if that is the only indicator being used, when the duty pump reaches the end of its life, there is no history of failure in customer service to justify replacing it. Instead, the station continues to operate without any standby. As time passes, the customer service will deteriorate, depending on the ability of the maintenance team to carry out repairs before the terminal reservoir empties. In real life, the operators will soon bring the problem to the attention of management, and hopefully the issue would be addressed proactively before customers suffer what could be a major disruption in supply.

These examples do not highlight failures in the regulatory indicators – they simply demonstrate that the regulator takes a very high-level overview of the state of each company. However, if the company is to meet the economic regulator's targets, it must have a far better awareness of the state of its assets than is provided by regulatory targets and indicators. In the language of total quality management, it needs to be able to identify the root causes of problems, not just the symptoms, even though the presence of the symptoms may be sufficient for regulatory assessments.

9.6   Asset assessment techniques
9.6.1   Asset inventory
Banyard and Bostock (1998) describe the development of a suite of computer programs that allowed them to define a long-term capital programme in order to optimise the delivery of all regulatory outputs. To do this, the starting point is the development of an asset register, which is a database containing all of the company's assets.

At this point, it is necessary to consider what level of detail is required. For example, it is of very limited benefit to record that there is a water treatment works at a certain location. What is required is to consider what assets collectively make up the waterworks. Most water treatment works will have been constructed in phases, and hence some assets will be older than others. Additionally, it needs to be recognised that while concrete tanks may have a life of 60–70 years, mechanical and electrical equipment such as pumps and compressors will only have a life expectancy of 20–25 years. Telemetry equipment will have an even shorter life. It becomes clear, therefore, that an asset register has to be established that will distinguish between these very different assets.

The easiest way to tackle this problem is decide at what level assets might be abandoned and replaced by newer assets once their useful life has expired. So, each pump would be separately listed, including details of its manufacturer, the date of installation, output and power consumption. But there would be no separate listing for the pump casing or pump impeller. If the pump impeller needed to be replaced, it would be a maintenance item paid from revenue, not capital. This issue will be revisited later in the chapter.

It is also helpful to record details of tank dimensions so that meaningful cost models can be developed. It is stressed here that all data must be capable of audit, so there can be no question over the integrity of the data, which, after all, are a foundation block of asset management. It is almost impossible to deliver effective asset management if there is uncertainty about which assets are actually being managed.

For pipe systems, the introduction of GIS systems has been a huge step forward. The ability to display the pipe network against an up-to-date map background has been invaluable and has also led to significant reassessment of the amount of pipe work actually in the ground. As with the above-ground assets, this needs to be supplemented with other information such as the pipe material, pipe internal diameter, date of laying, etc. A word of warning is appropriate at this point: it is essential that the data cannot only be displayed but can also be exported to other databases for manipulation with other information. Many of the early GIS systems were very restricted in allowing such export of data, and frequently required purpose-written software before manipulation was possible. This at times led to separate databases being developed, which frequently were out of step with one another. The modern GIS systems have avoided this problem.

The next step is to supplement the fundamental asset data with an assessment of the 'serviceability' of the assets.

9.6.2   Assessing serviceability

In April 2000 The Director General of OFWAT wrote to the Water Companies (MD161) raising concerns over the economic aspects of customer serviceability, this was the starting point for what was to become the Common Framework. Skipworth *et al.* (2002) distinguish between three major types of PIs:

- Condition PIs: used to measure the physical condition or state of an asset.
- Performance PIs: used to measure the behaviour of the asset(s) in terms of its performance.
- Serviceability PIs: used to measure the quality of service received by customers and the ability of the assets (and their operator) to maintain the quality of service.

The Water Research Centre (WRc) had produced condition grades for various water industry assets, and these were adopted by many companies. They complied with the preferred Ofwat grading of 1–5, with 1 being excellent and 5 being awful. However, there were no similar performance grades, and it was left to individual companies to develop these, resulting in a lack of consistency between companies.

An example of condition PIs are Ofwat's pipe condition grades:

- Condition grade 1: no failures, fully complies with modern standards.

- Condition grade 2: no significant failures (minimal impact on service performance), not quite consistent with modern standards.
- Condition grade 3: deterioration beginning to be reflected in service levels or increased operating costs.
- Condition grade 4: considerable corrosion affecting service performance, nearing end of useful life, frequent bursts.
- Condition grade 5: substantially derelict and source of service problems, no residual life.

As can be seen from the above, the condition grades are of a descriptive nature and involve implicit performance information which was often criticised in the past. Ofwat has moved away from this approach as far as its periodic reviews are concerned, but it still has value within companies for their day-to-day management, and is worth further exploration here.

First, there needs to be much greater clarity between condition and performance grades, and a more numerate approach is also helpful. Condition grades describe the physical condition of the asset: whether it is structurally sound, whether it is safe to operate (particularly electrical equipment), etc. For pipes, the number of bursts per kilometre per year would be one possible numerate assessment. Performance grades indicate whether or not the asset is capable of meeting the required output. So, the output from a pump measured in volume and head would be graded in terms of how far it fell short of requirements.

A simple example illustrates how this approach operates: consider a pipeline which can deliver the desired flow, but fails eight times per kilometre per year. Such a pipe would be rated 5 for condition and 1 for performance. In contrast to this, a new pipeline that had been badly designed and was incapable of delivering the flow required would be rated 1 for condition but 5 for performance.

It is now possible to examine a more complex problem to see how these condition and performance grades can assist a company in managing its asset base. Consider a situation where an existing pipeline is no longer capable of delivering the necessary flow because a new housing development is being constructed. The options for dealing with this problem might be:

1. replace with a new pipe
2. duplicate with a new pipe that would be sized to make up the deficit in capacity
3. renovate the existing pipe and add a second pipe only if necessary.

The problem facing any large company is that these options will be evaluated on the basis of individual subjective judgements, unless company-wide standards have been defined. However, the individual preferences of different members of staff are not a satisfactory way of controlling the investment in a huge asset base. Thus, standardised condition and performance grades are a major step towards objective decision-making. Further, some of the decision-making can be standardised across the company. The decision on whether

239

or not relining is economic is driven by a net present value (NPV) calculation comparing the NPV of a new main today with the NPV of relining now and replacing when the relined pipe is worn out. Once this calculation has been made, it can be simplified into a rule that says no pipe will be relined unless it has a certain life after relining, which can be linked with a condition grade. So, for example, it could be said that only pipes of condition grade 3 or better are to be relined.

Finally, there is a common misunderstanding that condition and performance grades in some way lead to a position whereby all assets are replaced when they reach the end of their book lives. This is incorrect. Assets should generally only be replaced when they are life expired, i.e. when they reach condition grade 5. They may be retired before that if upgrading performance is best achieved by early retirement, but this should be determined through discounted cash flow analysis (see Chapter 10), not subjective judgement. In any system there will be a mix of asset condition and performance, and this will constantly change as assets age.

9.6.3   Risk

The use of condition and performance grades has been criticised because it fails to take account of the risk of asset condition/performance failure, and to some extent this is a fair criticism; additionally they only gave a limited and rather subjective view of the likelihood of future asset failure and no indication of the consequence: a more sophisticated approach was needed. In the general case in the mid-1980s, Germanopolous et al. (1986) considered the case of a service reservoir serving a large town which had only 4 hours nominal storage, when the water service provider's standard was 18 hours. However, the town had never run out of water. Either the company standard was too high (nominal retentions are usually between 14 and 18 hours for design purposes) or there were other factors at work. Germanopolous et al. showed by use of statistical analysis that the town was extremely well protected because there were numerous independent feeds serving the reservoir, and it was highly unlikely that they would all fail at the same time. This example showed the dangers of simply taking company standards and failing to understand the precise situation. Mathematically, because of the number of feeds into the reservoir, the nominal capacity understated the resilience of the actual situation, or, put another way, the effective capacity of the reservoir was well above the water service provider's standard.

Unfortunately, this idea of incorporating 'risk' may lead to potentially damaging conclusions, and great care is needed. In the general case, risk of some failure is defined as a function of the likelihood and the consequence of that failure. In the context of asset management, this function is usually defined as a simple product of the likelihood and the consequence components of risk (despite the obvious drawback that small likelihood and large consequence can be represented by the same risk number, as in the case of a large likelihood and a small consequence). In the case of water distribution systems, consequence is usually equated with 'population affected' (or some similar measure). From here, it is only a very small step to concluding that money should be used to protect the large centres of population at the expense of small village communities, because that is the route to minimise risk. The fallacy in

Figure 9.1 An idealised trade-off curve between the NPV of asset management interventions and the service performance benefit (risk)

After Tynemarch: www.tynemarch.co.uk

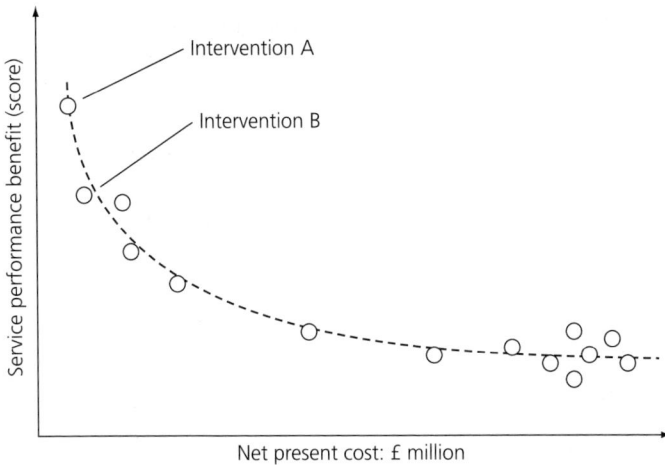

this approach is that the statutory requirements are normally for all customers to receive a defined minimum level of service. There is no defence to regulatory failure to argue that only a few hundred people were supplied with unwholesome water because the money had been used to give extra protection to a large city. However, if it is understood that there must be a minimum level of service for all customers in most aspects of a water utilities work, then risk analysis does provide valuable tools in deciding where any further funds can be best applied to obtain best value for money.

It is certainly possible to trawl operational staff for their views on which schemes should be pursued to deliver risk reduction. These schemes can then be evaluated to see what actual risk reduction they would bring, and then risk quantification methods applied to help distinguish between competing schemes and groups of schemes. The output would be similar to the idealised plot in Figure 9.1, and the utility is left to draw the cut-off point where it wishes, either in terms of funds available or residual risk (see Figure 9.1). Thus, the concept of risk is a valuable contribution to asset management, but its use must be fully understood and controlled accordingly (Cook *et al.*, 1999).

9.6.4 The Common Framework
Although the use of condition and performance grades had proved very useful, the regulator for England and Wales was concerned that it did not properly meet his needs, and sought an alternative approach. This resulted in UK Water Industry Research (UKWIR) – the industry's common-interest research body – commissioning Tynemarch to produce an alternative. The report was published in May 2002 (UKWIR, 2002).

It sought to build on the regulator's preferred four-stage process for assessing capital maintenance needs (capital monies used to replace worn out assets – not to be confused with maintenance of assets):

- historical analysis to identify levels of expenditure in previous years and levels of customer serviceability
- forward-looking analysis to determine future needs
- identification of any changes that make continuing at historical levels unreasonable
- a case being made for a revised level of capital maintenance.

The Common Framework (UKWIR, 2002) was then developed, incorporating the following:

- Capital maintenance should normally be justified on the basis of current and forecast risk analysis of asset failure, i.e. the likelihood and consequence of asset failure should be assessed. Consequences are to be expressed in terms of both service levels and costs.
- Service includes both service to customers and to the environment, to capture statutory obligations that are not represented by standard customer service indicators.
- A least-cost approach must be demonstrated for the capital expenditure/operational expenditures (CAPEX/OPEX) trade-offs.

The framework deliberately did not specify detailed methodology, as the regulator wished to encourage innovation and diversity of approach between companies. It has turned out to be a very comprehensive way of assessing capital maintenance needs, but is very data hungry. Unfortunately, much of the data required were not routinely collected before the publication of the framework, and this has hampered progress. It is possible to synthesise missing data using expertly facilitated teams of practitioners, but such approaches are not always fully accepted by regulators. Since the introduction of the Framework, data quality has improved and several computer based tools have been introduced, such as SEAMS and PIONEER to assist companies in the application of the Framework, which have allowed further improvements in the understanding of the asset base.

9.7   Asset interventions

In the asset management context, the term 'intervention' is typically used in a very broad sense, denoting any activity of capital or operational nature that is used to maintain and/or improve the condition, performance and/or serviceability of a single or group of assets (including water supply and distribution system as a whole).

A wide range of interventions is usually available. Within water distribution system asset management, interventions are often classified into infrastructure (pipe related) and non-infrastructure interventions (applied to other, typically above-ground, assets).

One of the generic intervention classifications often used is as follows (interventions shown in the order of increasing costs and impacts):

- routine asset maintenance (e.g. regular pipe cleaning using jetting, scouring or some other method)

- asset repair (e.g. a local pipe patch, such as a pipe repair to take care of a recent burst)
- asset refurbishment or rehabilitation (e.g. pipe relining)
- asset renewal (e.g. pipe replacement).

Another common classification of water distribution system asset interventions is into *capital* (CAPEX) and *operational* (OPEX) type interventions. The accounting conventions allow expenditure to be charged to CAPEX only if it creates an asset that will exist for more than 1 year. Creating an asset can mean extending the original book life of an existing asset, but not simply allowing it to meet its intended life.

In the above list, routine maintenance and asset repair would normally be OPEX, and the other two, CAPEX-type interventions. The decision of the source of funding will vary in detail, based on the accounting policies of individual companies. However, the fundamental issue is not based on the magnitude of the expenditure but whether or not it extends the life of the asset beyond the originally assumed book life. So, if the expenditure is simply to allow the asset to achieve its book life, it should be charged to revenue, and if it extends the book life it can be charged to capital, and the residual value of the asset in the balance sheet will be increased accordingly, together with its new book life. The scope for variation arises because, while most companies will have a minimum value below which all expenditure will be charged to revenue, it should be noted that the converse (i.e. a maximum value that can be charged to revenue or OPEX) never occurs, although many engineers seem to believe that it does.

Interventions can be applied to a single asset (e.g. pipe replacement) or groups of asset (e.g. flushing of a group of pipes to mitigate the risk of discoloured water). Sometimes the intervention covers large parts of a water distribution network (e.g. active leakage control by means of regular sounding, or systematic flushing of pipes to prevent discolouration events).

Each intervention comes with a cost attached to it. Developing accurate intervention cost models, especially the parametric ones, where the cost of an intervention (e.g. building of a new service reservoir) is a function of several parameters (e.g. reservoir volume, material, etc.), is not an easy task, and it often requires data that is not readily available (e.g. data on past engineering works and related costs, data on various technological details and related unit costs, etc.).

Different interventions have different levels of impact on asset condition, and therefore also generally on performance. This is true not just for interventions of different types (e.g. pipe cleaning will change the roughness coefficient of the pipe, i.e. hydraulic capacity only, while pipe replacement may also change the pipe diameter and material, thus also changing its structural performance) but also for the interventions of the same type (e.g. cement–mortar and epoxy are two pipe-relining techniques that will result in different hydraulic and structural characteristics of the relined pipe).

A single intervention is likely to have a simultaneous impact on future values of a number of performance and/or serviceability indicators, e.g. pipe replacement will affect the

243

hydraulic capacity, the burst rate, leakage, etc. Quantifying the impact of interventions on one or more asset/system PIs and, especially, serviceability indicators is a complex issue often requiring the development of specialised mathematical models. Some of these models may require running various physically based and computationally expensive system simulation models to obtain more-accurate predictions. For example, to quantify accurately an effect of replacing several pipes in a distribution system on the pressure at a critical point in the system, a hydraulic simulation model of the water distribution system should to be run with and without modified pipe characteristics, i.e. before and after interventions are applied. Even worse, in some cases, it is very difficult (if not impossible) to quantify the impact of some interventions on some assets and related indicators simply because accurate impact models do not exist (e.g. how to accurately quantify the impact of active leakage control on leakage reduction?). Some examples of consequence modelling approaches and tools are presented in Section 9.9.

Asset interventions, especially those of capital type, are often costly, and, as such, need to be timed well. This is especially important for the proactive type of asset management where assets should be replaced just before they come to the end of their useful life. This, of course, is easier said than done. To achieve an optimal mix of asset interventions over some future planning horizon, especially at the system level, some sort of asset management decision support tool should be used (see Section 9.9). The reason for this comes from the large number of possible solutions (i.e. combinations of different intervention types and their timing), which makes it a very difficult, if not impossible, job for the human brain.

9.8   Asset deterioration
9.8.1   General
Asset deterioration is one of the key drivers behind large investments required to manage water company assets. The UK water and infrastructure systems are particularly affected by this, as many pipes were buried under the ground a long time ago. The average age of water mains in the UK is approximately 45 years (Ofwat, 2008c), with some pipes being older than 100 years.

Deterioration in asset performance is usually a consequence of ageing, but can also result from other external factors/processes. For example, depending on the type of source water, the hydraulic performance of unlined ferrous pipes may reduce with age due to the internal corrosion process, leading to an increase in the absolute roughness of a pipe or, in some cases, a reduction in its effective internal diameter. However, even where a pipe is not subject to the above corrosion, its hydraulic performance may reduce in the future owing to a need for it to convey increased flow rates as a result of, for example, urbanisation and/or climate change. Therefore, asset deterioration can be expressed in terms of its condition deterioration but also in terms of its performance deterioration. Note that the two are not necessarily always fully correlated (Hall *et al.*, 2006).

Physical asset (i.e. condition) deterioration is a consequence of various mechanisms that may take place over time. These processes are often asset-specific and usually involve complex physical, chemical and/or bacteriological processes. An example of an asset deterioration process is the corrosion of metallic (i.e. ferrous) water pipes. Corrosion of these pipes is a process of ion release from the pipe into the water (internal corrosion) and/or the surrounding soil (external corrosion).

Quantifying and predicting asset deterioration, especially in terms of performance, is critical for successful asset planning and management. This was less important in the past when reactive-type asset management was used, i.e. where assets were replaced following a major failure. The proactive approach aims to replace the asset just before it comes to the end of its service life (e.g. just before the pipe starts bursting frequently in short periods of time). However, predicting asset deterioration is a difficult task. To illustrate this, examples of pipe deterioration models are presented in the next section.

9.8.2 Pipe deterioration modelling

Two general types of pipe deterioration models exist: physically based models and statistically based models. The former aim to describe the physical mechanisms underlying the pipe deterioration processes, while the latter aim to develop predictive models by applying various statistical analyses to the data available.

Bearing in mind the complexity of pipe deterioration processes (see the previous section), it is no surprise that very few physically based models have been developed so far. Most of these models are limited in scope, i.e. they typically apply to very specific types of pipes and conditions. An example of such a model is that developed by Rajani and Tesfamariam (2007), which aims to estimate the time to failure of cast-iron water mains.

In contrast to physically based models, quite a few statistically based pipe deterioration models have been developed (Kleiner and Rajani, 2001). These models can be further classified into models aiming to predict condition deterioration and those aiming to predict performance deterioration. In this context, the deterioration performance of a pipe is usually defined in terms of its burst rate (equal to the number of pipe bursts per unit time) or its burst frequency (burst rate divided by the pipe length). Regardless of the type of predictions made, almost all statistical models developed aim to link burst rate/ frequency to various pipe attributes and other explanatory factors. Examples of these factors (also known as 'covariates' in the statistical literature) include pipe material, age, diameter, external loading (traffic on the road above), quality of the workmanship during pipe installation, surrounding soil conditions and other factors (Boxall *et al.*, 2007).

According to Kleiner and Rajani (2001), all statistical pipe deterioration models can be broadly classified into deterministic, probabilistic multi-variate and probabilistic single-variate group-processing models.

The deterministic models are regression-type models that predict pipe break rates/ frequencies using two or three parameters (typically time/age, pipe material and diameter). The models can be further classified into time-exponential (Shamir and Howard, 1979; Walski and Pelliccia, 1982) and time-linear models (McMullen, 1982; Kettler and Goulter, 1985; Jacobs and Karney, 1994), depending on the shape of the curve used to approximate the deterioration period on the 'bath tub' curve (Figure 9.2).

Figure 9.2 The bath tub curve

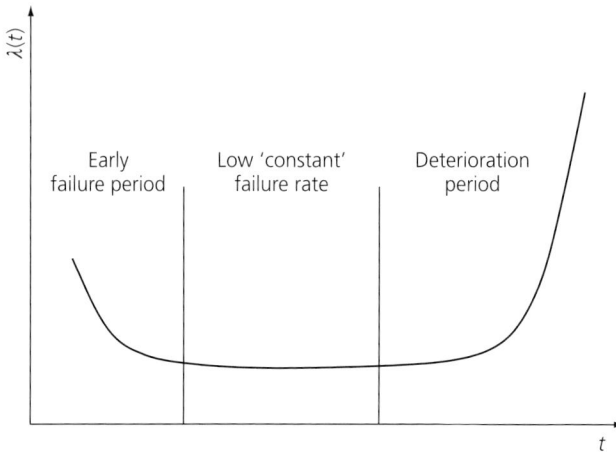

This curve is often used in reliability theory to depict the deterioration of an asset (or system) expressed as the failure rate of the asset λ with time t. Depending on the statistical technique used, the probabilistic multi-variate models can be further classified into proportional hazards models (Marks *et al.*, 1985; Andreou *et al.*, 1987), accelerated-lifetime models (Lei and Saegrov 1998) and time-dependent Poisson models (Constantine and Darroch, 1993). Probabilistic single-variate group-processing models can be classified into cohort survival models (Herz, 1996), Bayesian diagnostic models (Kulkarni *et al.*, 1986), semi-Markov process-based models (Gustafson and Clancy, 1999) and break-clustering models (Goulter and Kazemi, 1988). For details of these models and their respective pros and cons, see Kleiner and Rajani (2001).

Several authors identified the need to aggregate pipes into homogeneous groups in order to conduct more-effective analysis (Shamir and Howard, 1979; Lei and Saegrov, 1998). Shamir and Howard (1979) were the first to suggest that data groups ought to be considered as homogeneous with respect to the causes of failure. Pipe material, diameter and age, with or without additional factors such as soil types and/or land use above the pipes, have been widely adopted as grouping criteria to emphasise their influence on failure (Herz, 1996; Lei and Saegrov, 1998). While easier to develop and more accurate then the models aiming to make burst rate predictions at the single-pipe level, these models are less useful for generating detailed asset management plans. As a result, pipe level models have started to appear recently (Le Gat and Eisenbeis, 2000; Economou *et al.*, 2008).

The data used for the development of statistically based pipe deterioration models typically come from two sources:

1. the asset database, containing basic information about pipes (material, diameter, age, etc.)

2. the incident database, containing information about observed pipe burst events (i.e. incidents).

Three major problems associated with these data sets in the UK are as follows:

1. data sets have been recorded for relatively short periods of time (most of the UK water companies started collecting these data sets in the early 2000s)
2. both databases often have a lot of missing and/or incorrect information
3. in many cases, linking the two databases is not straightforward, as locations of bursts in the incident database tend to be recorded based on the nearby property rather than on the pipe itself.

In the remainder of this section, two pipe deterioration models are presented in more detail. The first model is based on the evolutionary polynomial regression (EPR) technique (Giustolisi and Savić, 2006; Berardi et al., 2008). The second model is the Bayesian-based zero-inflated non-homogeneous Poisson process (NHPP) model, developed by Economou et al. (2008).

9.8.2.1 Example 1: the EPR model
The EPR model is a deterministic model aiming at predicting the burst rate/frequency for a single pipe or a group of pipes as a function of a number of potential explanatory factors. The EPR method is essentially a smart regression approach. Unlike the conventional regression approach, where the exact mathematical model structure (e.g. a function linking the pipe burst rate and explanatory variables) has to be assumed before the regression, in EPR only the following, very general, polynomial form of this function (or similar) has to be assumed:

$$\mathbf{Y} = a_0 + \sum_{j=1}^{m} a_j (\mathbf{X}_1)^{\mathbf{ES}(j,1)} \dots (\mathbf{X}_k)^{\mathbf{ES}(j,k)} f\left((\mathbf{X}_1)^{\mathbf{ES}(j,k+1)}\right) \dots f\left((\mathbf{X}_k)^{\mathbf{ES}(j,2k)}\right) \qquad (9.1)$$

where $\mathbf{Y}$ = the predicted burst number/rate/frequency
$\quad \mathbf{X}_k$ = the kth explanatory variable
$\quad \mathbf{ES}$ = the matrix of unknown exponents
$\quad f$ = a function selected by the user
$\quad a_j$ = unknown polynomial regression coefficients (i.e. model parameters)
$\quad m$ = the number of polynomial terms (in addition to the bias term a_0).

The EPR model works in two main loops (Giustolisi and Savić, 2006):

1. In the external loop, EPR uses the multi-objective genetic algorithm (MOGA) to determine the unknown $\mathbf{ES}$ values.
2. In the internal loop, given the selected $\mathbf{ES}$ values, EPR uses the singular-value decomposition-based least-squares method to determine the values of the unknown polynomial regression coefficients a_j.

The EPR model starts its search with all possible explanatory factors, and automatically determines the significant ones, which are, typically, a very few of them. This is achieved by the means of **ES** values: every time a value of the exponent **ES**(j, k) becomes equal to zero, this means that the value of kth input variable $\mathbf{X}_k$ is insignificant, i.e. effectively deselected from the model shown in equation (9.1). This way, a lot of time is saved, as the user needs to run the EPR model only once, as opposed to the conventional regression approach.

In addition to the above, because the EPR model uses the MOGA method it generates a whole set of Pareto optimal solutions, ranging in complexity and accuracy of the burst prediction models identified. More importantly, all the models generated are fully transparent, as they are presented in the equation format, which makes it easier to inspect them for any potential illogical relationships, etc. An example of such an equation is

$$\text{BR} = 0.084904 \frac{A}{D^{1.5}} \tag{9.2}$$

where BR = the burst rate (number of bursts per year)
  A = the pipe age
  D = the pipe diameter.

The EPR method has already been successfully used to develop real-life pipe burst prediction models in the UK (Hall *et al.*, 2006), including a recent application to develop all-pipe predictive models for the entire networks of two water companies in the UK.

9.8.2.2 Example 2: the zero-inflated NHPP model

The zero-inflated NHPP model is a stochastic, Bayesian-based pipe burst prediction model (Economou *et al.*, 2008). It is based on the non-homogeneous Poisson process, which is flexible enough to capture the non-linear relationship of the failure rate (here, the pipe burst rate) with time, while simultaneously allowing for the inclusion of suitable pipe explanation factors (Loganathan *et al.*, 2002). Furthermore, the model was adjusted to account for the possible zero inflation that may well exist in pipe failure data due to the fact that many pipes never experience a break during the observation period. Finally, the model was calibrated using the Bayesian-based approach, which provides a natural framework for updating the model parameters, e.g. every 5 years when new business plans have to be submitted to the regulator (see Section 9.3). The Bayesian approach also provides an effective way for dealing with short observed data sets by using the relevant prior estimates of model parameters, which are based on engineering judgement rather than data. These priors can then be updated once the new observed data become available.

The zero-inflated NHPP model developed by Economou *et al.* (2008) uses the following function to approximate the deterioration period on the bath tub curve (see Figure 9.2):

$$\lambda(t, \mathbf{x}) = \theta t^{\theta-1} \, e^{\beta x} \tag{9.3}$$

where λ = the pipe burst rate
  $\theta > 0$ is the deterioration curve shape parameter
  $\mathbf{x} = (1, x_1, \ldots, x_q)$ is a vector of related explanatory variables
  $\boldsymbol{\beta} = (\beta_0, \beta_1, \ldots, \beta_q)$ is a vector of model parameters
  t = time.

Figure 9.3 Example of the zero-inflated NHPP model output

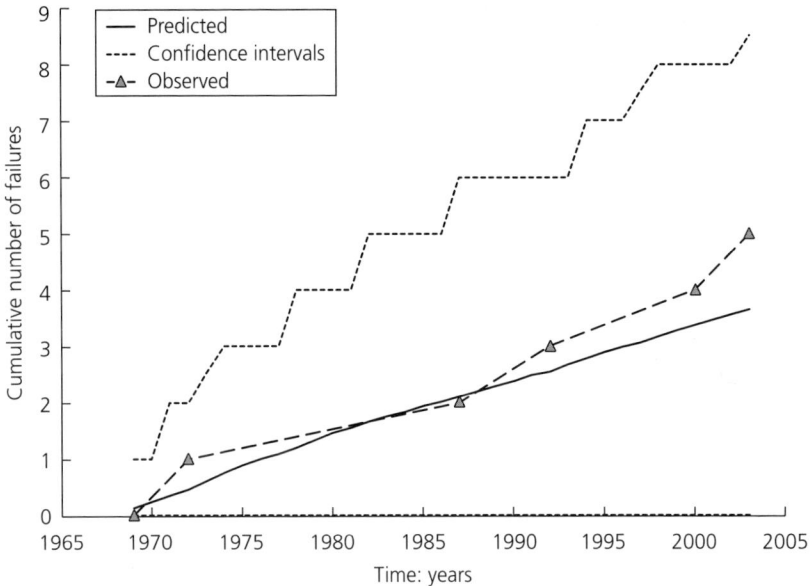

The values of θ and β are determined by using the aforementioned Bayesian-based calibration approach.

The zero-inflated Bayesian model makes pipe burst predictions at the single level, which is a significant advantage over many other models available nowadays. Being a stochastic model, it enables prediction of confidence intervals in addition to mean values. This is demonstrated in Figure 9.3, which depicts the zero-inflated NHPP model 'predicted' and 'observed' cumulative number of bursts over time for a single pipe in a large North American town. The calibration period is 1969–1998, and the validation period is 1999–2003.

9.9   Failure consequence evaluation
9.9.1   General
Section 9.8 addressed the process of asset deterioration and some of the modelling approaches that have been applied to predict the frequency of failure. In compiling the information needed to identify the actual 'risk' presented by assets (see Section 9.6.3), and hence to carry out the cost–benefit calculations which drive asset intervention decisions (as explained further in Section 9.10), it is useful to apply models which assess the consequence of failure.

The principal service impacts relevant to water infrastructure intervention decisions include the hydraulic service measures monitored by Ofwat (see Table 9.1), i.e. interruptions to supply and low pressure, together with discolouration. (Leakage is also a desirable addition to this list, although the complex relationships affecting deterioration and the difficulties in consistently modelling the impact of active leakage control on leakage levels hinder this.) Secondary impacts may also be included where their costs may be determined, e.g. the disruption arising to traffic caused by an unplanned burst repair, or the biological water quality risk created by supply interruption of long duration.

As referred to elsewhere in this chapter, the impact of inadequate pipe capacity on system pressures is best assessed by the use of a hydraulic model or system monitoring, or by a combination of both. In most cases, this type of deterioration (caused either by physical asset deterioration through corrosion or by increasing demand) is gradual, or is at least planned, and may be investigated selectively on an 'as-needed' basis linked to planned development or low-pressure monitoring data. Such capacity inadequacies are, of course, relevant to integrated asset management decisions. However, the less-predictable service impacts are those which arise from a sudden asset failure, predominantly a water main burst, and it is on this aspect that the rest of this section focuses.

9.9.2 Supply interruptions (DG3) and low pressure (DG2)

Regardless of the frequency with which a pipe is predicted to burst, its impact on customer service in terms of loss of supply depends on its position in the local network, i.e. connectivity, number and relative position of valves, etc.

Assessing the hydraulic impact that the failure of a pipe would have is ideally carried out using hydraulic models. Some hydraulic simulation packages offer a batch run facility which allows the model to be set up to run a suite of scenarios sequentially. By setting up these runs to represent the closure of each pipe in the system in turn, then running a simulation and logging the results (specifically identifying the nodes at which there is either no supply, or where pressure is inadequate relative to a target minimum), it is possible to assess the number of customers affected through the demand information used to build up the model. This allows a pipe-specific consequence to be established for each pipe which, when factored by the predicted failure frequency for the cohort to which the pipe belongs, provides a risk number for each asset in its current position (or at some future horizon, to reflect ongoing deterioration). If this is compared with the risk number associated with, for example, a replacement pipe with a much lower predicted failure frequency, the risk mitigation impact and its value may be determined as part of the cost–benefit evaluation.

When carrying out a large-scale selection of asset interventions, as would typically be the case for the preparation of water companies' business plans at the price review (and which companies are increasingly moving towards as part of their 'business as usual' processes), the practicability of using hydraulic models to cover all areas is, however, limited. Many companies (at least in the UK) do not have a comprehensive stock of hydraulic models, particularly not models which have been calibrated to the high standard necessary to determine the impact results with sufficient confidence; and even if they do, the time taken to run the models may be prohibitive. Alternative pragmatic approaches have therefore also been applied using GIS tracing, allowing these pipe-specific consequences to be determined in a relatively crude but nevertheless effective way.

GIS scripts have been written and applied by several water companies to determine the number of houses that would be affected as a result of pipe failure. There are three prerequisites for this approach to be feasible:

1. the GIS must contain connectivity information, to allow tracing along the appropriate pipes only
2. information must be present regarding the number of properties connected to each pipe
3. information regarding valve locations and settings must be comprehensive and accurate.

Two separate logical approaches have been taken:

1. Logic that tests whether each individual pipe is a 'sole feed' to an area of demand and, if it is, counts the number of properties connected downstream of that point. The sole feed test includes an assessment of the likely adequacy of alternative paths based on pipe size. This approach provides an indication of the initial impacts in the event of a catastrophic pipe failure. Its disadvantage is that, to determine what is 'downstream', there is a fourth prerequisite: the hydraulic direction of flow must be known, which, in looped systems particularly, can be difficult to determine consistently.
2. A second tracing analysis that returns the number of properties which would be isolated, assuming that valves are closed to allow the repair of each pipe in the system in turn. The number of properties is the sum of the properties contained within the isolated boundary, together with any clusters outside of the area but cut off from their 'source' as a result of the isolation. This does not require a knowledge of hydraulic flow direction, and is arguably a more-authentic representation of the real problem.

Figure 9.4 shows an example of DG3 consequence tracing from a pipe using the second approach.

The ability to determine the impacts of failure on low pressure is clearly limited without a hydraulic model. However, there is scope for refinement of the GIS-type approaches by developing more-sophisticated rules which, perhaps analysed against hydraulic model results using a technique such as neural networks, assess the likely extent of inadequate pressure service when particular combinations of failed and surviving pipe capacities apply.

9.9.3   Discolouration modelling
Models have also been developed which provide an indication of the risk that each pipe in a system poses to discolouration. Although not itself a hydraulic measure, discolouration is closely associated with hydraulic changes in a system when there is a burst or operational action, and therefore requires a hydraulic model.

Figure 9.4 Example of DG3 consequence tracing

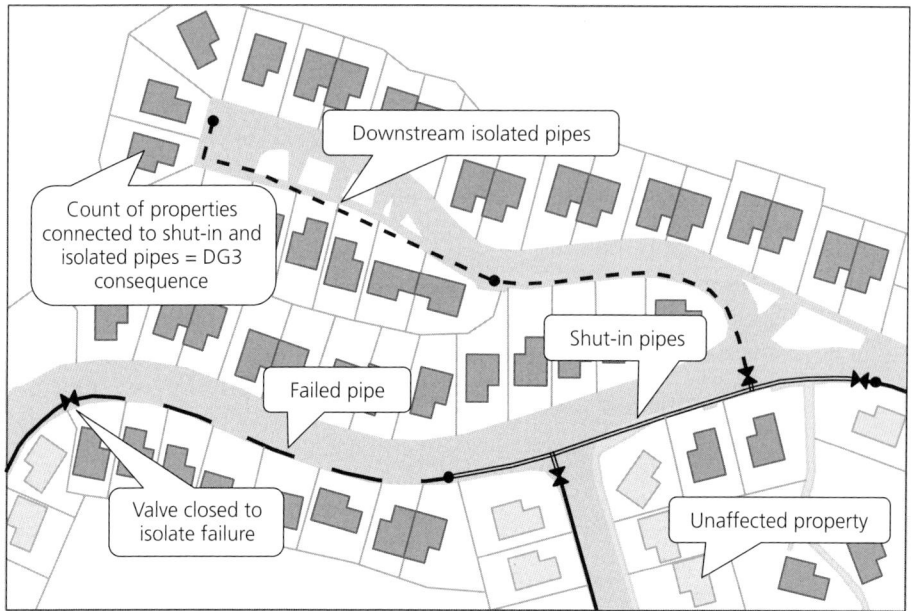

A modelling tool known as the discolouration risk model (DRM) has been developed (Dewis *et al.*, 2005), which, similar to the DG3 batch run approach described above, simulates the sequential failure of each pipe in a system in turn and its effects on the hydraulics of other pipes in the system. Each failure is assumed to comprise two phases: first, the application of an additional demand at the pipe to represent the water lost from the burst, and, second, the closure of the pipe for repair. A matrix is established summarising the impacts, with the risks being greatest in circumstances where a pipe which has excess capacity under normal conditions (and therefore has a low conditioning shear stress, i.e. self-cleaning condition, acting at the pipe wall) is forced to work hard to an extent which could cause discolouration (i.e. a mobilising shear stress is introduced, resulting in transportation of material accumulated at the pipe wall). Other factors reflect further considerations that could influence the magnitude of the risk, such as the source water type or the length of ferrous trunk mains upstream of the system.

Discolouration is a complex phenomenon, and the industry still has much to learn before being able to predict precisely its occurrence and scale of impact. However, the tools currently available provide an initial means of establishing a consequence value at the individual pipe level. The points made above about the impracticability of applying hydraulic models on a wholesale basis are, of course, equally applicable here, so some initial screening of water systems by readily available measures such as the number of discolouration complaints that have been received is usually applied to filter the small proportion of systems with the greatest apparent risk for more in-depth hydraulic analysis.

9.10   Whole-life-costing-based asset management
9.10.1   The concept
Whole-life costing (WLC) is a term that is often used in an engineering context, particularly civil engineering with respect to the provision of infrastructure. The Construction Research and Innovation Strategy Panel (CRISP, 2009) defines WLC as 'the systematic consideration of all relevant costs and revenues associated with the acquisition and ownership of an asset'. Therefore, the WLC concept considers all the costs incurred throughout the life of an asset, i.e. all the costs arising from the installation, provision, operation, maintenance, servicing and decommissioning of the asset. It is important to understand that these costs will be used as input data to economic analysis, i.e. they will be used as part of a discounted cash flow analysis, as explained in Chapter 10. They in no way replace such analysis.

Skipworth *et al.* (2002) developed a new water infrastructure asset management framework that makes use of the WLC concept (Figure 9.5). The costing framework takes into account both activity-based costs (e.g. due to interventions) and the aforementioned life cycle costs associated with assets and system as a whole. This way, costs (e.g. leakage costs) are linked to system performance (leakage) through the quantities that drive costs – the cost drivers (volume of leakage). Therefore, as the performance of a system changes with time as a result of deterioration and/or interventions, so the changes in costs are tracked.

Figure 9.5 WLC-based asset management framework
Based on Skipworth *et al.* (2002)

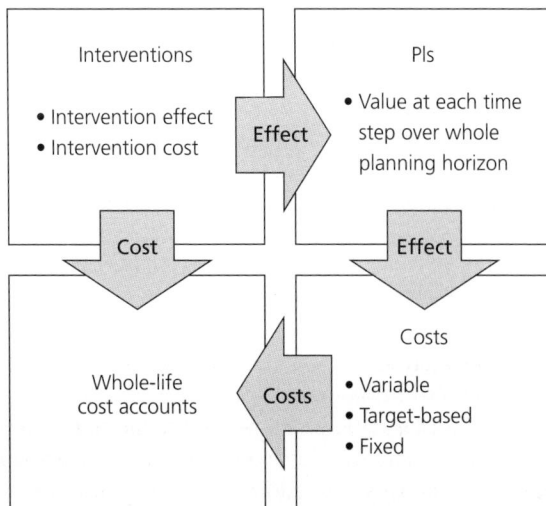

In the above framework, the optimal set of interventions to be applied to the analysed water distribution system (or part of it) over some long-term planning horizon (typically 20–50 years, although shorter time periods can be considered as well) is determined by choosing the set of interventions that will lead to the lowest NPV of the total whole-life costs over this planning horizon subject, to target levels of service and a number of regulatory, environmental and other constraints (all defined over the same planning horizon). The whole-life costs include the cost of interventions (both CAPEX and OPEX) but also any other costs arising in the asset management process (e.g. maintenance cost, regulatory penalties, etc.). Based on ownership, all the costs are classified into private (i.e. water company) and social (i.e. public/society and environmental) costs (Skipworth *et al.*, 2002). All the costs can also be classified into fixed or variable costs.

Every time an intervention occurs somewhere on the planning horizon (e.g. a pipe is replaced at some point in time), the associated intervention cost is accounted for, and the impact of this intervention on the relevant asset(s) and the overall system performance (e.g. minimum pressure at critical point) is quantified by using relevant impact models (e.g. a system hydraulic model). The estimated PI values are then compared with the corresponding targets, and, if target constraints are not met, penalty-type costs are added to the overall sum of costs. At the same time, if the intervention applied has an impact on some other PI(s)/cost(s) (e.g. the future pipe burst rate), the associated cost(s) (e.g. maintenance costs related to the number of pipe bursts) are evaluated too and added to the total whole-life cost of the analysed system. In addition to the above, deterioration of various system elements is modelled to take into account the deteriorating asset/system performance in periods of time without any interventions.

9.10.2 ExSoft decision support tool

The above concept is encapsulated in the ExSoft software tool. This decision support type tool has the following main modules:

- a whole-life cost-accounting module
- a network performance module
- a decision tool module.

The whole-life cost-accounting module stores and calculates all the whole life-costs mentioned in the previous section. It is an accounting-type module that attempts to identify, store and report all costs. More details about the whole-life cost-accounting approach adopted in ExSoft can be found in Chapter 10.

The network performance module is used to store the information about the analysed water distribution system and the associated performance models. This module uses a GIS to store the physical asset data (e.g. the pipe material, diameter, etc.; the installed capacity, head flow and other curves of a pump, etc.) but also the information on the connectivity of network elements. In addition to this, the network performance module stores all the performance (e.g. a pipe burst model) and other models (e.g. a hydraulic model) that are used to simulate different aspects of the performance of a network with time under changing conditions (e.g. the deterioration of assets, changes in demand and changes due to interventions). This way, the impacts of various interventions on the performance of the analysed system can be quantified.

The decision tool module performs the optimisation of interventions by minimising the whole-life asset management costs over the given planning horizon, subject to target system performance. It uses the optimisation technique called genetic algorithms (Goldberg, 1989), which is capable of solving large, complex, non-linear optimisation problems such as the asset management problem. The decision tool makes use of both WLC accounting and network performance modules, and generates as an outcome a list of interventions to be applied to the system assets (together with associated timing). In addition to this, the decision tool generates the optimal whole-life cost (Figure 9.6) and associated PI profiles over the planning horizon (Figure 9.7). The decision tool can also solve alternative optimisation problems, e.g. the problem where the optimal intervention selection is driven by the maximisation of the performance of a water distribution system subject to limited available budgets.

The ExSoft tool and the WLC-based methodology behind it have provided the basis for the development of a well-known commercial asset management tool, WiLCO (Engelhardt and Skipworth, 2005). This tool is one example of several that have now been used by major UK water companies and other infrastructure industries (transport, gas) to help manage their assets.

Figure 9.6 Whole-life cost profiles generated by ExSoft
From Skipworth *et al.* (2002)

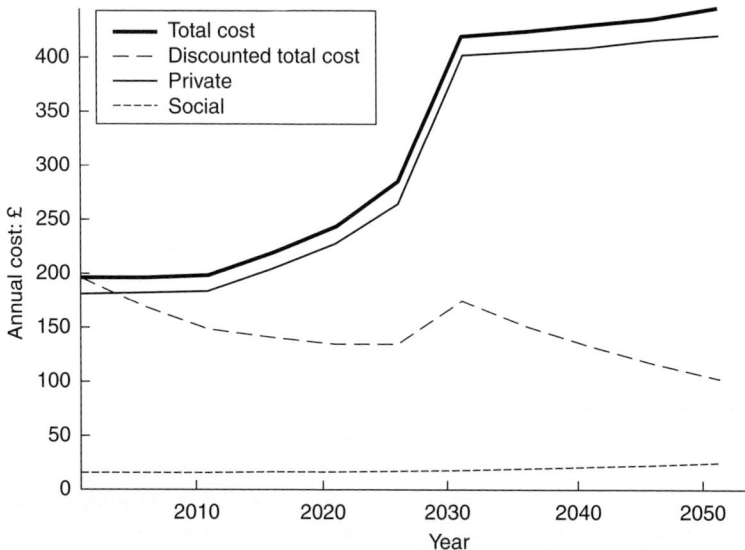

Figure 9.7 Leakage profile generated by ExSoft

From Skipworth *et al.* (2002)

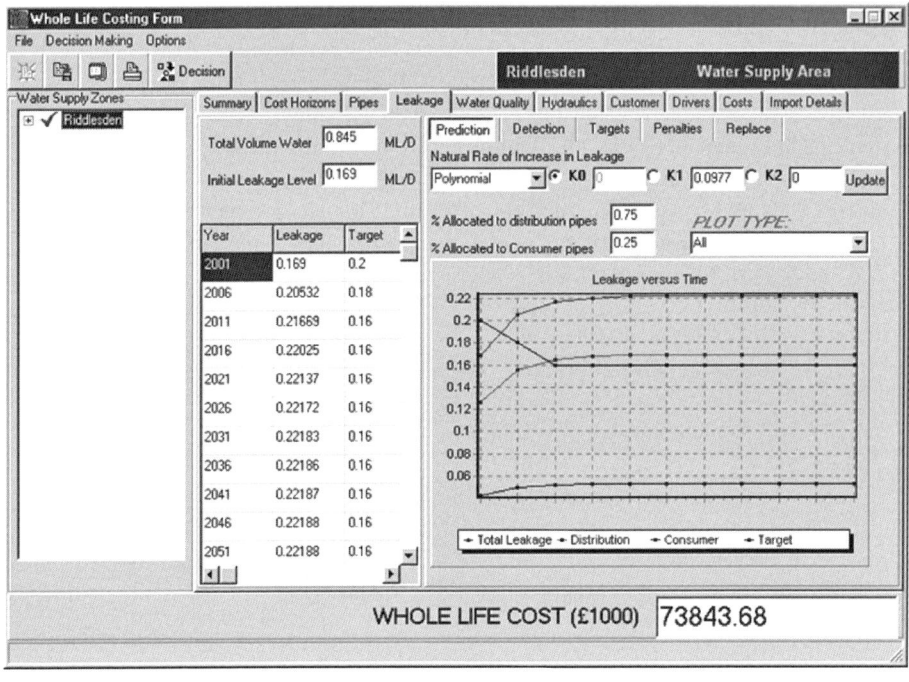

9.10.3 Other asset-management-modelling approaches and tools

Many other asset management approaches and tools exist nowadays. One of the first asset management methodologies developed for the optimal strategic investment planning in the water industry was the GAasset model and software (Miller *et al.*, 2001). Developed in the mid-1990s, this was probably the first asset management approach that suggested the use of an integrated and optimised approach for selecting the best future asset investment planning strategy, incorporating a WLC assessment over a predetermined planning period.

Another important asset management approach is the methodology developed as part of the large EU Fifth Framework Programme research project entitled 'Computer Aided Rehabilitation of Water Networks' or CARE-W (Saegrov, 2005). The overall aim of the project was to establish a framework for water network rehabilitation decision-making. The work on the CARE-W project resulted in a computer-based system that comprises a suite of tools, providing a cost-efficient system for the maintenance and repair of water distribution networks, with the aim of guaranteeing a security of water supply that meets social, health, economic and environmental requirements. The CARE-W system consists of software dealing with fundamental instruments for estimating the current and future conditions of water networks (i.e. PIs, the prediction of network failures and the calculation of water supply reliability) and routines for estimating long-term investment needs as well as selection and ranking of rehabilitation projects. These tools are integrated and are operated jointly in the GIS-based prototype software. The principal outputs obtained are a prioritised list of rehabilitation asset candidates and the associated short- and long-term strategic plans with project costs.

Many other tools exist that can be broadly classified as asset management tools. The main difference between these and the aforementioned tools is that these tools tend to address specific aspects of the asset management problem (e.g. water distribution system rehabilitation) by using a limited number of PIs and related asset interventions (e.g. CAPEX-type interventions only). Examples of commercial tools are the CapPlan software by MWH Soft (2009), the Darwin Designer software by Bentley Systems (2009) and the OptiRenewal software by Optimatics (2009). Many customised methodologies and software that were developed to solve specific real-life problems exist too (Walters *et al.*, 1999; Randall-Smith *et al.*, 2006). All these commercial tools have their origins in a number of research methodologies and associated prototype tools developed since the mid-1990s (e.g. Simpson *et al.*, 1994; Savić and Walters, 1997). A review of these methodologies, together with the relevant historical timeline, can be found in Lansey (2006).

Finally, note that some companies, especially the larger water companies in the UK, have developed their own asset management methodologies and tools, which are often customised to suit their own needs and, in particular, the data currently available to them. However, details of these models are not publicly available due to the commercial nature of business in the UK water sector.

9.11   Summary and future work

A number of issues related to the asset management of water supply and distribution systems have been addressed and presented in this chapter, including key asset management drivers, the regulatory framework, asset PIs, interventions and associated deterioration models, the WLC asset management framework and several real-life software tools.

Based on the material presented, it can be concluded that a lot of excellent work has already been done in the area of asset management of water supply and distribution systems. Having said this, a number of challenges remain to be addressed in the future. Some of the key challenges remaining are as follows (in no particular order):

■ The development of new technologies and methods resulting in assets and related water supply and distribution systems that are more cost-effective yet reliable, easier and safe to install and use, more energy-efficient and sustainable. Examples include new water treatment equipment and related structures, improved flow and pressure-regulating devices, etc.

■ The development of new technologies and methods for more cost-effective and energy-efficient observation of the condition and, especially, the performance of water supply and distribution assets and systems. Examples include more-energy-efficient equipment for real-time pressure and flow monitoring, new sensors for real-time water quality monitoring, equipment for the automatic detection of various anomalous events including pipe bursts/leaks, various equipment failures, discolouration events, etc.

■ The development of new technologies for more cost-effective and reliable asset interventions, including cleaning, repair, rehabilitation and renewal of pipes and other assets. An example would be the development of new burst/leakage repair methods which can be applied more quickly and efficiently than the existing ones.

■ The development of improved asset management models, including: (1) more-accurate asset performance models – in particular, better asset deterioration models that can make accurate long-term predictions at the single-asset level; (2) improved impact-type models that can more accurately predict the effects of complex interventions on single/grouped asset(s) and overall system performance, including more realistic hydraulic network models with modelled pressure-dependent demands and leakage, water quality models able to more accurately predict discolouration events and various other water quality changes in the network (e.g. concentration of disinfection by-products), models to more accurately estimate GHG emissions (i.e. carbon footprints), etc.; (3) new asset-management-modelling tools that require less parameter tuning, and generally less specialist expertise, which, in turn, would make these tools usable by a large number of smaller, typically publicly owned water utilities run by municipalities worldwide.

As can be seen from the above, there are plenty of challenges remaining for the future.

REFERENCES

Alegre, H., Baptista, J. M., Cabrera Jr, E., Cubillo, F., Duarte, P., Hirner, W., Merkel, W. and Parena, R. (2006) *Performance Indicators for Water Supply Services*, 2nd edn. London: IWA Publishing.

Andreou, S. A., Marks D. H. and Clark, R. M. (1987) A new methodology for modelling break failure patterns in deteriorating water distribution systems: theory. *Advances in Water Resources* 10: 2–10.

Banyard, J. K. and Bostock, J. W. (1998) Asset management – investment planning for utilities. *Proceedings of the Institution of Civil Engineers: Civil Engineering* 126(2): 65–72.

Bentley Systems (2009) DarwinDesigner software overview. www.bentley.com/en-US/Products/WaterCAD/Darwin-Designer.htm [accessed 15.09.09].

Berardi, L., Kapelan, Z., Giustolisi, O. and Savić, D. A. (2008) Development of pipe deterioration models for water distribution systems using EPR. *Journal of Hydroinformatics* 10(2): 113–126.

Boxall, J. B., O'Hagan, A., Pooladsaz, S., Saul, A. J. and Unwin, D. M. (2007) Estimation of burst rates in water distribution mains. *Proceedings of the Institution of Civil Engineers: Water Management* 160(2): 73–82.

Constantine, A. G. and Darroch, J. N. (1993) Pipeline reliability. In: Osaki, S. and Murthy, D. N. P. (eds), *Stochastic Models in Engineering Technology and Management: Proceedings of the 1st Australia–Japan Workshop*. Singapore: World Scientific.

Cook, S., Fowler, M. and Banyard, J. (1999) Quantitative risk analysis for ranking security of supply investment schemes. In: Powell, R. and Hindi, K. S. (eds), *Computing and Control in the Water Industry*. Baldock: Research Studies Press.

CRISP (Construction Research and Innovation Strategy Panel) (2009) Whole life cost forum. www.wlcf.org.uk [accessed 11.09.09].

Dewis, N. and Randall-Smith, M. (2005) Discolouration risk modelling. In: *Proceedings on the 8th International Conference on Computing and Control for the Water Industry*, Vol. 2, pp. 223–228. Exeter.

Economou, T., Kapelan, Z. and Bailey, T. (2008) A zero-inflated Bayesian model for the prediction of water pipe bursts. In: *Proceedings of the 10th International Water Distribution System Analysis Conference*. Kruger National Park.

Engelhardt, M. O. and Skipworth, P. J. (2005) WiLCO – state of the art decision support. In: *Proceedings of the 8th International Conference on Computing and Control in the Water Industry*. Exeter.

Germanopolous, G., Jowitt, P. W. and Lumbers, J. P. (1986) Assessing the reliability of supply and level of service for water distribution systems. *Proceedings of the Institution of Civil Engineers* 80(2): 413–428

Giustolisi, O. and Savić, D. A. (2006) A symbolic data-driven technique based on evolutionary polynomial regression. *Journal of Hydroinformatics* 8(3): 207–222.

Goldberg, D. E. (1989) *Genetic Algorithms in Search, Optimisation and Machine Learning*. Reading, MA: Addison-Wesley.

Goulter, I. C. and Kazemi, A. (1988) Spatial and temporal groupings of water main pipe breakage in Winnipeg. *Canadian Journal of Civil Engineering* 15(1): 91–97.

Gustafson, J. M. and Clancy, D. V. (1999) Modelling the occurrence of breaks in cast iron water mains using methods of survival analysis. In: *Proceedings of the AWWA Annual Conference*. Chicago, IL.

Hall, M., Kapelan, Z., Long, R. and Savić, D. (2006) *Deterioration Rates of Sewers*. London: UKWIR.

Herz, R. K. (1996) Ageing processes and rehabilitation needs of drinking water distribution networks. *Journal of Water SRT – Aqua* 45(5): 221–231.

Institute of Asset Management (2009) Asset management definition. www.theiam.org [accessed 26.11.09].

Jacobs, P. and Karney, B. (1994) GIS development with application to cast iron water main breakage rate. In: *Proceedings of the 2nd International Conference on Water Pipeline Systems*. Edinburgh: BHR Group.

Kettler, A. J. and Goulter, I. C. (1985) An analysis of pipe breakage in urban water distribution networks. *Canadian Journal of Civil Engineering* 12: 286–293.

Kleiner, Y. and Rajani, B. B. (2001) Comprehensive review of structural deterioration of water mains: statistical models. *Urban Water* 3(3): 131–150.

Kulkarni, R. B., Golabi, K. and Chuang, J. (1986) *Analytical Techniques for Selection of Repair-or-replace Options for Cast Iron Gas Piping Systems – Phase I*. Chicago, IL Research Institute.

Lansey, K. E. (2006) The evolution of optimizing water distribution system applications. In: *Proceedings of the 8th International Water Distribution Systems Analysis Symposium*. Cincinnati, OH.

Le Gat, Y. and Eisenbeis, P. (2000) Using maintenance records for forecast failures in water networks. *Urban Water* 3: 173–181.

Lei, J. and Saegrov, S. (1998) Statistical approach for describing lifetimes of water mains – case Trondheim municipality, Norway. In: *Proceedings of the IAWQ 19th Biennal International Conference on Water Quality*, pp. 21–26. Vancouver.

Loganathan, G. V., Park, S. and Sherali, H. D. (2002) Threshold break rate for pipeline replacement in water distribution systems. *Journal of Water Resources Planning and Management, ASCE* 128(4): 271–279.

McMullen, L. D. (1982) Advanced concepts in soil evaluation for exterior pipeline corrosion. In: *Proceeding of the AWWA Annual Conference*. Miami.

Marks, H. D. *et al.* (1985) *Predicting Urban Water Distribution Maintenance Strategies: A Case Study of New Haven, Connecticut*. Washington, DC: Environmental Protection Agency.

Miller, I., Kapelan, Z. and Savić, D. A. (2001) GAasset: fast optimisation tool for strategic investment planning in the water industry. In: *4th International Conference on Water Pipeline Systems*. York.

MWH Soft (2009) CapPlan water software overview. www.mwhsoft.com/page/p_product/CapPlanwater/capplanwater_overview.htm [accessed 15.09.09].

Ofwat (2008a) *Financial Performance and Expenditure of the Water Companies in England and Wales 2007–08*. Birmingham: Ofwat.

Ofwat (2008b) *Service and Delivery – Performance of the Water Companies in England and Wales 2007–08*. Birmingham: Ofwat.

Ofwat (2008c) *Service and Delivery – Performance of the Water Companies in England and Wales 2007–08 – Supporting Information*. Birmingham: Ofwat.

Ofwat (2009) Regulating the companies: Ofwat's role. www.ofwat.gov.uk/legacy/aptrix/ofwat/publish.nsf/Content/RegulatingCompanies.html [accessed 25.08.09].

Optimatics (2009) OptiRenewal software overview. www.optimatics.com/go/software/optirenewal/optirenewal [accessed 15.09.09].

Rajani, B. and Tesfamariam, S. (2007) Estimating time to failure of cast-iron water mains. *Proceedings of the ICE: Water Management* 160(2), 83–88.

Randall-Smith, M., Rogers, C., Keedwell, E., Diduch, R. and Kapelan, Z. (2006) Optimized design of the City of Ottawa water network: a genetic algorithm case study. In: *Proceedings of the 8th Annual Water Distribution System Analysis Symposium*. Cincinnati, OH.

Savić, D. A. and Walters, G. A. (1997) Genetic algorithms for least-cost design of water distribution networks. *Journal of Water Resources Planning and Management, ASCE* 123(2): 67–77.

Shamir, U. and Howard, C. D. D. (1979) An analytic approach to scheduling pipe replacement. *Journal of the American Water Works Association* 71(5): 248–258.

Simpson, A., Dandy, G. and Murphy, L. (1994) Genetic algorithms compared to other techniques for pipe optimisation. *Journal of Water Resources Planning and Management, ASCE* 120(4): 423–443.

Skipworth, P. J., Engelhardt, M. O., Cashman, A., Savić, D. A., Saul, A. J. and Walters, G. A. (2002) *Whole Life Costing for Water Distribution Network Management*. London: Thomas Telford.

Saegrov, S. (2005) *CARE-W – Computer Aided Rehabilitation for Water Networks*. London: IWA.

UKWIR (2002) *Capital Maintenance Planning: A Common Framework*. Report No. 02/RG/05/3. Birmingham: UK Water Industry Research.

Walski, T. M. and Pelliccia, A. (1982) Economic analysis of water main breaks. *Journal of the American Water Works Association* 74(3): 140–147.

Walters, G. A., Savić, D. A., Thurley, R. W. F., Halhal, D., Kapelan, Z. and Atkinson, R. (1999) Optimal design of water systems using genetic algorithms: some recent developments. In: Powell, R. and Hindi, K. S. (eds), *Proceedings of the Computing and Control for the Water Industry*, pp. 337–344. Baldock: Research Studies Press.

REFERENCED LEGISLATION

The Drinking Water Directive. Council Directive 98/83/EC of 3 November 1998 on the quality of water intended for human consumption. *Official Journal of the European Communities* L330: 32–54.

The Water Act 1989. London: HMSO.

The Water Act 2003. London: TSO.

The Water Industry Act 1991. London: HMSO.

The Water Supply (Water Quality) Regulations 2000. London: TSO.

Water Distribution Systems
ISBN: 978-0-7277-4112-7

ICE Publishing: All rights reserved
doi: 10.1680/wds.41127.263

ice

Institution of Civil Engineers

publishing

Chapter 10
Finance and project appraisal

Adrian Cashman University of the West Indies, Barbados
John K. Banyard Independent consultant, Warwick, UK

10.1   Introduction

No two engineering projects are the same, although there may indeed be similarities and commonalities. They are unique in their composition, location, size and purpose, and, as such, there will be uncertainty over their performance and effectiveness. Therefore, all projects to some extent or another require appraisal. How much time and effort (the degree of detail and expertise) put into an appraisal varies with the nature of the project under consideration. Thus, the purchase of a table for an office will not be subjected to the same degree of scrutiny as, say, the acquisition of a new corporate headquarters. But behind these two examples there are some basic facts that are needed before proceeding. First, who or what body is making the decisions, and what is their interest? The answer to this can have a significant impact on the type and scope of an appraisal as well as on the level of detail required. It has a direct influence on the information required and to be presented. For example, the answer will influence whether a financial analysis is required or an economic one. It could also determine if the project appraisal includes some form of multi-criteria analysis (MCA). Understanding the environment in which a project appraisal is to be conducted is fundamental to understanding its purpose, and the shaping and presentation of the results.

Knowing the audience and purpose will assist the engineer in identifying the processes to be followed and the requirements that are to be met. In almost all cases there will be requirements that the process of decision-making should be formalised and follow accepted methodologies. The formal process and methodologies adopted should demonstrate that a rational and documented basis on which a decision was made has been followed. This is important for quality control and auditing purposes at a later date, should this become necessary. Quite apart from reasons of audit and quality control, it is just good practice to keep a proper track of the work that has been undertaken.

This chapter on project appraisal techniques first seeks to address why projects should be examined from an economic and financial standpoint. Having been satisfied that there are good reasons, the basic concepts on which project appraisal techniques are based are introduced to the reader. Next, the question of where the necessary funding for projects comes from is considered, before discussing the various alternative techniques that are available and the circumstances under which they might be used. Discounting is a fundamental part of many techniques, and so the principles that are behind it are introduced, followed by the presentation of evaluation techniques along with examples of how they can be applied. The chapter ends with three shorter sections that look at some additional considerations that the reader might bear in mind when carrying out project appraisals.

10.2 Why project appraisal?

Why should there be concerns about how projects are appraised and decisions made? One very basic reason is that projects require the input of scarce resources in order to be realised. Therefore, a legitimate question is to ask whether the proposed project constitutes the best use of those scarce resources. The question is even more pertinent when there is more than one course of action available to achieve the desired result. Under these circumstances, choices have to be made, and the grounds on which they are made explicit.

Some cultures in the past, such as the Greeks and Romans, sometimes informed their choice of course of action by (among other things) recourse to auguries and the interpretation of portents. While such methods may not appear rational to us in the 21st century, they were regarded as acceptable and explicit at the time. In the 19th century, project selection for developments such as railways and water systems relied heavily on engineering considerations. Today, it is expected that projects and options should be assessed in a structured way, making use of comparisons based on a decision analysis technique. This would include some form of economic appraisal. Economic appraisal can be thought of as the counterpart of and complementary to a technical appraisal: both consider performance relative to a benchmark requirement and often relative to other alternative options or courses of action.

Drummond *et al.* (2006) identify three reasons for adopting a systematic approach as to whether or not to commit resources. First, without a systematic approach it is difficult to clearly recognise what are the relevant alternatives available to achieve a desired objective. Secondly, the viewpoint assumed in an analysis is important. In other words, from whose perspective should a project be considered and appraised? Lastly, a systematic approach is an important means of reducing uncertainty, although not an infallible one. Some writers have suggested that as options for appraisal have to be developed, they do not from the start exist in detail. Thus, one of the functions of an appraisal process is to identify areas of uncertainty or lack of information, and by doing so gradually improve the level of knowledge and detail of a project proposal. A systematic approach identifies what data are required to refine the appraisal, and in doing so improve the performance of the project by avoiding problems.

In the past, the development of options was seen as being very much a technical activity, to be undertaken by specialists in the area. This was on the presumption that with their experience and expertise they would 'know best' what would work and what would not. This view has been increasingly questioned. Requirements for stakeholder involvement and degrees of public participation have in some cases broadened the pool of knowledge involved in the generation of options. And this has been seen as a good thing, although it is by no means widespread. The idea behind this is that broadening the range of participants brings different types of knowledge to the process. In doing so, more

Figure 10.1 Comparison of projects

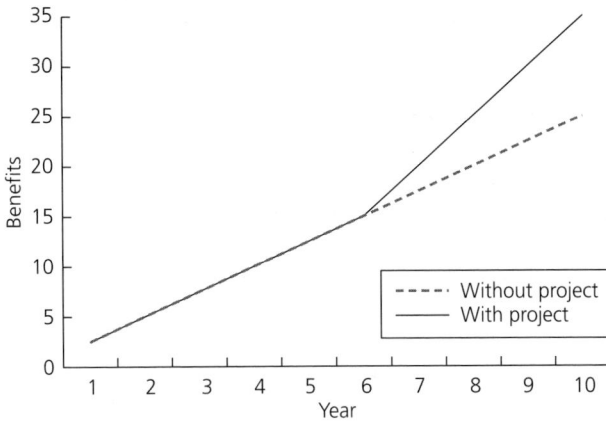

potential options are generated – all of which should be treated seriously and equally. This, it is argued, leads to better projects and decision-making.

Appraisal implies the exercising of judgement, usually against some norm – in some cases this may be a 'do nothing' or 'business as usual' option. So, proposed projects or alternatives are measured against each other and/or against whatever is the appropriate baseline. This point is illustrated in Figure 10.1, where the 'with project' benefits would be compared with the 'without project' alternative – business as usual. Appraisal as a process is also a way of assessing the degree of confidence in the preferred option that emerges from the assessment. It can serve to highlight key uncertainties and the potential risks that might be associated with the project.

Project appraisal is therefore not just something that informs decision-makers: it is a method of communication that at best provides a way of sharing information and understanding between diverse parties. In other words, it is not an arcane tool for a small clique of people who can understand the intricacies of the process but, rather, it should be your tool for clear communication.

10.2.1 Value for money
The overall objective of economic appraisal is to look at value for money by considering the associated costs and consequences of a course of action. It tries to answer the question 'Is this the most effective and beneficial use for the scarce resources, compared with the other options or opportunities available?' In order to do this, it is necessary to have ways of being able to compare alternatives on an equal basis using the same measure for all the alternatives, including 'do nothing'. By far the most common means of doing this is to denominate all costs, expenditures, benefits and impacts in monetary terms.

That said, it also needs to be remembered that decisions are very often not made purely on economic grounds. Economic appraisal of a project must be seen as an input into a decision-making process, a fact that tends to be overlooked. It can at times lead to a degree of dismay over the decisions that are eventually made. The economic appraisal of a project is not an end in itself, in the same way that a technical appraisal is not an end in itself. Both are part of a larger picture. There are other questions that need to be considered. For example, what are the various categories of unknown factors and risk associated with a proposal? How these might impact on a project proposal should be evaluated, especially if there are associated cost consequences. These must be taken into consideration if the affordability of a project is to be properly determined, especially if there will be costs associated with mitigation measures. All things considered, the objective with the appraisal of a project is to be satisfied that a course of action is chosen that has a (required) degree of certainty that it will realise the largest set of net benefits as compared with other alternatives.

In addition to this, there are other considerations, such as distributional issues: who bears the costs and the benefits and what is their relationship to the project proposed? It is often the case that costs are borne by one set of stakeholders while the benefits are enjoyed by others. This gives rise to equity considerations, which project appraisal techniques such as cost–benefit analysis (CBA) are not that well equipped to handle without modification. To some extent, the reasons why or by whom an appraisal is required will go some way to determining such considerations, and therefore the completeness of the information that will be required to be generated.

An appraisal is the means of communicating ideas and describing how far a set of proposals meets required objectives. This involves a series of sequential steps: the identification of the alternatives and a determination of their impacts; measurement of the resources required and quantification of the impacts; determination of the value of the inputs, the outcomes and the impacts; and comparison of the alternatives in terms of their costs and consequences. It is good practice to include sensitivity analysis and test the robustness of your assumptions (Green, 2003). It is clear from much of what has been said that the appraisal of a proposal requires the gathering of data and the generation of relevant information that has to be brought together as a first step. Preparation is a key element, as there must be a reasonably good idea of what is to be appraised before it can be undertaken. This is not to say that no appraisal can take place before all the facts are known. Appraisal in one form or another is an ongoing part of the development of a proposal; it is an iterative process. But when significant decisions have to be taken, then a more rigorous approach is required, and for this the information requirements are more demanding.

10.2.2 Communication

Whatever the outcome of a project appraisal, the results have to be communicated to an audience. This should be done in a way that is clear and comprehensible to *all* parties and not just those familiar with the intricacies of economic or technical analysis. Regardless of the intended audience, a good practice is to assume that whoever is reading the appraisal knows very little about appraisal techniques. Thus, the presentation of the appraisal should be set out in clear and simple steps. Tables and diagrams should be used as much as possible, as these are often easier to understand than blocks of text.

This chapter will look at some of the basic terms and concepts that underpin project appraisal and the techniques that are used. Included in this will be a short description of public–private partnerships and the role of private finance initiatives – these are included as some of the appraisal techniques that will be discussed are inherent in them. What constitute costs, expenditures and benefits are central to any appraisal that deals with the comparative assessment of value for money. Discount rates and inflation often constitute a much debated issue, often revolving around the question of what is the 'right' discount rate to adopt under a particular set of circumstances. Thus, the project appraisal must address this issue. Various evaluation techniques are discussed, some of which make use of discount rates as part of their assessments. Finally, questions of affordability, willingness to pay, social and environmental costs, and the boundaries of any analysis will be touched upon.

The overall aim of the chapter is to introduce to and assist the reader with how to go about appraising projects and their alternatives in a systematic and consistent manner. It assumes some basic prior knowledge of economics, but it is not intended to be an economic text. Undertaking a project appraisal may be time-consuming, but it does not have to be difficult. A good project appraisal requires a blending of knowledge of how things work and what happens to them, and understanding that there are associated costs.

10.3 Basic concepts
10.3.1 Economic or financial appraisal?
Up to this point, reference has been made to the economic appraisal of a project or proposal. But there is an important distinction to be made before proceeding further. Economic appraisal is not the same as financial appraisal, and so it is necessary to be clear as to which one of the two is to be conducted. Financial appraisal considers the expenditures and revenues associated with a project from the point of view of the entity or project participants, and it looks at the money needed to finance a project. Economic analysis considers the various effects of a project from the point of view of its impact on the wider (national) economy. In other words, economic analysis takes a broader view than does a financial analysis. The commonality between the two is that both are conducted in monetary terms. Furthermore, the two are complementary to each other: for a project to be economically sound it must also be financially sound. If not, then the economic benefits associated with the project will not materialise.

Economic analysis takes on board societal aspects and includes them in the evaluation by comparing the financial value of resources used and the financial value of social, environmental and economic impacts with the extra benefits that would be generated for the economy as a whole. The measurement of positive and negative impacts is included through the assessment of willingness to pay for units of increased consumption and willingness to accept compensation for foregone units of consumption of goods and services. Willingness to pay or willingness to accept are used in place of actual market prices in order to account for those impacts that are either not marketed or at best incompletely marketed, such as biodiversity. It also allows for those situations where government policies and interventions distort market prices or where imperfect competition also impacts on prices. This is, relatively speaking, of greater importance in developing countries or where the size of the project relative to the size of the economic sector is such that the project would have no discernible impact on prices.

From the short discussion given of the difference between financial and economic appraisal, it might appear that the choice of which one to apply is straightforward. If a private company, then it is financial analysis. But in the case of the water industry, things are not so straightforward. Water service providers (WSPs) undertake a wide range of projects, many of which require some form of evaluation in order to assist decision-makers. Some projects will be of such a nature that they have a significant 'public good' element to them. For these projects, which have the potential to impact on society, the economy or the environment, it is often the case that the government believes that it should have a say in how such projects are evaluated.

Over time, the boundary between whether a project should be evaluated from a financial or from an economic standpoint has become blurred. Infrastructure projects are subject to a range of assessment requirements by authorities, requirements that have accumulated over a number of years. One of the reasons for the imposition of requirements is that infrastructure projects have impacts on society and the environment which might not otherwise be taken into consideration due to 'deficiencies in the market'. For this reason, directives, regulations and policies have been put in place to ensure that consideration is given to some of the wider impacts that water sector projects among others, might have on a wider environment (Box 10.1). Whether this is fair or reasonable is not for debate here; that is another issue.

The point is, is that in the majority of cases a purely financial evaluation will not be adequate. But on the other hand, it will not be the case that a full economic appraisal

---

Box 10.1 What is an 'economic level of leakage'

Ofwat defined the economic level of leakage as 'the level of leakage at which it would cost more to make further reductions in leakage than to produce water from another source'. In 2005 the House of Lords' Science and Technology Committee asked if this was the most appropriate concept to apply. In its opinion, it focused exclusively on the relative costs, paying insufficient attention to the environmental impact that the development of additional resources might have.

In other words, it was questioning whether the application of a financial analysis that excluded costs associated with environmental impacts was the proper approach to employ. Its recommendation was that something more than a financial evaluation was required.

How would you go about this?

For more about the economic level of leakage, see Section 8.3.1.

---

will be required either. In other words, any project appraisal must recognise what the various regulatory requirements are that have to be complied with. This will be touched on again later in the chapter. In some cases, it may even be prudent to go beyond that which is strictly required.

10.3.2   Expenditures and costs

The question of what expenditures and costs should be considered in an appraisal is very closely linked to the discussion of financial and economic appraisal. As with financial and economic appraisal, there can be a tendency to view costs and expenditure as being more or less the same thing. But they are not. The difference between them can be understood by asking the question: 'Who incurs a cost?' The answer to that question, together with the regulatory guidance as to what other factors are required to be included, will indicate the *boundaries* of the appraisal.

Expenditures are those monies which are spent as payment for the cost of acquiring goods and services needed for a given purpose. The purpose is the realisation of the project. In other words, it is only those costs associated with the difference between the without and the with project situation that can legitimately be considered. Put another way, it is only the expenditures on the extra use of resources, goods and services necessary to realise the corresponding benefits from the project that can be included. And these expenditures are incurred by the party or parties who are responsible for the realisation of the project.

Not all costs need to be included, especially when they are associated with what economists refer to as *externalities* (Box 10.2). For example, the construction of a new facility might lead to people living nearby developing respiratory complaints. Those persons would incur costs, but these may not be required to be considered in the appraisal of the facility; it might be that it is not known at the time that such effects could happen. Expenditures are involved because they are associated with the use of resources, and they have a cost in monetary terms: doctors' bills, medication, loss of earnings. The question is whether or not they can or should be accounted for in the appraisal. Application of the 'polluter pays' principle suggests that they should be, or there may be regulatory or policy guidance on such matters. On the other hand, regulators are cognisant of the adverse impact that too broad an interpretation of relevant costs might have on prices. Thus, there is a balance to be struck over what expenditures and costs might

---

Box 10.2 Externalities

An externality occurs when the action of one party directly or indirectly changes the options available to a third party and the generating party does not have to account for that change. Therefore, any associated cost is not reflected in the prices charged for any goods or services provided.

Can you think of some externalities associated with a wastewater treatment works?

---

be relevant, especially those associated with externalities and whether society as a whole or the beneficiaries of a project should bear the costs.

10.3.2.1 Identification of expenditures and costs

In any project there is a need to account for expenditures, to identify them and categorise the nature of the expenditure. Traditionally, the water industry, like others where infrastructure is a significant component, distinguishes between capital and operational expenditures, especially for accounting purposes. Other distinctions that are made include direct and indirect costs and internal and external costs – which have been touched on above. Within any one of these there will be a number of subdivisions that categorise costs according to a schema. Different industries and organisations will often have their own schema, developed to meet their particular reporting and cost control needs. The advent of accounting software has led to a certain standardisation. Also, regulatory bodies, e.g. Ofwat, have their own reporting requirements.

In project appraisal there are two questions that should be asked that can inform the identification and quantification of expenditure. First, what would be the most appropriate categorisation? Secondly, what level of detail is appropriate for the appraisal? The answers depend to a certain extent on what stage in a project cycle the appraisal is being carried out. In the initial concept stage, a few broad cost categories would suffice, as the level of cost information available and the relative accuracy of the figures will not be of the highest. As a project progresses and more information becomes available, more details concerning the proposals are tied down, and then the granularity of expenditure information becomes much finer. A word of caution, though: information on expenditures is to be used to inform decisions, and not to be an end in itself.

SYSTEM COSTS

If the project being considered is part of a larger system, the question might arise as to whether to include or exclude expenditures on the larger system. Here the test would be that if the expected benefits of a project cannot be realised without expenditures on the larger system, then one should include those expenditures. The cost boundary must include all expenditures required to achieve the expected benefits.

SUNK COSTS

In contrast to the systems costs described above, where the project uses existing facilities but does not require any 'upgrading' to realise the benefits, then the prior expenditures should not be included. They are regarded as 'sunk' costs, and have no bearing on the project being appraised. The only proviso that there is to this would be if there were opportunity costs associated with the prior project. In other words, using the prior project would foreclose on other potentially more beneficial opportunities. Under these circumstances it would be the opportunity costs that would need to be considered: the benefits that have been foregone.

WORKING CAPITAL

Working capital is usually defined as net current assets, which would include bank balances and cash in hand. In terms of a project appraisal, it would be the additional amount of capital required to run a project created by investment in fixed assets. And it would be capital required until such time as an increase in net cash flow removed the need for the additional financing.

TRANSFER PAYMENTS

Transfer payments are one-way payments for which no money, goods or services are received in exchange. They can affect the distribution of costs and benefits among project beneficiaries, as they transfer responsibility for resources from one party to another without altering the availability of the resource as a whole. It has a cost effect. Taxes, duties, royalties and franchise may be considered as forms of transfer payments. On the other hand, if a tax is levied on an output which is included in the market price, this should not be accounted for in a financial analysis, as the producer does not benefit from the tax.

CAPITAL COSTS

The expenditure required to put in place a facility that, when commissioned (normally embedded in a physical asset), is capable of producing goods or services is normally referred to as the capital cost. It could include the purchase of land, purchase of equipment, construction of structures and so on. They may not be limited to initial expenditures, as the purchase of equipment can be seen as a capital expenditure. Typically, capital costs are thought of as one-time costs, although the expenditure to pay for them may be spread over a period of time. Capital costs are fixed, and are therefore independent of the level of output.

However, the creation of a physical asset through forms of construction or other processes will be achieved through the employment of human resources, use of machinery and so on – all of which would be seen as operating expenditures on the part of the parties creating the facility. In other words, what for one entity may be seen as a capital cost could be seen as an operating cost by another. It depends on whose perspective you are viewing the expenditures from. Equally, the value of the asset may not reflect the expenditures required to create it, which gets us into the realm of book values, which will be left to the accountants to deal with and are not discussed in this chapter.

Therefore, what counts as a capital expenditure can be open to interpretation. Note also that capital cost is *not* the same as the cost of capital.

OPERATIONAL COSTS

While the capital cost is seen as one-off expenditures, operational costs are regarded as recurrent expenditures required to maintain the flow of goods and services. They may include the cost of employment of personnel, payment for services, insurances, maintenance costs, consumables and much more. Operational costs may be further divided into direct and indirect costs as well as fixed and variable costs.

Direct costs are those that can be directly assigned to the production of goods or services, and are very often seen as variable costs. In other words, the level of usage of resources is

271

correlated with the level of production or output. But direct costs and variable costs are not synonymous: the cost of running a machine used in production is not necessarily a direct cost but is likely to be a variable or semi-variable cost. Indirect costs are those not directly attributable to any given function. They may also be fixed or variable. An example of an indirect cost might be the machine used in the production mentioned above, especially if it can be used for more than one purpose. Indirect costs are sometime called overheads.

Variable costs are those that vary with the intensity of operation or level of output, while fixed cost are said not to vary with the level of output. Overhead costs representing the cost of resources used by an organisation to be able to continue to perform its functions are usually taken to be fixed costs.

The idea that there are fixed operational costs is a valid assumption in the short run, where production can be varied within certain limits without requiring any change in overhead or other of your 'fixed' costs. However, any change in the level of operation outside of those limits will result in a change of 'fixed' costs. Therefore, in the long run, so-called fixed costs can also be thought of as also being variable. This point lies at the heart of activity-based costing (ABC) (Innes and Mitchell, 1995; Emblemsvag, 2001). ABC works from the presumption that there are no fixed or indirect costs. It shifts the focus of attention towards the activities that enable production, rather than the production itself. It is activities that consume resources and drive overhead costs; thus, it seeks to convert indirect/overhead costs into direct costs associated with carrying out activities that then result in the production of the goods or services.

While there is consensus that ABC is a more accurate way of costing goods and services, one of the drawbacks is the complexity of the task of applying it. There are only a few examples of it being used in the water sector.

OPPORTUNITY COSTS
This is the value of the next best alternative, which will be foregone if the proposed project goes ahead. Identifying opportunity costs can be an important aspect of project appraisal when there are alternatives that are mutually exclusive and where there are choices over what projects can be embarked upon. Identifying the opportunity costs involves an assessment of the costs and likely returns for the available alternative options. To an extent, the objective that is to be satisfied must also be considered. In other words, if the objective were to maximise income as opposed to, say, reducing leakage in a distribution system, then that will influence the range of opportunities available. Thus, the consideration of opportunity costs can be seen as being part of the financial appraisal of alternative ways to achieve a particular set of goals.

10.3.3 Cash flow
Cash flow is what it says it is – the flow of monies associated with a project or enterprise into or out of that project. In the case of a project, the cash flow would be taken over the life of the project. It can be taken as historical cash flows or projected future cash flows. A

consideration of net cash flows, that is, the difference between what is coming in and what is going out, can be used to determine if there is an imbalance that would need some form of financing to bridge any deficits.

In the case of water services, the potential income is often highly regulated and set by governments and not by sales in a market. Under these circumstances, the choice of project may have little influence on the income generation. Nevertheless, net cash flow is still an important consideration for the reason given above relating to the need for working capital.

10.3.4 Project impacts

Impacts from the proposed project and the alternatives will have a bearing on the flow of benefits as well as of any negative consequences. The direct consequences of project implementation can be divided into investment effects and exploitation effects. Investment effects would be the consequences the project would have for other potential investment opportunities. In other words, what are the opportunity costs associated with the project that would be foregone if you undertake it? Does proceeding with the proposed project preclude other courses of action? If so, the costs associated with this should be accounted for. Exploitation effects can be thought of as the consequences that would arise as a direct result of going ahead with the project: these could be environmental, social or economic.

Apart from the direct impacts, are there any indirect consequences or multiplier effects, whether associated with investment in the project or the impacts? Again, you can approach this from the point of view of the entity undertaking the project and also from the societal point of view – do external impacts that the project has have to be formally account for? In this respect, techniques such as environmental impact assessments (EIAs) and life-cycle analysis (see below), among others, play an important supportive role in identifying what the impacts are and their possible extent. In most instances, there are formal requirements for carrying out EIAs. EIAs and related techniques can play an important role in not only identifying the potential impacts and benefits but also the associated areas of risk that the project may be subject to or give rise to.

In this respect, you need to identify which parties should be included in any analysis, because this will help you understand the impacts of a project. On the one hand, there are those parties which would be investing in the project in various ways, not necessarily financially; this could be direct investment/involvement or having to invest or incur forms of expenditure as a result of the proposed project. Then, in terms of the effects of the project, there should be considered those parties which would be undertaking new activities and what results from them, as a result of the realisation of the project. And there are those whose level of activity is substantially modified by the project: this may be related to costs savings on the part of the project proposers, for example. In a financial analysis, these are some of the factors that would need to be identified and closely examined for possible inclusion in an appraisal of a project and its alternatives.

10.3.5 Economic efficiency

As has been seen, CBA draws on the idea of obtaining the best possible return for the use of scarce resources. This being the case, the concept of economic efficiency becomes useful as a way of thinking about choices that might be made. It has been observed that the criterion of efficiency is an attractive one, especially to economists because it has little ethical content (Perman *et al.*, 2003). Economic efficiency is something of an umbrella term, as it covers alternative criteria such as:

- allocative efficiency – where the available resources are allocated in such a way as to maximise the benefit derived from their use
- productive efficiency – when production is achieved at the lowest cost possible
- distributive efficiency – when the goods and services that are produced are obtained by those who have the greatest need for them
- Pareto efficiency – where any change in allocation would make one party better off without making other parties worse off.

The underlying assumption for these concepts is 'all other things being equal'.

While the various concepts of economic efficiency are important, there is seldom a direct connection made between economic efficiency and CBA, even though how a CBA is carried out can be used to deduce something about the economic efficiency of a project. Furthermore, it can also be noted that economic efficiency can be looked at either from the private or societal perspective. Efficiency can be broken down into two interrelated parts: Should a project that produces goods and services go ahead? If so, what should be the level of provision? The first question only asks if the benefits outweigh the costs for any given project. The second asks how the benefits could be maximised; in other words, what are the various alternatives and which one of these would satisfy the appropriate efficiency condition?

10.4 Where does the money come from?

When carrying out any form of project appraisal it is always good practice to understand where the project promoter will obtain the financing (the money) from to pay for the project. Any form of financing brings its own set of constraints, and may impact on access to further monies. Understanding how a project is to be financed will make dialogue between the project promoter (usually the employer under the contract, or the client for the consulting engineers) and those executing the project very much more meaningful. First and foremost, it is important for all parties involved, and especially engineers and contractors, to understand that there are few, if any, clients who have access to unlimited funds. In reality, there never were unlimited funds: it was simply that, in the past, the consequences of cost over-runs were less well publicised than is now the case.

In the water sector, project promoters come in a variety of forms, depending on the country or part of the world being considered. In no particular order, the project promoter could be a government (national or sub-national), through a ministry, department or through a statutory corporation; or the project promoter could be a private company that has some form of management contract or franchise for the operation and maintenance of water services (as in France) or that owns the water infrastructure (as in England and Wales). Each operates under different constraints when it comes to financing.

Projects promoted and paid for by governments will be considered first. It is often thought that governments can always raise money by increasing taxes or by printing more money. Both of these assumptions are, to a certain extent, correct. But they bring with them unwanted consequences; for example, potential loss of electoral support in the first case, and inflation in the second. Governments do borrow considerable sums of money both from their own citizens and from the international money markets. They do this through a variety of financial instruments, including national savings products. The financial instrument that is perhaps most important is the issuing of government bonds. Here, the government issues a promise to potential investors that they will receive a predetermined rate of interest for a predetermined time period. At the end of the period they will have their original deposit returned to them. These bonds are often referred to as gilts, because originally there was a gilt edge to the documents, and the interest that is paid on them (usually at 6-monthly intervals) is referred to as the coupon. They are normally viewed as a safe form of investment, because the interest rate is known, be it a fixed rate or one linked to inflation, and hence the annual return is predictable. There is actually a market in these instruments after they have been issued, but this complexity does not need to be considered here.

However, government bonds are neither risk-free nor offer a source of unlimited finance to any government. It is not unknown for governments to not honour the bonds they have issued. Perhaps the most obvious case of failure to honour the promise of financial returns is the example of war bonds, issued by the losing side in a conflict. The winners are most unlikely to feel any obligation towards the supporters of the losing side. Other examples abound where national economic difficulties have meant that governments have been unable or unwilling to honour the bonds. Investors are always concerned about the ability of any government to be able to repay their debts when they become due, and all government debt around the world is carefully monitored by the investment community. As the level of annual borrowing increases as a percentage of GDP (gross domestic product), so the risk is perceived to increase, and hence the lenders require higher rates of interest before they will buy the bonds. The impact of this can be seen in the financial problems experienced by the government of Greece in the spring of 2010, where concerns over the level of debt and debt servicing impacted on the ability to service the sovereign debt, and as the concern over the level of debt spread to other countries, the euro itself came under pressure. This balance of risk and reward tends to be a major factor in controlling the level of government borrowing in most developed countries.

It is probably true that few engineering projects will in themselves limit the monies that any government can borrow. However, the cumulative effect of problems stemming from increasing the government debt beyond levels perceived as reasonable or prudent have most certainly led to the development of vehicles such as the public finance initiative

(PFI). The PFI is in effect the outsourcing of government debt, whereby the government or nationalised industry asks contractors to borrow the money to finance a group of projects, such as a hospital building programme, on its behalf. The government then undertakes to pay an agreed annual charge for the construction and maintenance of the hospital over an agreed period of up to 30 years. Unless there are tax advantages for specific classes of work, this is usually a more expensive way of paying for projects (for a variety of reasons), but it has the advantage of keeping the debt off the government accounts. It thus preserves lower government borrowing rates, at least in the short term.

It should be said that not all government borrowing is used to finance engineering projects: it can be used for a multitude of purposes, but it is important to understand that no government has access to unlimited borrowing, and high levels of borrowing incur high interest costs.

Generally, governments insist on directly funding the expenditure of their own departments and the expenditure of nationalised industries. Thus, the defence budget will come from government monies, as will, generally, education and health service expenditure, although, as noted above, the UK's National Health Service now utilises a form of PFI to allow hospital construction programmes that would otherwise be beyond the government purse. This model of financing directly from government still predominates in many parts of the world where water services are provided directly by a government department.

As far as nationalised industries are concerned, the government will normally require them to borrow money from the government, and, generally, the rate it will charge them will be above the rate that the government itself incurs. In special cases, it may allow this borrowing from government to be 'topped up' with initiatives from the private sector, but seldom through direct borrowing, as this would have to appear on the government accounts. Municipalities and statutory bodies are generally treated in the same way as nationalised industries in the UK, and are restricted to borrowing only from the UK Treasury. However, this is not always the case around the world. In the USA, for example, it is common for municipalities to raise money by issuing their own bonds, and, as shall be seen in a later section, this can have an impact on the optimal phasing of engineering projects. Under certain circumstances, such bodies (including governments) may also borrow money from banks and other lending agencies, such as international or regional development banks. This borrowing comes with conditions and, often, higher interest rates. Recently, a great deal of international attention has been given to ways in which water services can be financed, pointing out that there needs to be, among other actions, sustainable cost recovery and the development of local capital markets (Winpenny, 2003; OECD, 2004).

This now leads to the question of funding for private companies, usually referred to as public companies. What is being referred to are companies that are independent of government ownership (hence private, in one sense of the word), but may be subscribed to by the public (hence public companies), or owned by private individuals (hence private, in a very narrow sense of the word). In all cases, companies will have shareholders, who have invested their monies in the company in the anticipation of receiving a reward. Sometimes the reward will be an increase in the value of the shares as the company expands, and sometimes it will be a dividend which is paid out of profit. It is unusual for all profit to be paid out as dividend, as normally some will be retained for capital investment. In the case of 'not-for-profit' bodies, any surplus has to be either distributed or re-invested. The shareholder investment is usually referred to as 'equity'.

In addition to this, the company will be able to borrow monies, either directly from banks, or by issuing its own corporate bonds, which are similar to government bonds but are only backed by the company. For completeness, there are other instruments such as preference shares, but, again, these are beyond the scope of this brief summary. What is important is that the company's creditworthiness, and hence the rates at which it can access finance, is determined by its financial performance, and ratios such as the value of equity to the value of debt are crucial in determining these issues. In addition, most borrowing from banks will be governed by covenants which clearly set minimum or maximum values for ratios which, if exceeded, allow the bank to immediately call in the debt. In most cases, the bank would not put the firm into liquidation but would insist on renegotiating the loans at higher interest rates. The money to pay these higher rates would come from monies that would otherwise have been paid to shareholders in dividends, and, as result, the shareholders will themselves be seeking explanations.

Because these financial ratios, together with the overall performance of the company, are so important to the financial well-being of the company, there are organisations, such as Moody and Standard & Poor, who specialise in providing assessments of the credit-worthiness of companies. This leads to the term 'investment grade', whereby companies with a sufficiently high rating are viewed as safe for investment by pension funds, insurance companies, trusts, etc. Conversely, companies that fail to achieve this standard have their bonds referred to as 'junk bonds' – this does not mean that they are worthless but, rather, that they are high-risk investments, offering a high return but with an increasing likelihood that the company may not be able to pay either the interest or the capital at some time in the future.

It is essential that appraisals contain realistic estimates of cost, as significant cost over-runs are unlikely to be greeted kindly by promoters. To reinforce this point, it worth remembering that, in 2009, the State of California was on the verge of bankruptcy (though not as a result of engineering costs). This demonstrates that even very large economies (California, if it were an independent state, would be between the seventh and 10th largest economy in the world) can run out of money. And, of course, in 2008, Iceland, Hungary, the Baltic States and others had to be supported financially by the International Monetary Fund (IMF). In 2010, Greece had to seek support from the IMF and the European Central Bank in order to, among other things, lower the cost of borrowing necessary to service its debt. Other examples from the past include IMF assistance to countries in Latin America (e.g. Honduras and Haiti) and Africa (e.g. Tanzania and Zambia), among others.

Closer to home, there are two recent examples of publicly quoted companies getting into terminal difficulty because of seriously flawed engineering appraisals.

The first is the Channel Tunnel, which started with an estimate, including contingencies, of about £2.6 billion (1985 price base). By the time it was finished, the cost had risen to £4.7 billion (1985 prices). As a result, the revenues from users of the tunnel were inadequate to pay the interest on the loans. Prices could not be increased because ferry operators offered economic alternatives, and, eventually, the whole financial structure of the company had to be changed, to the detriment of the shareholders, whose investments were heavily diluted by the new structure.

The second example is the demise of Railtrack, which went into the modernisation of the West Coast Mainline on the basis of an estimate of £2.5 billion, and finished up with a project costing almost six times that figure (National Audit Office Report, 2006). This was a major factor in the decision of the incoming Labour government to renationalise the company. Again, it was the shareholders who lost their investment.

Engineers cannot expect to have their views and advice respected by the financial community when they make errors of this magnitude. While these are two exceptional cases, they do illustrate the sometimes quite devastating consequences of inadequate appraisal.

10.5   Techniques and alternatives

The focus of this chapter is on project appraisal, specifically looking at costs and benefits. So far, the elements that might inform an appraisal have been discussed. At some stage, a decision has to be made as to how to operationalise the appraisal. In other words, what technique should be used to inform decision-making? A variety of techniques could be used, depending on the circumstances. Table 10.1 shows three forms of appraisal, each

Table 10.1 Measurement of costs and consequences

Type of appraisal	Measurement or valuation of costs	Identification of consequences	Measurement or valuation of consequences
Least-cost analysis	Monetary units	Achieved to the same degree	None
Cost-effectiveness analysis (CEA)	Monetary units	Single effect of interest, common to alternatives but achieved to different degrees	Natural units, e.g. dissolved oxygen level, leakage reduction
Cost–benefit analysis (CBA)	Monetary units	Single or multiple effects, not necessarily common to alternatives	Monetary units

After Drummond *et al.* (2006)

of which deals with costs but treats consequences in different ways. The more commonly used techniques include least-cost analysis, CEA and CBA. The latter is probably the best known and most widely used. Indeed, CBA is a commonly used appraisal tool, and one which the US government, the European Union, the World Bank and other regional and international agencies routinely use to appraise projects. Although the aims of each of these techniques may be different, as evidenced by their descriptive titles, they do share some commonalities. The end point is the achievement of a goal for which there are different alternative options, and you need to decide how to go about taking into account the costs associated with the achievement of the goal(s).

In the following section, the three techniques mentioned above are briefly discussed. In addition, some other possible approaches to project appraisal are considered that have emerged over the last decade or so. The additional approaches considered are life-cycle analysis, whole-life costing (WLC) and the application of private finance initiatives. Each of these four approaches to appraisal can and, in some cases, do draw on the techniques already mentioned. They differ in the manner in which they are used, and they may be thought of as alternative ways of identifying the range of costs and consequences associated with alternative courses of action. Each, in turn, is the subject of a growing body of literature detailing their underpinnings and assumptions as well as case studies of their application.

10.5.1 Least-cost analysis

Least-cost analysis focuses on the determination of the project alternative with the least cost, given that the alternatives are mutually exclusive and technically feasible options. To be mutually exclusive, the alternatives must be capable of producing the same outcome or output of a specified standard or quality. If there are differences between alternatives in terms of their output, then these would have to be normalised in order to compare alternatives on the same basis. For projects where the benefits are in the form of a single commodity, such as water supply, and where the potential income stream is not affected by the choice of alternative, then a least-cost approach can be considered. Such an example might be the extension of a water supply which could be secured either through leakage reduction or by incorporating new sources.

In such a case, if incorporating a new source provides more supply than the leakage reduction, then in order to compare them equally, the relative benefits achieved would have to be normalised. This could be done by valuing the foregone incremental benefits from incorporating the new source, and to add that as a cost to the leakage reduction option. In doing so, it would ensure that the two options are essentially equivalent to each other. The alternative with the lowest present value of costs would then be the least-cost option. Note that benefits do not enter the calculation, where they are the same for both alternatives under these assumptions.

10.5.2 Cost-effectiveness analysis

CEA usually considers a single measure of output such as the level of leakage reduced or the level of dissolved oxygen in a river. The alternatives that are considered may be able to achieve different levels of reduction in the chosen parameter, but at differing costs. The results are expressed in the form of a cost-effectiveness ratio, which is used to compare options. In other words, cost-effectiveness looks for the best way to achieve a given end result. The most cost-effective alternative will be that which achieves the desired outcome at the least cost. When the budget is limited, this may be a useful way of deciding which alternative should be adopted where there is a limited range of options. The difference between cost-effectiveness and least-cost methods lies in the way that benefits (consequences outcomes) are handled, as can be seen from Table 10.1.

The main difficulty in using CEA is that it can only really handle one measure of output. Where there are multiple outputs, unless some aggregate measure can be derived, the method becomes difficult to apply. Even if such an analysis were to be applied to each of the outcomes, engineers would be faced with the task of deciding which output is the more important and how they should be ranked. This introduces an element of subjectivity, which may not be seen as helpful.

10.5.3 Cost–benefit analysis

In CBA, benefits (consequences) are treated in the same way as the costs by assigning a monetary value to them. The broad objective is to determine whether the outcomes from an alternative justify the costs incurred in realising the benefits. The problem becomes one of ensuring that all the benefits or dis-benefits arising from the proposed project can be identified and costed. In some cases, the difficulty lies in identifying the benefits and the externalities associated with projects alternatives. As such, CBA has a broader scope than least-cost analysis or CEA. The latter address mainly questions of productive efficiency while CBA helps to inform questions of allocative efficiency to determine if a goal is worth achieving, given alternative uses of the resources.

CBA is used in a wide variety of applications, although it was developed for use in infrastructure projects. It assesses the net impact of a project or alternative project, and thus is used to inform decisions where budgets are limited. Among the strengths of the techniques is that it is able to consider externalities and other factors that distort prices, allowing market imperfections to be explicitly considered. This allows the techniques to go beyond simple financial effects for the private investor and to consider projects from the point of view of society or of the public. In principle, CBA could be used to evaluate programmes and policies. The aim of CBA is to determine if a project is desirable from the point of view of the private investor and that of social welfare, taking into consideration all the costs and all the benefits associated with the proposed project. It does so by summing the discounted costs and benefits of project alternatives and comparing them. The process of discounting will be discussed below.

The purpose of the project and the nature of the parties responsible for the project will have an influence on determining the extent of the CBA. Where the purpose of a project is to promote local development and is the responsibility of a public body, then the CBA should go beyond just the financial analysis. After all, what is important is the extent to which the project will be able to meet the goal of local development and at what overall cost. A further consideration is the extent to which the carrying out of a project will affect the local 'market': the availability of goods and prices for goods and services. In regions such as Europe, it would be the exception where a project is of such a scale as to have any noticeable impact on the market, and therefore such impacts can be ignored. This, though, will not be the case in other geographic settings. Under these circumstances, aspects such as the impact on wages or displaced consumption would have to be factored in. Many of the international lending agencies, such as the World Bank, Asian Development Bank, the European Development Fund and others, have developed their own guides to carrying out CBA. There are also specialised or standardised methodologies that have been developed by various agencies for use under particular circumstances, for example the evaluation of flood alleviation or coastal protection schemes by the Department of Environment Food and Rural Affairs in the UK.

In cases where there is the private provision of services to the public (sometimes referred to as public goods) such as water supply and sanitation, the question of what benefits should be included and whether the appraisal should go beyond that of a financial analysis becomes somewhat blurred. In certain instances, this question will be answered in part by the requirements of government agencies, regulatory authorities and their requirements as to what should be considered or included. It might be argued that some of the evaluation or decisions with respect to benefits have already been taken. For example, the disposal of sewage waste by long-sea outfall is no longer an option in Britain, one reason among many being that there is a presumption that the amount the public is willing to pay to improve the marine environment outweighs any additional costs associated with the alternative means of disposal of the waste on land. Thus, any cost–benefit appraisal of a waste water treatment works can take this as a given and need not include it in the calculation of costs and benefits associated with the various options considered.

The extent to which CBA can incorporate environmental and social costs and benefits is a matter of ongoing debate. It is related more to the ability to identify the relevant social and environmental facets to be included and how to go about valuing them than it is to the ability of the methodology to handle them. For some, the very idea of placing a value on the goods and services that the environment supplies is questionable. Leaving that point aside, with environmental impacts considered – and they should be – then not only should the benefits outweigh the costs but they should outweigh the costs plus the environmental cost of the impacts of the project. It might also be said that the environmental cost could also mean external costs, as unpriced environmental damages are externalities associated with the project. The question of how to arrive at environmental costs and, by the same token, environmental benefits is a subject area all of its own. Suffice it to say, this is an area in which techniques such as EIAs, life-cycle analysis (see the following section) and environmental economics all have an important role to play.

10.5.4 Life-cycle analysis

Life-cycle analysis is based on the idea that a detailed examination of the life cycle of a project and all the goods and services that are required to bring it to completion and keep it operating should be carried out. The goal is to compare the full range of social and environmental damages that will arise, and to use this as a means of choosing that which imposes the least burden on the environment. It takes into account aspects such as the use of raw materials, energy consumption and the amount and type of waste produced. Life-cycle analysis has found most use in being applied to the production of goods and services. It is unusual to be able to use the approach to consider the impact of the use of those goods and services, as normally this information is not available. Theoretically, it provides a way of assessing raw material production, manufacture, distribution, use and disposal, and intermediary steps. It has been used to optimise environmental performance, either of a single product or a company, but is much more difficult due to the informational requirements to apply to projects – as part of an evaluation.

The procedures of life-cycle assessment have been systematised by the International Organisation for Standardisation (ISO) through its incorporation into environmental management standards, ISO 14000. Life-cycle analysis is a potentially powerful tool which can assist regulators to formulate environmental legislation, help manufacturers analyse their processes and improve their products, and perhaps enable consumers to make more-informed choices. There are four main phases in carrying out an analysis:

- The goal and scope of the assessment have to be agreed upon as well as the methods for assessing environmental impacts.
- An inventory of relevant data has to be drawn up, some of which may be produced through modelling, focusing on the inputs and outputs from the process considered. In some cases this can include life-cycle costing in parallel with the environmental life cycle.
- The impacts have to be assessed, and, where possible, the impacts normalised so that they can be compared.
- The results have to be interpreted, and subjected to an independent review.

In recent years, this process has been much facilitated by the development of various software tools and packages.

One of the most commonly identified problems with the application of life-cycle analysis is that of where to draw the boundaries around an analysis. It is usual to ignore second-generation impacts, such as the energy required to extract aggregate used in concrete to build the wastewater treatment works to treat the sewage to environmental standards. While life-cycle analysis is a way of finding out about the environmental footprint and converting this into some form of common metric for analysis, comparisons of alternatives are more challenging. As a technique within project appraisal, it is probably best employed as part of an EIA.

10.5.5 Whole-life costing

WLC (Skipworth *et al.*, 2002) is a term that is predominantly though not exclusively used in an engineering context with respect to the provision or procurement of service infrastructure. It has been advocated as an alternative decision-making approach that is more appropriate when considering alternative means of achieving a set project objective and the balance between the initial capital investment and subsequent operational costs over the whole life of the project and the associated service infrastructure. A distinctive feature is that, as an approach, it moves away from the perspective of projects as being one-time exercises, to considering a stream of potential interventions over a period of time. As such, it is of particular use in the water industry.

> The *whole life cost* of construction in a conventional sense ... should include such factors as initial construction cost, operating, maintenance and repair charges and an allowance for demolition.
>
> (Chenery, 1984)

In other words, it combines capital and operational costs over the life of the asset infrastructure. This has implications for investment decision-making when compared with alternative capital-budgeting decision-making approaches. The adoption of WLC approaches has been underpinned by developments such as best value in local government, building and constructed assets (ISO/DIS 15686), private finance initiatives and public–private partnerships. The US Environmental Protection Agency (EPA) (1996) provides a comprehensive treatment of WLC and the range of costs that can be considered for inclusion. These include the initial capital as well as all up-front costs, operational, maintenance, rehabilitation, repair costs and, where relevant, the decommissioning costs. It is not just internal or direct costs that have to be considered but also overhead costs as well. A significant feature of WLC is its attempt to minimise the distinction between capital maintenance and operating costs.

In the context of the provision of water services, WLC has been proposed and adopted as an integral part of decision support applied to the maintenance and rehabilitation of distributed networks such as water mains. In the case of highly regulated privatised WSPs, not only are the costs of service provided important but also the quality aspects. Generally, the application of WLC to network distribution management aims to achieve the lowest network service provision operating cost, when all costs are considered, while meeting statutory standards and regulatory performance requirements. One of the key differences between WLC application within the utilities sector as against the built environment lies in the stage of the service provision cycle. Generally, the service infrastructure already exists, and thus the focus is not on its provision but on maintaining the serviceability of the assets. Within this context, the physical condition of the service infrastructure and the effects of deterioration on serviceability assume a greater significance. This adds extra importance to the timing dimension of management decisions and interventions, as these impact not just on the net cash flow generated but also on the relative benefits of an intervention *vis-à-vis* its impact on serviceability and performance. Thus, in applying a WLC approach to ageing assets, knowledge of the physical performance and the consequences of that performance are important (Box 10.3). This is due to the linkages between such performance and the activities and demand for resources that operating and maintaining the levels of serviceability require.

Box 10.3 Performance modelling

How well a system such as a water distribution system performs depends on a number of factors. If it is already in place, then factors such as age, materials used, usage, etc., will have an important influence. If it is a new system, then there will be certain expectations as to how it will perform over its useful life and the level of maintenance it will require. At the same time, how well a system performs and what has to be done to maintain or improve that performance come with a cost. These costs will vary over time, and will depend on the type of interventions made (e.g. repair or replacement), while the type of intervention will also affect future performance as well as running costs. A growing trend is to couple performance modelling with associated costs, and to optimise overall performance subject to constraints, such as the minimisation of overall costs. The performance model is a mathematical representation of the key interactions and characteristics of the system, and varying the parameters allows the impact of alternative decisions regarding design, management or other interventions to be assessed.

10.6   Discounting

Where the time over which a proposed project will come into being and operate is in excess of at least 1 year and where the costs incurred and the benefits that accrue do so at different points in that time horizon, then the concept of discounting becomes a central consideration in the evaluation of alternatives. The vast majority of people would rather have a benefit sooner rather than later. This idea lies at the heart of discounting, and is referred to by economists as a *positive rate of time preference*, for a variety of reasons. First, because there are varying degrees of uncertainty as to what tomorrow might bring, there tends to be a preference to enjoy benefits today rather than wait until later. Then, there is the general idea that people will be better off in the future than they are today, thus money today is worth more than the same amount in the future because people will be richer – it is hoped. In other words, a given amount of money has more value today than at some time in the future. This idea provides the basis for discounting, and can be applied not just to costs and expenditures but also to other goods and services that are not so easily traded or for which markets exist. Put the other way around, it can be said that an individual would have to be offered some form of incentive to defer consumption until a later point in time: the payment of interest related to the scale of benefit they expect to receive.

If individuals think like this, and society is an aggregation of individuals, then it should follow that society should also make choices about the value of projects or investments on a similar basis. But the question is, should society adopt the same set of preferences as an individual? In other words, should society discount the future to the same degree that private individuals (or entities) do? Many economists argue that there should be a difference between them. One distinction that has been made is that individuals will express a different set of preferences depending on whether they are considering something from their own individual perspective as, say, a customer or from their perspective as a citizen concerned about what may happen in the future. A further argument that has been advanced is that individuals often underestimate the benefits of future consumption and that there is no good reason to carry this short-sightedness over into the social context. It seems, therefore, that there is consensus that there is and should be a difference between the rate at which a private individual views future consumption and the way society views the future. One of the implications of this is that projects that are undertaken on behalf of society (by a government) and which are intended to benefit society should not be discounted at the same rate as projects undertaken by private bodies for their own benefit. Quite what the discount rate should be is controversial, and is a subject of much discussion. A very good example of such a discussion is that surrounding the Stern Report (Stern, 2006), where there is much ongoing debate over what should be an appropriate discount rate when considering measures to address climate change.

If there were a perfect capital market it would aggregate individual savers' and investors' preferences, and it would be capable of producing a supply of funds to and demand for funds from the market such that supply would equal demand at the social discount rate. In such a world, the question of which discount rate to use would not arise as there would be just one rate. But markets are far from perfect. There is no such thing as perfect information about 'the market': there is a mix of different assets with respect to how risky they are, their liquidity, time span, tax regimes and other factors, which has already been touched upon. The result is that there are on offer a range of different interest rates that reflect the different mix of assets. The question once again is: 'Which interest (discount) rate should be used in a particular set of circumstances, especially when carrying out project appraisal?' In some cases, interest rates are determined by monetary authorities such as central banks, which may be setting rates in order to achieve objectives other than to do with matching supply with demand. For example, the UK Treasury's Green Book (HM Treasury, 2003) in setting out guidance for the evaluation and appraisal of government investments proposes that a discount rate of 3.5% be adopted with factors such as risk, optimism bias and the cost of variability dealt with separately and explicitly. Even within governments, rates may be set that reflect political pressures and goals rather than the workings of the market. The Stern Report adopted a discount rate of 1.4%, and noted that this is close to long-term yield on government bonds (gilts).

As might be expected, there are theoretical approaches to analysing discount rates which seek to separate out the pure rate of time preference from the influence of the rate of growth of consumption over time. Conventional thinking is that, on public projects, explicit allowance should be made for the cost of bearing tax, externalities and forms of risk associated with the mix of assets. Therefore, the 'correct' rate should be the risk-free market interest rate. This is taken to be the interest rate on government bonds – on the assumption that there is little or no risk attached to government investments. Some refer to this as the *pure rate of time preference*. With respect to investments in the private sector, it can be observed that rates of return are considerably higher than on, say, government bonds: but should these be used? Economists agree that it is the 'real' rates and not the nominal rates that should be used. 'Real' rates are nominal

Table 10.2 Present value of 100 at various discount rates

Discount rate: %	Time horizon in years				
	5	10	20	50	100
1	95.15	90.53	81.95	60.80	36.97
2	90.57	82.03	67.30	37.15	13.80
4	82.19	67.56	45.64	14.07	1.98
6	74.73	55.84	31.18	5.43	0.29
8	68.06	46.32	21.45	2.13	0.05
10	62.09	38.55	14.86	0.85	0.01

rates adjusted for inflation. If there is only one interest rate on offer in the private sector, say 5% per year, and inflation is 2% per year, then the real interest rate would be 5−2 = 3%. This also has implications for costs and benefits and how they are to be handled. The use of a real interest rate implies that the effects of inflation on costs and benefits should be ignored. All costs and benefits should be accounted for in constant money terms, and the use of any historic cost data should be adjusted accordingly.

Another question that can arise is whether, for long-term projects, those that operate and deliver benefits over an extended period of time, the discount rate employed should be uniform. One of the reasons for asking this question is the impact that the use of a discount rate has on future values (Table 10.2). Table 10.2 illustrates that future values are of increasingly less importance, especially as the discount rate increases. This aspect can lead to some types of projects being disadvantaged when compared with others. Take, for example, two alternatives that achieve the same stream of benefits, but one has a high initial cost but low recurrent costs and the other a lower initial cost but higher recurrent costs. The effect of discounting, other things being equal, will favour the option with the lower initial cost. It is for this reason that in certain instances a decreasing discount rate may be applied to future values. This approach has been adopted by the UK Treasury in its 2003 Green Book, where a declining long-term discount rate was set out (Table 10.3). The length of time over which a project is appraised can also have an important influence on the choice between alternatives.

Table 10.3 Declining long-term discount rate

Period: years	Discount rate: %
0–30	3.50
31–75	3.00
76–125	2.50
126–200	2.00
201–300	1.50
>301	1.00

In cases where the time horizon may be long and there are associated environmental impacts as a result of the project, the proper time horizon for the appraisal is the time at which the impacts of the project cease. It is not the time at which the project ceases to serve its original purpose. The examples of nuclear power stations and the effects and mitigations of climate change come immediately to mind.

In practice, the choice of discount rate requires the exercise of judgement and the inclusion of sensitivity analysis to explore the effects of different rates on outcomes. Even then there should be scope for debate and disagreement. The former senior vice president and chief economist of the World Bank, Joseph Stiglitz, once said that the choice of discount rate:

> depends on a number of factors, and indeed I have argued that it might vary from project to project depending, for instance, on the distributional consequences of a project. These results may be frustrating for those who seek simple answers, but such are not to be found. The decision on the appropriate rate of discount thus inevitably entails judgments.
>
> (Stiglitz, 1994)

10.7   Evaluation techniques

There are a number of evaluation techniques that can be applied, depending on what objectives a project proposer has. This indicates that there is no 'right' evaluation technique but rather a range of possible options. All of the techniques look at the initial investment required against a project set of returns on that investment, and because they are projections they will necessarily involve varying degrees of uncertainty. Uncertainty is an additional aspect that has to be dealt with explicitly to inform decision-making. All of the techniques try to assist in answering the question 'Should the project go ahead?' And in some cases they also help to answer the further question 'Is this option better than another one?' There is one technique that ignores the time value of money: the payback approach. Then there are those techniques that are based on the application of the time value of money: discounted cash flow, net present value (NPV), the internal rate of return (IRR) and the cost–benefit ratio. Finally, there are those techniques in which the last-mentioned approaches are combined with other non-monetary factors, referred to as multi-criteria approaches.

10.7.1 Payback period

Payback is probably the simplest appraisal tool. It involves a calculation to estimate the time that would be required for the projected cash flow to cover the initial capital cost or investment required to establish a project. The strengths of this technique are its simplicity and that it gives a quick indication of the ability of a project to pay for itself. It is particularly useful where project time spans are relatively short and where there is a comparison to be made between two or more possible alternatives. Table 10.4 and Figure 10.2 illustrate a simple example of this technique. It can be seen from this simple illustration that the payback period is 3 years and approximately 4 months.

Table 10.4 Payback period

Year	Capital investment	Income	Cumulative cash flow
0	650	0	−650
1	0	100	−550
2	50	200	−400
3	0	300	−100
4	0	350	250
5	0	400	650

The payback method is a popular method of appraisal in some quarters, but it has a major limitation: it just concentrates on payback and ignores other aspects. It pays no attention to what happens after payback, and, more importantly, it ignores the time value of money by assuming that money tomorrow has the same value as money today, implying a zero discount rate.

10.7.2   Discounted cash flow

Discounted cash flow looks at the expected cash flow associated with a project that has been forecast into the future over some finite period of time. To this is then applied the principle of the time value of money – that money today is of more value than money tomorrow or at some other point in the future. Essentially, what is being done is to ensure that the cash flow amounts are equivalent to each other no matter at what point in time they are realised. The way of achieving this is to 'discount' the future amounts in such a way as to ensure that all amounts are equivalent to each other, not in terms of total amount but in terms of the value placed on each of the single units that make up each total amount. In order to do this there has to be some reference point in time around which the costs are to be compared or related. The usual practice is to take the 'present' as the reference point, but this does not have to be the case.

Figure 10.2 Payback period

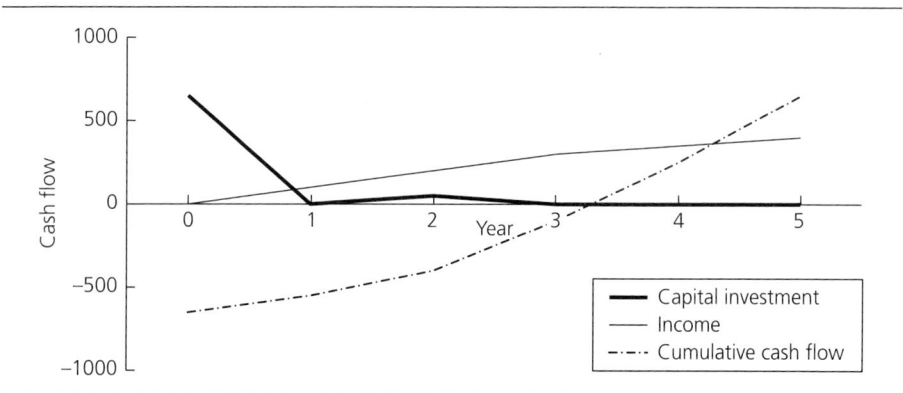

Table 10.5 Discounted cash flow for $r=4\%$

Year	Income	Discounted amount
0	0	0
1	100	96
2	200	185
3	300	267
4	350	299
5	400	329

Whatever reference point in time is chosen, it has to be explicitly stated. The reasons for this are that the present is itself not static: what is the present today will not be the present tomorrow. Discounting also offers a means of comparing the results of similar projects that may have been evaluated at some time in the past, by enabling the results to be compared based on the same values.

Let us assume that it has been determined that, for a particular project, the appropriate discount rate is 4% and the cash flow is the same as that given in Table 10.4. But, before that, the way discounting works will be looked at.

Consider a case where there is an amount of money to invest. Given an initial amount invested at the start of the project, A, and an interest rate being offered on investments of $r\%$ per annum, then at the end of 1 year, the investment would be worth the initial sum (A) plus the interest earned, which would be $A \times r\%$. Thus, the total would be $A + A \times r\%$, which can be written as

$$B = A \times (1+r) \tag{10.1}$$

If this sum were to be left for another year to earn further interest (and assuming that the interest rate is unchanged), then the initial investment would be worth

$$C = A \times (1+r) \times (1+r) \text{ or } C = A \times (1+r)^2 \tag{10.2}$$

Discounting is the reciprocal of this 'compounding'. This can be illustrated by rearranging the above equations. So, the amount at the end of the first year would be $B \times 1/(1+r) =$ the initial amount invested, A. Similarly, for 2 years the amount at the end of the second year would be $C \times 1/(1+r)^2 =$ the initial amount invested, A. And so on for subsequent years. Applying this to the income stream shown in Table 10.4 gives the results in Table 10.5 and Figure 10.3.

Put another way, this represents the present value of the cash flow.

10.7.3 Net present value
The discounted cash flow only looks at one part of the proposed project – the benefits arising from the project, represented in this case by the revenue generated from its

Figure 10.3 Discounted cash flow

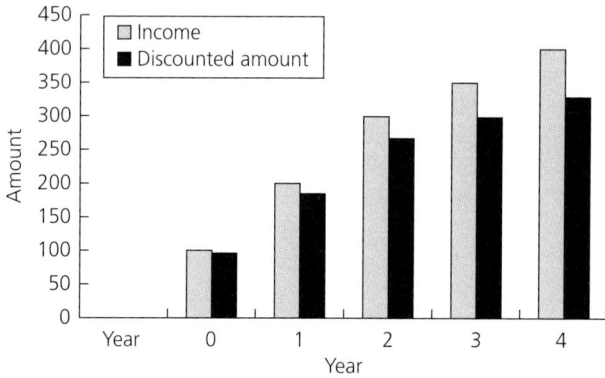

operations. It does not look at the investment side. The NPV is the present value of the benefits generated by a project less the present value of the investments required to realise the function and operation of the project. Using the amounts in Table 10.4 and discounting them all to year 0, the NPV can be found (Table 10.6).

The NPV of this project over a 5 year lifespan is 479 currency units, at a discount rate of 4%. If at the end of the period some of the project components still had a value that could be realised, then that would have to be factored in as a benefit: it represents an income for the project. If, on the other hand, no benefit in the form of the value of the project infrastructure can be realised, then the investment made must be regarded as a sunk cost, and should not be factored into the calculation.

If the funding for a project involves a loan that is repayable over a period of time, then the cost of servicing the loan and the schedule of payments has to be included in the cash flow.

Table 10.6 Net present values

Year	Capital investment	Present value of investments	Income	Present value of income	Cumulative cash flow	Net discounted cumulative cash flow
0	650	650	0	0	−650	−650
1	0	0	100	96	−550	−554
2	50	46	200	185	−400	−415
3	0	0	300	267	−100	−148
4	0	0	350	299	250	151
5	0	0	400	329	650	479
Totals		696		1176		479

The above can be represented in the form of the expression

$$\text{NPV} = \sum_{t=0}^{t=T} \frac{\text{benefits} - \text{costs}}{(1+r)^t} \qquad (10.3)$$

where r = the discount rate
t = the life of the project

The NPV takes the future projected cash flows into account, and it gives absolute values rather than percentages, which can in some cases be misleading.

10.7.4 Internal rate of return

The calculation of an IRR is an extension of the NPV technique. The IRR is that discount rate which when applied to the costs and the benefits is such that the NPV is zero. The IRR can be determined through trail and error calculation (or using the solver function in a spreadsheet). For example, in the case of the expenditures presented in Table 10.4, the calculation would show that the required discount rate to set the NPV to zero (i.e. the present value of investments equals the discounted income cash flow) is 22.2%.

The point of calculating the IRR is that it allows a comparison to be made with the cost of capital for the project. In this case, the discount rate (= the cost of capital) has been given as 4%. This project yields an IRR value of 22.2%, which is much greater than the 4%. In other words, the project gives a much higher return than just leaving the money in the bank or investing it in some form of savings: it is, on the basis of the information presented here, a good investment.

The IRR takes the NPV approach a step further, but can sometimes be difficult to grasp. Because the IRR deals with percentages, it ignores absolute values; for example, a 22% return may sound better than a 10% return, but not if the 22% is on 1000 currency units and the 10% is on 1 000 000 currency units.

10.7.5 Application of CBA

To all intent and purpose the determination of the NPV is the CBA. The result obtained has required considering all the costs incurred and an evaluation of all the benefits that would accrue to the project. The process has yielded a single number, the NPV, and, as will be seen below, this can be used for comparative purposes. But the question arises as to whether this is the only metric that can be derived from the figures produced. Clearly, there are other measures that can be deduced from the figures. The most obvious, and one quite commonly used, is to look at the ratio of the benefits to the costs. Benefits can be taken as gross benefits or, alternatively, as net benefits after having deducted operational costs from the stream of benefits. Care must be taken not to double account for costs, so that, if net benefits are used, the operating costs are not also included in the costs, only the initial investment costs.

Using the example in Table 10.6 at a 4% discount rate, the present value of the benefits is 1176 currency units and the present value of the costs is 696 currency units. The figures give a benefit–cost ratio of 1176/696 = 1.69.

More is said about the application of CBA as a means of comparison in the next section. For the moment, it can be noted that essentially CBA is a method of evaluating the comparative economic efficiency of options. Because it seeks to reduce everything to value in monetary terms, it is poor at being able to handle different concepts of value and relative value, what different people might consider as being of importance and therefore more 'valued'. Indeed, much of the criticism of CBA comes down to the reduction of decision-making to just one of monetary value.

10.7.6 Comparing projects

So far, only single projects have been looked at. All of the techniques outlined above can and are used as a basis for comparing between options.

- In the case of payback, the decision question would be: 'Of the options considered, which one has the shortest payback period?'
- In the case of a discounted cash flow, the question would be: 'Of the options considered, which one yields the highest positive cumulative cash flow?'
- In the case of NPV, the question is similar to that of the discounted cash flow: 'Which yields the highest NPV?'
- And in the case of IRR: 'Which option yields the highest rate of return?'

The answers to all of these questions are of great interest to decision-makers, and can provide the basis for making informed decisions. An example will illustrate the point. Take the case where there are two alternative projects that, due to the availability of funds, are mutually exclusive: if one is undertaken, then the other cannot proceed. The case is illustrated in Tables 10.7 and 10.8.

The results are somewhat contradictory: on the basis of NPV, project A would be favoured, while on the basis of IRR and the benefit–cost ratio, project B would be chosen. Where projects are mutually exclusive in terms of their benefits, then the project with the greatest NPV should be chosen. However, if there are constraints over the availability of funding, then it makes sense to invest in the one that would

Table 10.7 Comparison of projects

Year	Project A		Project B	
	Cash flow	Present value of cash flow	Cash flow	Present value of cash flow
0	−500 000	−500 000	−300 000	−300 000
1	170 000	163 462	50 000	48 077
2	200 000	184 911	100 000	92 456
3	200 000	177 799	150 000	133 349
4	150 000	128 221	200 000	170 961
5	250 000	205 482	250 000	205 482

These figures can be used to compare the two projects

Table 10.8 Comparison of projects using NPV,
IRR and the benefit–cost ratio

Technique	Project A	Project B
NPV	359 874	350 325
IRR	21%	25%
Benefit–cost ratio	1.72	2.17

give the highest returns – the highest benefit–cost ratio and, possibly the highest IRR – because in some instances there can be more than one IRR. This can occur if the sign of the annual net benefits changes a number of times.

Given that the results of comparison can be ambiguous and that in trying to determine what the relevant costs and benefits are there will be uncertainty and errors, it makes sense to subject the appraisal to a sensitivity analysis. In this, some of what might be regarded as the key assumptions on which the appraisal is based need to be tested to see how sensitive the results are to their variation. Another way of looking at this is as a way of seeing how robust the assumptions and the outcomes are. If the results suggest that there are one or more parameters that account for variability in the outcomes, then perhaps it would be prudent, if possible, to try to make sure that the most reliable value is determined for that parameter. In other words, it could suggest where further investigation might be required. On the other hand, this may not be as useful as it first sounds, as it might well be that the source of sensitivity was already known, and all that has been done is to demonstrate that fact, which may not be helpful.

Sensitivity analysis might, however, indicate whether the uncertainties are such as to lead to spurious choices. The benefit–cost ratio does give us some indication as to how much things would have to change so as to make a given alternative marginal. The higher the benefit–cost ratio, the more robust the alternative to uncertainties, unless the uncertainties are mutually reinforcing. Green (2003) suggests that the benefit–cost ratio is an indication of the confidence that the 'do something' option is preferable to the 'do nothing' option (Figure 10.4).

One method of carrying out a sensitivity analysis is to select a number of key variables. The value each of the chosen variables should systematically be altered by the same amount, say 10%. The effect on the overall answer can then be observed. If the answer varies by less than 10%, the variable is not too significant, if by more than 10%, then it could well be significant. As indicated previously, it suggests where effort might be required to narrow uncertainties. And, taken together with the results from CBA, it can indicate just how much uncertainty can be allowed for in evaluating the outcomes.

10.7.7 Multi-criteria analysis

MCA differs from CBA in that provides a way of considering more than one decision criteria. It can be used when some of the impacts or benefits of project alternatives

Figure 10.4 Benefit–cost ratio and confidence in the project
Reproduced with permission from Green (2003). © John Wiley & Sons Ltd

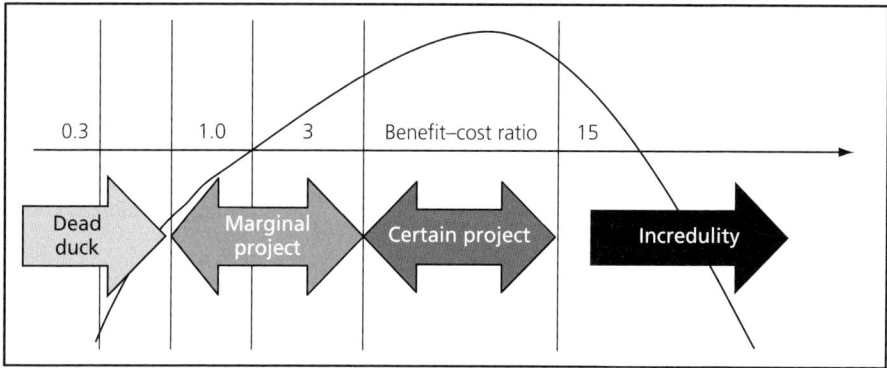

cannot be assigned a monetary value but they might have an importance in the decision-making process. It is an approach that assumes that different objectives can be expressed with respect to some common denominator by means of trade-offs, so that the gains or losses associated with one objective can be traded off against gains or losses in another objective. The role of decision support is then to facilitate us in the process of evaluation. It may well be that there is no solution that optimises all the criteria, and, therefore, the results are used to find compromises between different options. The technique is of use where there is a need to handle a lot of complex information in a consistent way.

MCA is a structured approach to determine what the preferences are for options that satisfy several explicit sets of objectives which have been identified to varying degrees. The objectives that are to be met have to be specified and their attributes identified. It is not necessary that these are in monetary terms: often, they are based on analysis using scoring or weighting of qualitative criteria. It is more important that there are measurable criteria to assess the extent to which the objectives are satisfied. This means that it is possible to combine a number of different criteria or categories within an overall framework, i.e. environmental and social criteria as well as costs and benefits. There is the clear idea that decision-making is informed by more than just costs and incomes. It means that MCA can be used where social and environmental impacts cannot be given a monetary value, as would be required within a CBA. The methodology does have a number of approaches that have been used to aggregate data on individual criteria to provide indicators of the overall performance of alternatives (aggregation, lexicographic, graphical, consensus maximising and concordance). Examples of MCA methodologies include ELECTRE (Figueira et al., 2005), ANP (Saaty, 2005), MAGIQ (McCaffrey, 2005) and NSFDSS (Tam et al., 2006). MCA can be used to identify a preferred alternative, to rank alternatives, to short-list alternatives for further appraisal or to weed out the acceptable from the unacceptable.

One strength of MCA is that it is said to be more representative of decision problems, as it takes into account more than one objective – unlike CBA. However, one of the difficulties with it is how the preferences of different groups or individuals are accommodated within its framework. Different stakeholders may have differing perceptions as to priorities, and thus it may be difficult to come up with a single solution. Under these circumstances, MCA is a means of uncovering those preferences and a starting point for resolving different understandings, making it part of a negotiation process rather than an end in itself. The value then lies partly in the process of constructing the problem to be analysed within the MCA by indicating why factors are important and how they interact. Hence, used in conjunction with other approaches and undertaken from different perspectives, MCA, together with stakeholder consultations, can inform how differences between individual, social and environmental preferences can be negotiated.

In formulating an MCA the objectives that are to be satisfied have to be identified, and frequently this can be proceeded with in a hierarchical fashion. Broad objectives can be broken down into lower-level objectives, allowing a better chance of being able to assess them. Different forms of multi-criteria evaluation are available, with the most suitable method dependent on the nature of the decision to be taken.

A basic feature of MCA is the performance matrix, which draws together the various alternatives on one axis and the performance criteria on the other. The individual performance assessments are usually, but not always, numerical. In some cases this may be all that is required, and decision-makers assess the extent to which the objectives have been satisfied by inspection, though there are dangers with this (e.g. incorrect assumptions). More usually, some further processing is carried out. The two simplest MCA methodologies are ranking and rating. Ranking gives each objective a rank depending on its perceived degree of importance relative to the decision that is to be made. These can then be ranked in order of the priority given to them. Rating takes this a step further by giving each objective a score, between 0 and 100, say. The scores for all the objectives must add up to 100, meaning that if one objective is scored high, then another must be lower. The next stage that can be applied is weighting: numerical weights are given to each criterion that reflect the relative importance assigned to that criterion. It is the process of deciding what the relative weighting should be that causes problems, and is often where stakeholder input is required. Using the scoring and the weighting, a simple weighted average of scores is calculated. The use of weighted averages assumes that the preferences are mutually independent of each other, an assumption that sometimes needs to be verified. If preferences are mutually independent, then a linear additive evaluation may be applied in which the scores are multiplied by their weights, and combined to get an overall score. Most MCA approaches use this, and an example is given in Tables 10.9 and 10.10 of a sewage system with a wastewater treatment works located on an estuary. There are three options to be evaluated, and each option has a collection system, treatment facility and disposal system.

Table 10.9 shows in the first column the criteria, followed by the subcomponents for each of the three main criteria. Each of these has been assigned a weighting on a scale of 1–5, with 5 denoting high importance and 1 denoting low importance. In the following

Table 10.9 Multi-criteria analysis

Criteria	Attributes	Weights	Option A			Option B			Option C		
			Collection	Treatment	Disposal	Collection	Treatment	Disposal	Collection	Treatment	Disposal
Environmental	Marine impacts	5	50	25	80	30	55	10	20	20	10
	Odour	4	40	50	10	40	30	30	20	20	60
	Biodiversity	3	33	50	60	34	40	35	33	10	5
Social	Disruption	5	60	50	10	20	30	30	20	20	60
	Public acceptability	5	20	20	10	40	40	70	40	40	20
	Land use conflicts	4	10	20	10	10	60	45	80	20	45
Operational	Reliability	4	75	5	80	15	5	10	10	90	10
	Flexibility	3	30	35	30	35	35	35	35	30	35
	Complexity	3	65	35	10	15	30	5	20	35	85

Table 10.10 Results of MCA

	Option A	Option B	Option C
Environmental	1604	1202	794
Social	1010	1610	1580
Operational	1255	585	1160
Total score	3869	3397	3534

columns the ratings for each of the subcomponents of the criteria have been assigned. If the ratings for the three collection options are summed across for, say, marine impact, they add up to 100. The higher the rating applied, the higher the beneficial impact associated with it. In Table 10.10, the results of the multiplication of the weights and the ratings are given for each of the criteria and then summed, to produce an overall score for each option. For this hypothetical case, it would appear that option A would be preferred, as it has the best overall performance. Notice that in this example the costs of the various options have not been included.

MCA procedures differ between each other in terms of how the basic information in the performance matrix is processed and the degree of sophistication used. These more complex approaches include analytical hierarchy process, pair-wise comparison, multi-attribute utility theory, outranking methods and fuzzy sets, but a discussion of these is outside the scope of this chapter.

In the hypothetical example of the sewage system given above, a CBA has been carried out, the results of which are given in Table 10.11.

How, then, can both sets of results be used to inform decision-making, as in terms of NPV, option C would be preferred, while option A gives the highest benefit–cost ratio? One way would be to combine the two within an MCA, and assign weights to the results from the MCA given in Table 10.10 and the NPV and benefit–cost ratio given in Table 10.11. This is shown in Table 10.12, and in this example, equal weighting is given to all three criteria. On this basis, option A would seem to be the preferred option.

Table 10.11 Results of CBA

	Option A	Option B	Option C
Capital cost	208 850 000	301 500 000	286 200 000
Operation and maintenance	77 550 000	120 500 000	132 250 000
Benefits	813 600 000	878 700 000	950 850 000
NPV	527 200 000	456 700 000	532 400 000
Benefit–cost ratio	2.84	2.08	2.27

Table 10.12 Combined CBA and MCA

	Option A	Option B	Option C
NPV	527 200 000	456 700 000	532 400 000
NPV ranking	2	3	1
Benefit–cost ratio	2.84	2.08	2.27
Benefit–cost ratio ranking	1	3	2
MCA score	3 869	3 397	3 534
MCA ranking	1	3	2
Unweighted score	4	9	5

Of course, the process of identification of criteria and the relative weightings and rankings assigned could well be the outcome of a stakeholder engagement process as well, as a means of communication between parties.

This example shows, albeit somewhat crudely, that it is possible to incorporate the results of a CBA in an MCA. Note, however, that the reverse is not true. In this respect, an MCA has more flexibility and can be used as a communication tool as well as a means of informing decision-making. As with CBA, there are numerous handbooks and manuals detailing how MCA can be applied. Indeed, there is even a *Journal of Multi-Criteria Analysis* (see the references for more sources).

10.8   A few practical points

At this point, most of the different methods of project appraisal that engineers will encounter have been presented. The importance of understanding where funding actually comes from has been discussed, the limitations that various promoters will face have been set out, and 'discounting' has been explored. However, few if any engineers will experience all of these different approaches within their working lives, and may well be asking themselves why their employer or client restricts itself to simple financial appraisal when more sophisticated techniques are clearly available.

10.8.1   Mandatory projects

First, it should be noted that engineers working on large 'one-off' projects, particularly those funded by international bodies such as the World Bank, will be expected to use the more complex techniques, and the use of these techniques will be led by specialists. The techniques will also be valid for major 'one-off' investments, such as new raw water reservoirs, by WSPs, nationalised industries and municipalities. This section, though, is more concerned with routine investment decisions such as reinforcement of distribution systems, sewerage systems and the construction of water or wastewater treatment plants: indeed, the investments which make up the majority of the work in a water undertaking's capital programme.

This issue has already been touched on by pointing out that, in the UK, the Environment Agency (EA) has decreed that long sea outfalls for the disposal of sewage in coastal regions are no longer acceptable, and so this option can no longer be considered. This is one aspect of an economic analysis that has been settled by the government through the EA. More correctly, it has actually been settled by EU legislation which decrees the standard of cleanliness required for bathing beaches, and the EA has the responsibility to ensure compliance in the UK. The EA has viewed that legislation, and concluded that the UK cannot comply by adopting long sea outfalls, and that at least secondary treatment is required. It is worth remembering that other countries that discharge to the Mediterranean and the North Sea have already reached this conclusion and have adopted higher standards of treatment to meet EU standards.

This impact of legislation goes much further. With distribution systems, the UK economic regulator Ofwat has set service standards that require a minimum water pressure at the stop tap of 15 m. Engineers working in the USA will be familiar with a similar restriction, although expressed differently, which sets minimum fire-fighting flows in distribution systems. With sewage works, the EA is charged with achieving the EU water quality standards within rivers. It therefore sets discharge consents, which can be appealed against, but, effectively, the river water has to achieve a pre-determined quality. So, in effect, a large part of what would be the economic analysis has already been determined by legislation and regulation. What the engineer has to do is find the most economic way of meeting these standards, and that generally reduces matters to a financial appraisal.

Reference has already been made to a House of Lords recommendation that the permitted level of leakage in the UK should take account of environmental as well as economic matters. While this is understandable, it introduces a degree of subjective assessments. For some, the value of the environment is beyond measure. For others, the exact opposite is the case, and the environment counts for little against the convenience for man. Of course, these are two ends of the spectrum, but there is a wide range for opinions, and therefore in cases such as what should be included and excluded from a leakage calculation, further guidance is required. For example, the UK government, through Ofwat, has tried to put a price on the carbon footprint, through a mechanism known as 'shadow pricing'. This has recently been replaced by a new concept, the 'non-traded price of carbon', which has approximately doubled the value of carbon. Doubtless, further changes will occur in the future, but these are simply attempts by the government to impose what it believes reflects acceptable or aspirational norms which should be used uniformly in project appraisal.

The consequences of this approach of defining much of what would traditionally be a major part of economic appraisal are very important. It effectively defines the form that appraisals will take. Since most of the variables are now defined, the task is to find ways of meeting a defined objective. Furthermore, the charging mechanisms for water are uniform within any water company or municipality area: they do not usually differentiate between high- and low-cost sources, but are averaged over the whole area served. This means that the concept of an investment producing different levels of return is not valid, and so it is the appraisals that determine the least cost for meeting specified outputs that are preferable. Effectively, this will be an NPV analysis, where the option with the lowest NPV will be selected. This assumes that all options meet the required outputs; options that do not should be rejected before the financial analysis is undertaken. So, although engineers will be correct in their belief that they are carrying out 'financial appraisal' in terms of the mechanics, in actual fact they will be taking account of economic factors through adherence to legislative and regulatory standards.

As has been hinted at already, there is often an argument that this type of analysis does not pay adequate attention to future costs. However, the NPV analysis must by its very nature include all capital plus operating and maintenance costs. For government bodies, nationalised industries and municipalities, the discount rate will be set by the government (the Treasury). For private WSP and other non-governmental bodies, it will be set by the company, usually requiring board approval, and will always be above the bank rate. It will be above the prevailing bank rate since a company will not be able to borrow money at the bank rate. In fact, it should be not less than the weighted average cost of capital, which is the average cost of borrowing and the dividend to shareholders. The details of fixing discount rates are beyond the scope of this book, nonetheless they do need to be considered a little further to understand the analysis of a second type of investment decision.

10.8.2 Optional projects

In the previous section, projects that the water undertakings are bound by law and regulation to carry out, and where different projects will not change the income stream (charges from customers), have been considered. But there are other projects put forward which are optional, and which do generate an income stream. An example might be the installation of a small hydroelectric scheme at a raw water reservoir, or a wind turbine on land owned by the utility. Any company will have a number of such opportunities, but it has no obligation to undertake any of them. It will therefore need to be satisfied that the project is worth investing in. To do this, it will normally use the IRR. One benefit of this rather complex financial analysis is that it lends itself to being an effective filter for such proposals, because the board can set a so-called 'hurdle rate', meaning that if the IRR does not exceed the hurdle rate, the proposal will not be implemented. If the hurdle rate is exceeded, then the highest IRR option is likely to be preferred, but not always. The total capital cost may also play a part – money is a scarce resource. However, if no option beats the hurdle rate, the proposal will fail. In setting the hurdle rate, the board will have to consider a number of issues. One dominant issue is that it will be using funds that could be applied to other optional projects, and therefore shareholders will expect a good return from any such investment. This alone means that the hurdle rate will usually be in excess of the discount rate adopted by the same entity for the NPV analysis of mandatory schemes. The board may also take the view that it should build in some form of risk premium against things going wrong, such as costs escalating, or the predicted income stream being less than forecast, and, again, this can be dealt with on a generic basis by simply raising the hurdle rate.

Many engineers find it difficult to accept that there should be different discount rates for projects, but in simple terms they are necessary to differentiate between obligatory projects and those where the project is using funds which could be used for other purposes, or indeed where the risks entailed in the project do not have to be taken. The approach to such schemes in both governmental bodies and nationalised industries depends largely on the availability of funds from the Treasury, rather than the commercial judgement of the board in a privatised utility.

10.8.3 Phasing of projects

There is one further issue that needs to be considered, and that is the optimum phasing of projects. In other words, what is the design horizon for any project?

It is tempting to believe that economies of scale will always dictate that long planning horizons should be adopted, and certainly this view has had its adherents in the water industry for many years, with those in the UK usually citing the benefits of the legacy of our Victorian predecessors. However, it is not a valid argument when viewed logically; indeed, this approach has sometimes been referred to as 'gold plating'. From the standpoint of an economist, the overprovision of facilities is a waste of financial resources at both a national level and in terms of an organisation's finances. It is very difficult to predict the future, particularly in terms of demand, and once a decision has been taken, there is only one scenario that will allow it to be proved correct, and any number which will prove it to be in error. The longer the planning horizon, the greater the degree of uncertainty and the more likely it is to be shown that the original assumptions made were wrong.

Ignore, for the moment, the possibility that the assumptions underpinning a proposed project could be incorrect, and consider the simple case of a pumping station which is planned for a 25 year future demand, where the demand is increasing linearly over that period. While pumps could be installed now that would meet the ultimate demand, that would mean they would be running at their theoretical duty just as they reached the end of their design life. Might it not be better, therefore, to install pumps that could meet the forecast in 12 years time? And then, when they are running at maximum capacity, consider what size of pumps would be required to supplement the output. This then leaves the alternatives of building the pumping station with space for extra pumps, or simply building a pumping station that can deal with forecast demand for the next 12 years.

To answer these questions, discounting techniques and, in particular, NPV analysis can be used. By carrying out an NPV analysis of the different options, the future value of money is taken account of, and the most attractive option determined. It is, of course, necessary to use different planning horizons, but generally it will be found that, for water industry projects, the optimum is between 10 and 15 years, but only for projects that can be phased sensibly. Clearly, waterworks and sewage works fall into that category, while other major works such as raw water reservoirs generally do not.

It is interesting that some years ago in the USA it was possible for municipalities to access federal funding for some water projects. This was cheap, and sometimes even in the form of a grant, with the result that 25 year planning horizons were the norm. When this source of funding ceased, and municipalities had to raise their own finance through issuing bonds, the same planning horizons were retained, often justified by the cost of issuing bonds, meaning that reducing planning horizons was not cost-effective. However, on analysis it could be shown that this argument was incorrect, and the discounted cash flow analysis still supported shorter planning horizons.

10.9 Additional aspects

There are other aspects which have a bearing on project appraisal but which might not in themselves be included in, for example, an NPV or CBA. These could include issues such as affordability, willingness to pay, and social and environmental costs. In the case of affordability, it is not the ability of the investors or project proposer to afford a project – it is assumed that they can, or can at least raise the necessary finance through various financial instruments if the project can be shown to be a good investment, as has already discussed: affordability here refers to the ability of recipients to pay for the benefits, goods or services that would be provided by a project. Affordability is usually a distributional and equity issue. In the case of water services, it may have a direct bearing on the design of how such goods and services are to be paid for. On the other hand, it might be argued that matters of equity and distribution are not the concerns of a service provider but rather of the welfare system. In this regard, the status of the service provider and the general national policies will have a bearing as to whether this is a significant issue that should be incorporated into an appraisal or not. In cases where service provision is through some form of public service, then affordability may be of somewhat more importance.

When determining the benefits that might arise from the implementation of a project, willingness to pay for those benefits may be a significant factor. The problem is that just because it is called 'willingness to pay' does not necessarily mean that the amounts are actually realised, unless it is used to inform the payment mechanism for those services. In the case of a WSP that receives government support, that support would, in turn, have to be recovered from some form of taxation. In this case, the willingness to pay could be translated into a willingness to pay higher taxes. In any event, this would not enter a financial analysis, as no direct transfer takes place, but it would be included in an economic analysis, as there is a net transfer within the economy as a result of the project. For projects that yield environmental and social benefits, the determination of willingness to pay would form an important part of the evaluation.

Just as social and environmental benefits can be important, so too are social and environmental costs that might arise through damage to parts of the environment, pollution, loss of biodiversity and other causes. The decision as to whether these are to be included will depend on who is carrying out an evaluation and what the legal and regulatory requirements are. As has already discussed above, such considerations rely heavily on the prevailing institutional arrangements within which water services operate. It is more than likely that environmental damages and other associated costs would have to be included in some form or other. It is to what extent and how this stacks up against other, often political and developmental, requirements that are of importance – aspects which engineers should be aware of, as they can exercise little control over them. How to go about the determination of environmental costs is also a subject in itself. As has been indicated, usually the regulatory bodies provide guidance on good practice. It is also true that requirements in this respect can change over time. An example of this is again the determination of the economic level of leakage. Not so long ago, the approach adopted by the UK water industry was criticised by the House of Lords. The Lords felt, along with some other bodies, that the limited inclusion of environmental factors had the perverse effect of allowing higher levels of leakage to take place than was socially acceptable. As a result, it was recommended that the regulatory bodies reconsider the methodology for determining the economic level of leakage, taking into account 'sustainability' factors. Prudence might therefore suggest a generous interpretation of the range of social and environmental factors that might be included in an appraisal.

10.9.1 Boundaries

The boundaries of an appraisal can be geographical as well as temporal and, as has been seen in the case of time span, the choice can have an influence on the outcome of an appraisal. Longer time spans together with low discount rates will tend to favour those alternatives that have higher initial costs but lower recurrent costs (e.g. operation and maintenance). It has also been pointed out that the appropriate time span should be that when the costs associated with the implementation of a project cease – in the case of a nuclear power plant this might be a very long time indeed, and might not be either feasible or acceptable to all parties.

In the application of WLC to pipe network systems, it has been argued that there is no design or useful life that is applicable. The reason for this is that any intervention effectively alters the system and its useful life. In this case, the approach has been to use a period of analysis which is of the order of 50 years. This is a period considered to be long enough to allow the effects of interventions to manifest themselves. With life-cycle analysis, the time span is taken as however long the period is 'from cradle to grave'.

Therefore, the appropriate time span depends on a variety of factors, and it should be made explicit within any appraisal on what basis the time span used has been chosen.

With respect to the geographical boundary, this should reflect the predominant area over which the proposed project will have an impact. This may well extend beyond the immediate surroundings of the project site, especially where there could be spill-over effects as a result of a project. Again, the choice will vary from project to project.

10.10 Final words

The appraisal of projects or alternatives is as much about the exercise of judgement as it is about the use and application of appraisal techniques. Each project is by its very nature unique, and presents new challenges. Those who carry out appraisals should have a blend of technical skills, an understanding of economics and accounting, and an appreciation for the environment and social issues. Increasingly, appraisals are going beyond being an end in themselves that supply answers which decision-makers take up. The process of carrying out an appraisal becomes an integral part of communication and shared understanding that can be used to inform better decision-making. Decisions are not always made on the basis of a formal appraisal, as there are other subjective factors that can influence choices and outcomes. One of the tasks of project appraisal is to try to make explicit what the consequences associated with the available options are.

10.11 Worked examples

The examples apply the discounted cash flow to find NPVs.

10.11.1 Example 1: the effect of operating costs on the choice of project

A client has three alternative options that would each deliver the same level of service.

- Option A would be to construct the works at a cost of £12 million, and the associated operating costs would be £0.250 million per year.
- Option B would be to construct the works at a cost of £7 million, and the associated operating costs would be £0.500 million per year.
- Option C would be to construct the works at a cost of £6 million plus upgrades costing £0.670 million every 5 years, and the associated operating costs would be £0.500 million per year.

Given that the evaluation horizon is 20 years and the discount rate to be used is 8%, which of the three options should be chosen? What would the effect of altering the discount rate to 10%?

See Table 10.13 for the results of the appraisal.

10.11.2 Example 2: the effect of phasing

A client has to build a new sewage treatment works, and is presented with two proposed options.

- Option A: build the whole works now at a cost of £24 million over a 3 year period.
- Option B: build half of the required works now at a cost of £15 million and the remainder in 12 year's time at a cost of £18 million. The construction period is 2 years for each period.

The discount rate is 8%.

Which option would you recommend? What would be the effect of reducing the discount rate to 5%?

See Table 10.14 for the results of the appraisal.

Table 10.13 Appraisal for example 1

Year	Option A				Option B				Option C			
	Capital cost	Operating cost	Discount factor	NPV	Capital cost	Operating cost	Discount factor	NPV	Capital cost	Operating cost	Discount factor	NPV
0	12.000		1.000	12.000	7.000		1.000	7.000	6.000		1.000	6.000
1		0.25	0.926	0.231		0.5	0.926	0.463		0.5	0.926	0.463
2		0.25	0.857	0.214		0.5	0.857	0.429		0.5	0.857	0.429
3		0.25	0.794	0.198		0.5	0.794	0.397		0.5	0.794	0.397
4		0.25	0.735	0.184		0.5	0.735	0.368		0.5	0.735	0.368
5		0.25	0.681	0.170		0.5	0.681	0.340	0.670	0.5	0.681	0.796
6		0.25	0.630	0.158		0.5	0.630	0.315		0.5	0.630	0.315
7		0.25	0.583	0.146		0.5	0.583	0.292		0.5	0.583	0.292
8		0.25	0.540	0.135		0.5	0.540	0.270		0.5	0.540	0.270
9		0.25	0.500	0.125		0.5	0.500	0.250		0.5	0.500	0.250
10		0.25	0.463	0.116		0.5	0.463	0.232	0.670	0.5	0.463	0.542
11		0.25	0.429	0.107		0.5	0.429	0.214		0.5	0.429	0.214
12		0.25	0.397	0.099		0.5	0.397	0.199		0.5	0.397	0.199
13		0.25	0.368	0.092		0.5	0.368	0.184		0.5	0.368	0.184
14		0.25	0.340	0.085		0.5	0.340	0.170		0.5	0.340	0.170
15		0.25	0.315	0.079		0.5	0.315	0.158	0.670	0.5	0.315	0.369
16		0.25	0.292	0.073		0.5	0.292	0.146		0.5	0.292	0.146
17		0.25	0.270	0.068		0.5	0.270	0.135		0.5	0.270	0.135
18		0.25	0.250	0.063		0.5	0.250	0.125		0.5	0.250	0.125
19		0.25	0.232	0.058		0.5	0.232	0.116		0.5	0.232	0.116
20		0.25	0.215	0.054		0.5	0.215	0.107	0.670	0.5	0.215	0.251
Totals	12.000	5.000		14.455	7.000	10.000	0.329	11.909	8.680	10.000		12.030
Discount rate	8%											

Table 10.14 Appraisal for example 2

Year	Discount factor	Option 1		Option 2	
		Capital cost	NPV	Capital cost	NPV
0	1.000	8.000	8.000	7.500	7.500
1	0.926	8.000	7.407	7.500	6.944
2	0.857	8.000	6.859	0.000	0.000
3	0.794	0.000	0.000	0.000	0.000
4	0.735	0.000	0.000	0.000	0.000
5	0.681	0.000	0.000	0.000	0.000
6	0.630	0.000	0.000	0.000	0.000
7	0.583	0.000	0.000	0.000	0.000
8	0.540	0.000	0.000	0.000	0.000
9	0.500	0.000	0.000	0.000	0.000
10	0.463	0.000	0.000	0.000	0.000
11	0.429	0.000	0.000	9.000	3.860
12	0.397	0.000	0.000	9.000	3.574
Total		24.000	22.266	33.000	21.878
Discount rate		8%			

10.12 Further information

The European Commission has an online resource for the evaluation of socio-economic development, providing guides and source books for practitioners. It includes sections on CEA and CBA. See http://ec.europa.eu/regional_policy/sources/docgener/evaluation/evalsed/index_en.htm.

The World Bank IFC is working on a cost–benefit tool to facilitate project reviews prior to approval, benchmarking/cross-project comparisons, and project monitoring during project implementation. See www.ifc.org/ifcext/rmas.nsf/Content/CostBenefitAnalysis.

REFERENCES

Chenery, C. (1984) Whole life cost of construction – informal discussion. *Proceedings of the Institution of Civil Engineers* 76: 822–825.
Drummond, M. F., Sculpher, M. J., Torrance, G. W., O'Brian, B. J. and Stoddart, G. L. (2006) *Methods for the Economic Evaluation of Health Care Programmes*, 3rd edn. Oxford: Oxford University Press.
Emblemsvag, J. (2001) Activity-based life cycle costing. *Managerial Auditing Journal* 16(1): 17–27.
EPA (1996) *An Introduction to Environmental Accounting as a Business Tool: Key Concepts and Terms*. Washington, DC: Environmental Protection Agency.
Figueira, J., Greco, S. and Ehrgott, M. (2005) *Multiple Criteria Decision Analysis: State of the Art Surveys*. New York: Springer Science and Business Media.

Green, C. (2003) *Handbook of Water Economics Principle and Practice*. Chichester: Wiley.

HM Treasury (2003) *The Green Book: Appraisal and Evaluation in Government. Treasury Guidance*. London: TSO.

Innes, J. and Mitchell, F. (1995) A survey of activity based costing in the UK's large companies. *Management Accounting Review* 6(2): 137–153.

McCaffrey, J. (2005) The analytical hierarchy process. *Microsoft Developer's Network Magazine* 20(6): 139–144.

National Audit Office (2006) *The Modernisation of the West Coast Main Line: NAO Report HC 22 2006–2007*. London: TSO.

OECD (2004) *Financing Water and Environmental Infrastructure for All: Some Key Issues*. Paris: OECD.

Perman, R., Ma, Y., McGilvray, J. and Common, M. (2003) *Natural Resources and Environmental Economics*, 3rd edn. Harlow: Pearson Education.

Saaty, T. (2005) *Theory and Application of the Analytic Network Process: Decision Making with Benefits, Opportunities, Costs and Risks*. Pittsburg, PA: RWS Publications.

Skipworth, P., Engelhardt, M., Cashman, A., Savić, D., Saul, A. and Walters, G. (2002) *Whole Life Cost Accounting for Water Distribution Network Management*. London: Thomas Telford.

Stern, N. (2006) *The Economics of Climate Change: Stern Review*. London: TSO. www.sternreview.org.uk [accessed 01.09.10].

Stiglitz, J. E. (1994) Discount rates: the rate of discount for benefit–cost analysis and the theory of second best. In: Layard, R. and Glasiter, S. (eds), *Cost–benefit Analysis*, 2nd edn, pp. 116–159. Cambridge: Cambridge University Press.

Tam, C., Tong, T. and Chui, G. (2006) Comparing non-structural fuzzy decision support system and analytical hierarchy process in decision-making for construction problems. *European Journal of Operational Research* 174(2): 1317–1324.

Winpenny, J. (2003) *Financing Water for All. Report of the World Panel on Financing Water Infrastructure*. Marseille: World Water Council. www.financingwaterforall.org [accessed 01.09.10].

REFERENCED STANDARDS

ISO 14000 Series: Environmental Management Systems. Geneva: ISO.

ISO 15686: Buildings and Constructed Assets – Service Life Planning. Geneva: ISO.

Water Distribution Systems
ISBN: 978-0-7277-4112-7

ICE Publishing: All rights reserved
doi: 10.1680/wds.41127.309

ice
Institution of Civil Engineers

publishing

Chapter 11
Sustainability and climate change

Paul Jowitt SISTech Ltd, Heriot-Watt University, Edinburgh, UK
Adrian Johnson MWH, Edinburgh, UK

11.1   Introduction
11.1.1   Sustainable water supplies – worth fighting for?

The art of engineering has traditionally been based on a successful combination of the rigorous application of technical rationality and engineering science, its practical application, and the driving force of an economic imperative. That is no longer enough. There are other phenomenological, social and environmental impacts, and factors that need to be anticipated and resolved. This is particularly true for the development and management of water resources and associated water supply systems, where conflicts often arise, and which cannot be resolved by scientific analysis alone. Even where the facts are accepted, their interpretation and consequences are often disputed. The sustainable development of water resources depends on the combination of sound science and effective conflict resolution in the face of divergent objectives and the need to balance costs and benefits between a number of different stakeholders.

This chapter is not concerned with particular techniques – which are adequately described elsewhere in this book – but with some of the larger global and systems level issues that need to be tackled at the larger scale.

Increasingly, real and emerging water management problems have impacts that either need to account for (non-commensurate) socio-enviro-economic effects, or which accommodate large-scale heterogeneity, and sometimes both. They require trans-disciplinary and/or large-scale systems models which cannot be resolved by scientific rationality alone but require the exercise of choice, the interaction of a diverse set of stakeholders and a much wider understanding of impacts.

The primacy of science, engineering and technology as the key drivers to fulfil an economic imperative is giving way to the primacy of socio-enviro-economic drivers supported by scientific and engineering knowledge of what is possible and, most critically, what might be desirable.

11.1.2   The fundamentals

At the most simple level, a sustainable water supply system is one that supplies anticipated (and reasonable) demands over a sensible time horizon without degradation of the source of the supply or other elements of the system's environment. This requires

Figure 11.1 Letter printed in *The Times* newspaper, 14 June 1991

Water shortages
From Mr R. Grant Paton

Sir, Amid the constant concern over water shortages, there appears to be a fallacy. Admittedly water is consumed in increasingly large quantities, but surely the world enjoys a closed system. Certainly human consumption of water is quickly returned whence it cames. Where therefore does all our water go?

I am, yours faithfully,

R. GRANT PATON

that the physical boundaries of the water supply system have to be drawn wide enough to ensure that $\sum (\text{inputs} - \text{outputs}) = 0$ over time.

At the global scale, the governing activity is the hydrological cycle, in which the supply system is an artificial component, but a component nonetheless. And just as with the hydrological cycle, the water supply system is rarely in a steady state. Water demand varies diurnally and seasonally. The availability of water at the source also varies, and it may be that supply and demand are sometimes out of balance. For the most part, these imbalances can be smoothed out by reservoir storage at the source, storage at the water treatment plant, storage within the distribution systems (service reservoirs), storage at the point of use (cisterns) or transfers from an adjacent system.

Occasionally, the imbalance may lead to water restrictions on some or all users. Inevitably, this causes adverse comment from the users, some of which often finds its way into the media. A letter to *The Times* on 14 June 1991 (Figure 11.1) shows one such example. What the writer of this letter has failed to realise is that elements of the hydrological cycle are rate limited. Aquifers can only be replenished at a certain rate; river channels have a finite capacity; and evaporation rates from the oceans are limited. And with the water supply and distribution system itself, the pipe work has a finite capacity, pumps have limited outputs and water treatment plants have a maximum throughput.

A water distribution system can fail in one of three ways:

1. source failure – insufficient water at the point of supply
2. demand failure – insufficient capacity in the treatment or distribution system to meet transient demands
3. system failure – a localised failure of a component of the treatment or distribution system, such as a pipe or pump.

In terms of sustainability, these factors extend the boundaries of the water distribution system to include its physical assets, their maintenance and renewal, and, therefore, the need for a sufficient and stable system of financing it, irrespective of the precise nature of ownership.

The boundaries of the water distribution system therefore extend to the socio-economic realm, to include the relationship between consumers (domestic and business) and the water service provider (WSP) responsible for operating, maintaining and investing in the water supply system. This relationship is usually transacted through water charges, which brings with it the establishment of appropriate and equitable principles of charging. The sustainability issues affecting a water distribution and supply system therefore very quickly encompass the environment, society, economics and finance, and business.

11.1.3 The need for change
So, the focus of engineering development, including water supply engineering, has changed from technical and cost efficiency to having to address multiple and often divergent social, environmental and economic objectives.

A range of drivers have emerged from socio-political constructs of sustainable development, including the needs for social inclusion and social justice, to avoid environmental degradation and, most recently, to address projected climate change. The problem is that decision-making, whether at policy level or at the level of individual (water supply) projects, can become mired in an amalgam of conflicting objectives. The continued rise of carbon emissions from the water industry in the face of continued investment to improve service to customers and ensure environmental protection is a pertinent example of this problem.

A strong framework for assessing sustainability will assist in agreeing objectives, making choices that are socially acceptable and incorporate effective measurement of progress in achieving agreed objectives.

While setting objectives and success criteria is important, such a framework also needs to recognise and address problems in the existing development paradigms. Some of these issues are explored in this chapter, specifically:

- End-of-pipe approaches may deliver certainty of outcome to narrowly defined 'engineering' problems, but this will not deliver an overall affordable, carbon-efficient water service that meets the needs of customers and the environment.
- The balance between water as a basic human right and as an economic good does not properly reflect social need or encourage the right customer behaviour.
- The dominant use of net present value to determine whole-life costs is inconsistent with achieving future resilience and equity, which must be inherent in a sustainable approach.
- Given the complex interaction of natural and social factors, insufficient attention is being given to understanding the variety of risks to water supply, with the consequences of more frequent and severe failures and loss of water security.

It is argued that what is needed is the embedding of sustainability principles within a strategic systems approach that encourages innovation through all stages of the asset life cycle, to address the various competing pressures, at local and regional levels, and to deliver a transformed, secure, yet affordable water supply service.

Earlier chapters are largely concerned with techniques applied by individual engineers as they go about their project work; by contrast, this chapter looks at larger system level issues that need to be tackled to ensure sustainable outcomes. It is recognised that in many cases it will be beyond the purview of the individual practitioner to influence these issues, but collectively we all have a responsibility to work within our spheres of influence – whether at project or policy level – to help challenge the status quo.

11.2 Sustainability drivers

Sustainable development has become one of the key axioms of modern times, not just of committed environmentalists but widely adopted across the political, business and social spectra. What are the roots of sustainable development? It is most closely associated with the 1992 Rio Summit (United Nations Conference on Environment and Development (UNCED), 'The Earth Summit') and encapsulated in Gro Harlem Brundtland's definition (World Commission on Environment and Development (WCED), 1987) as 'Development that meets the needs of the present without compromising the ability of future generations to meet their own needs'.

The roots of the Brundtland definition (WCED, 1987) can be traced back at least to Jeremy Bentham's (1789) dictat that we should seek 'the greatest good for the greatest number'. And even the notion of intergenerational equity – so central to the Brundtland definition – was made explicit by American politician and forester Gifford Pinchot's (1947) addition of the phrase 'for the longest time' to Bentham's imperative.

But in more recent times, the concept of sustainable development has its roots in works such as Garret Hardin's paper 'Tragedy of the commons' (Hardin, 1968), the Club of Rome's *Limits to Growth* report (Meadows *et al.*, 1972) and the UN's Conference on the Human Environment, held in Stockholm in 1972. It was through the WCED in 1987 and its report *Our Common Future* (WCED, 1987) that it gathered momentum as the means to bridge the gulf between advocates and critics of progress through economic growth. Since then, 'sustainable development' has been adopted in countless hues as a philosophy and a policy tool by individual nations and agencies.

But it is the Brundtland definition (WCED, 1987) that has come to embody what is, in essence, a very simple idea: ensuring a better quality of life for everyone, now and for generations to come.

The four key objectives which underpin sustainable development are often taken to be:

1. social progress which recognises the needs of everyone
2. effective protection of the environment
3. prudent use of natural resources
4. maintenance of high and stable levels of economic growth and employment.

For convenience, these are often reduced to the three headings of:

1. social inclusion
2. environmental protection
3. economic well-being.

But the classification of the dimensions of sustainable development does not prescribe precisely how a particular set of policy or decision options are adjudged. It simply defines the need to adopt a holistic approach and identify the broad criteria that should be used to discriminate between particular choices.

While much has been written on what the terms 'sustainable development' and 'sustainability' really mean, if nothing else, it has become a critique of the status quo (Gibson *et al.*, 2005). The WCED and subsequent UN environment conferences, the reports of the Intergovernmental Panel on Climate Change (2007) and the Millennium Ecosystem Assessment (2005) collectively reinforced the conclusion that present practices of human development are not sustainable and need to change.

There is no single 'one size fits all' model, and to be practical it must be tailored to suit the level and sector of application.

In its 2005 strategy (Defra, 2005), the UK government set out some shared principles for achieving sustainable development at a national level, which underpin a set of priority areas for action and a basket of sustainability indicators – that is, a group of measures by which progress in meeting the objectives is monitored. In short, to live within environmental limits and achieve a just society, we will need a sustainable economy, good governance and sound science (Figure 11.2).

Similar principles can be set for water supply. As discussed above, the endeavour to abstract, treat and distribute water for society's use is an intrinsic part of the natural water cycle, albeit now grossly modified by centuries of engineering.

The EU Water Framework Directive (2000), which requires the development and implementation of river basin management plans to achieve 'good status' of water bodies, has the potential to transform the management of water at a catchment level. Rather than prescribing standards, the Directive encourages the involvement of interested parties to determine objectives for individual water bodies. The achievement of these objectives depends on deploying measures – such as reducing abstraction for public water supply to restore river flows – that are deemed to be cost-effective both in terms of financial cost and wider environmental cost. Less-stringent objectives may be admitted (in the short term at least) if the costs of achieving good status (or good potential in the case of modified water bodies) are judged to be disproportionate.

The UK government has also set out its vision for a sustainable water sector for England in *Future Water* (Defra, 2008) (Box 11.1). The vision reflects the need to take a balanced approach to meeting the needs of society and the environment; progress

313

Figure 11.2 Shared principles for sustainable development

Based on Defra (2005). © Crown Copyright, reproduced with permission of the Controller of Her Majesty's Stationery Office

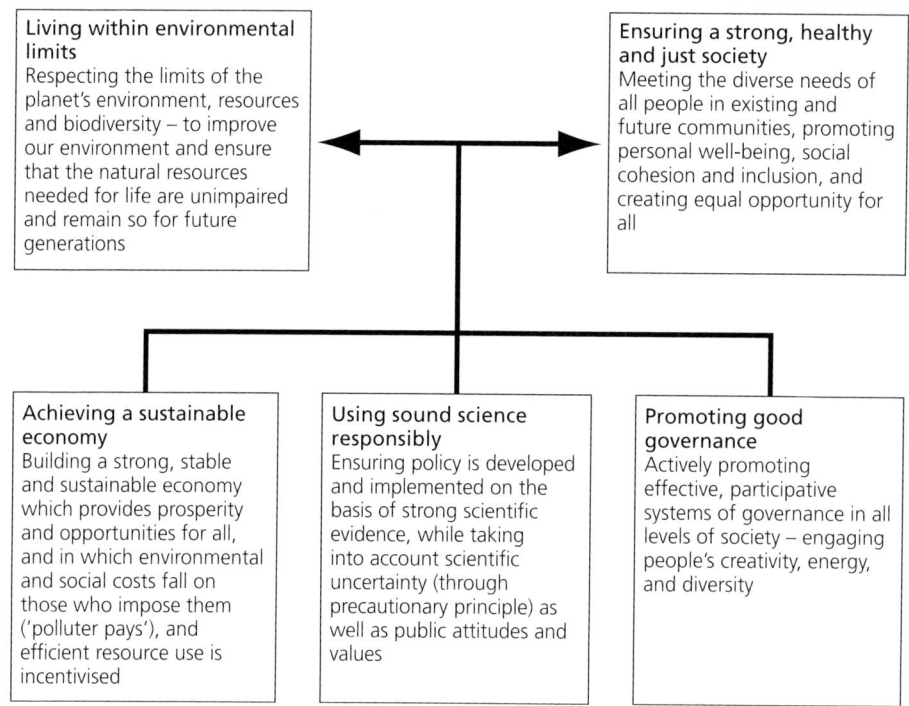

Living within environmental limits Respecting the limits of the planet's environment, resources and biodiversity – to improve our environment and ensure that the natural resources needed for life are unimpaired and remain so for future generations	**Ensuring a strong, healthy and just society** Meeting the diverse needs of all people in existing and future communities, promoting personal well-being, social cohesion and inclusion, and creating equal opportunity for all

Achieving a sustainable economy Building a strong, stable and sustainable economy which provides prosperity and opportunities for all, and in which environmental and social costs fall on those who impose them ('polluter pays'), and efficient resource use is incentivised	**Using sound science responsibly** Ensuring policy is developed and implemented on the basis of strong scientific evidence, while taking into account scientific uncertainty (through precautionary principle) as well as public attitudes and values	**Promoting good governance** Actively promoting effective, participative systems of governance in all levels of society – engaging people's creativity, energy, and diversity

---

Box 11.1 *Future Water* – a vision for sustainable water supply and improved water environment

'*Future Water* sets out how we want the water sector to look by 2030, and some of the steps we will need to take to get there. It is a vision where rivers, canals, lakes and seas have improved for people and wildlife, with benefits for angling, boating and other recreational activities, and where we continue to provide excellent quality drinking water. It is a vision of a sector that values and protects its water resources; that delivers water to customers through fair, affordable and cost-reflective charges; where flood risk is addressed with markedly greater understanding and use of good surface water management; and where the water industry has cut its greenhouse gas emissions. The vision shows a sector that is resilient to climate change, with its likelihood of more frequent droughts as well as floods, and to population growth, with forward planning fully in tune with these adaptation challenges.'

From Defra (2008), p. 8. © Crown Copyright, reproduced with the permission of the Controller of Her Majesty's Stationery Office

in achieving its objectives is being measured through the use of a suite of measurable indicators.

11.3   Climate change

In recent years, the huge amount of scientific research and modelling, in particular reporting of the Intergovernmental Panel on Climate Change (IPCC, 2007), together with other socio-economic research such as that by Lord Stern on the economics of climate change (Stern, 2006), has meant that climate change and its potential impacts have come to dominate the global environmental sustainability debate. Figure 11.3 is a representation of the Keeling curve showing the increase in atmospheric concentration of carbon dioxide (CO_2) over the last 50 years.

According to MacKay (2008), 'The consensus of the best climate models seems to be that doubling the CO_2 concentration would have roughly the same effect as increasing the intensity of the sun by 2%, and would bump up the global mean temperature by something like 3°C.' A 3°C temperature rise will lead to very significant impacts such as the release of methane from permafrost.

While global consensus on the extent and speed of future climate change and the allocation of responsibility for doing something about it remains elusive, humanity is

Figure 11.3 History of atmospheric CO_2 concentrations measured at Mauna Loa, Hawaii

Figure created by R. A. Rohde from NOAA ESRL published data (ftp://ftp.cmdl.noaa.gov/ccg/co2/trends/co2_mm_mlo.txt) and incorporated into the Global Warming Art project (www.globalwarmingart.com). Licensed for use under the Creative Commons Attribution ShareAlike 3.0 License

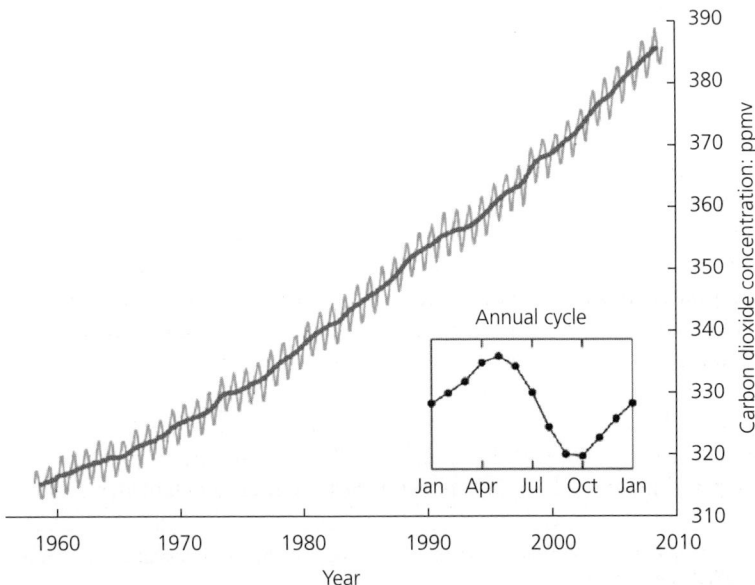

undeniably increasing the CO_2 concentration in the atmosphere and in the surface oceans through burning of fossil fuels and other activities.

Despite the lack of a global consensus, countries across the world are making commitments, devising mechanisms for reducing greenhouse gas (GHG) emissions and making plans to adapt to its impacts. Through its Climate Change Act (2008), the UK has committed to ambitious targets for reducing GHG emissions: an overall reduction of 80% on 1990 levels by 2050 plus interim targets. Public bodies and statutory undertakers are also required to assess risks from climate change to their operations and to report their plans for adapting to those risks.

As far as the water sector is concerned, environment agencies, water suppliers and their regulators are all now engaged in assessing the implications of climate change and how this could, and indeed should, affect the way in which services are provided in future.

According to Defra (2008), the UK water industry contributes less than 1% to the UK's inventory of GHG emissions, but 'there is a real risk that this will rise with water demand and more ambitious standards for water quality in the natural environment' (p. 11). As shown by Figure 11.4, the historic trend has been for continued increases in emissions. Given the continued need for water suppliers to invest to meet new legislation, deal

Figure 11.4 The carbon reduction challenge for UK water companies (AMP, asset management plan)

Adapted from Ainger (2010)

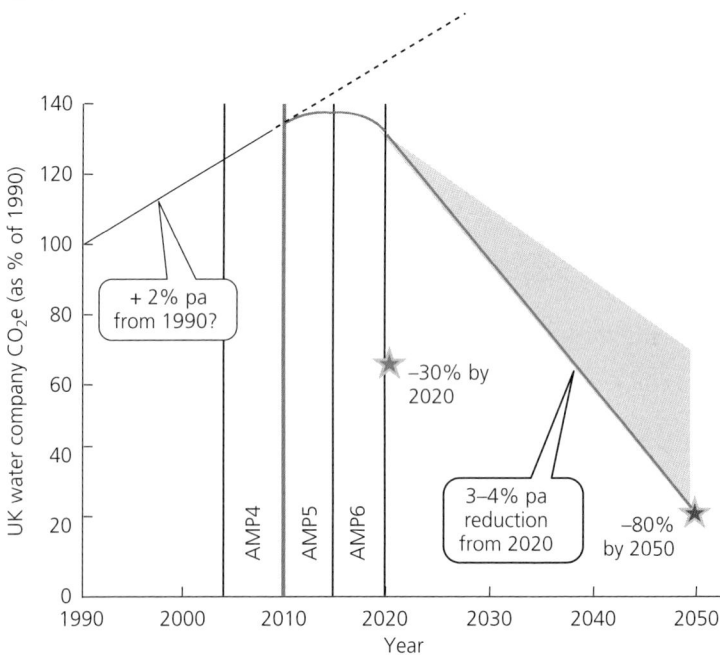

with growth in demand and to maintain deteriorating assets, the industry faces a formidable challenge to play its part in helping to reduce emissions.

But reducing emissions is only half of the story: water supply systems are of course particularly vulnerable to the impacts of climate change. In terms of assessing and adapting to such impacts, there remains significant uncertainty both within climate change projections and their application to water resources management for public water supply. Publication of the UKCP09 (Jenkins *et al.*, 2009) climate projections provides opportunity to understand climate change risks better. At the highest level, impacts on operations and assets could arise from:

- changes in temperature (annual and seasonal)
- change in precipitation patterns and evaporation rates (annual and seasonal)
- changes in the frequency and severity of extreme precipitation events
- changes in the duration, intensity and frequency of droughts
- changes in storm tracks (depressions)
- changes in sea level.

In some cases, climate change will directly affect water operations and/or assets (e.g. temperature changes affecting the efficiency of treatment processes), whereas in others it will lead to changes in human behaviour or environmental conditions that, in turn, will affect operations and/or assets (e.g. changes in population leading to changes in water demand). So, there is chain of potential first-, second-, third-, etc., order effects, which need to be considered (UKWIR, 2007; Water UK and MWH, 2007).

Second- and third-order effects could include changes in parameters such as:

- water demand (domestic and non-domestic)
- river flows and groundwater levels
- river flood levels
- algal blooms and odour nuisance
- changes in soil moisture
- efficiency of biological processes
- water quality
- saline intrusion.

These types of effects have the potential to variously impact water operations and assets in terms of:

- operational efficiency
- performance in terms of water quality compliance and flooding frequency/severity
- asset deterioration
- asset failure.

Of course, there may also be interactions between different effects on individual assets or groups of assets, which also need to be systematically evaluated.

317

Figure 11.5 Carbon emissions from the water cycle

Data from Environment Agency (2009)

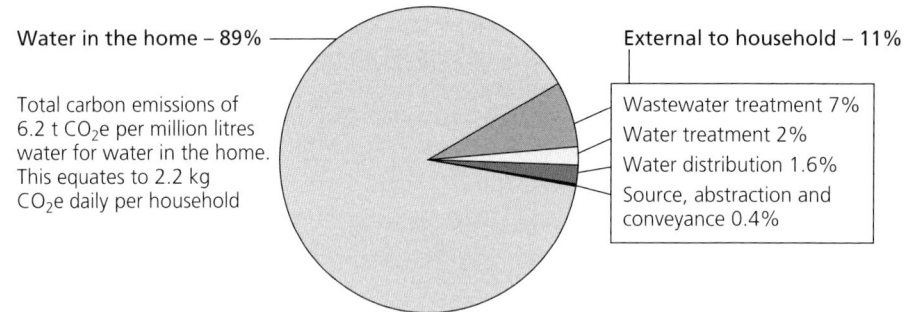

Water in the home – 89%

Total carbon emissions of
6.2 t CO_2e per million litres
water for water in the home.
This equates to 2.2 kg
CO_2e daily per household

External to household – 11%

Wastewater treatment 7%
Water treatment 2%
Water distribution 1.6%
Source, abstraction and
conveyance 0.4%

The twin challenges of reducing carbon emissions and maintaining resilience to climate change impacts, while meeting society's need for a wholesome and affordable water supply, demand a long-term view involving what future services and the assets to support them look like and working out how to get there. The UK Environment Agency (2009) has set out a range of objectives for next 20–40 years in its water resources strategy *Water for People and Environment* for addressing climate change, while protecting water resources and ensuring water is properly valued. As shown in Figure 11.5, nearly 90% of the emissions from the human water cycle arise from the use of water in the home (mainly from water heating for cooking and washing), and are additional to those arising from the operational activities of the water service providers. Thus, if a major shift in the way services are provided could be achieved – specifically to better manage the demand and use of water – this would have the most significant impact.

11.4 Assessing the environmental, social and economic sustainability of water supply projects

While vision documents such as Defra's (2008) *Future Water* set policy and advocate a range of measures for sustainable outcomes, including addressing climate change, the danger is that decision-making becomes mired in an amalgam of competing objectives. What is needed for infrastructure projects is a structured framework for 'sustainability assessment' that practitioners and decision-makers can use to assess alternative courses of action against a range of objective attributes and stakeholder choice criteria.

While there are now many variants of sustainability assessment, a successful approach will ensure that:

■ relevant physical, economic, social, cultural, environmental, etc., factors are identified at an early stage of project development, and analysed effectively
■ a set of sustainability objectives is adopted which reflect both the primary need for water development and the relevant national/local policies, plans and issues of concern to stakeholders

Figure 11.6 Typical stages involved in sustainability assessment

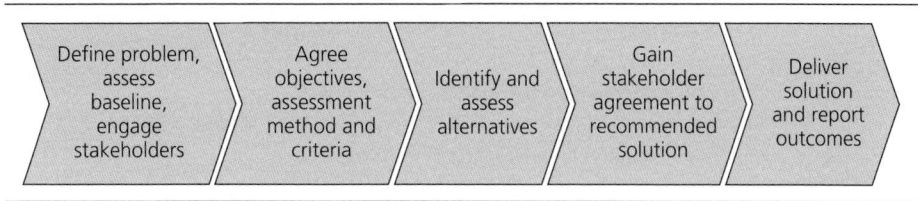

| Define problem, assess baseline, engage stakeholders | Agree objectives, assessment method and criteria | Identify and assess alternatives | Gain stakeholder agreement to recommended solution | Deliver solution and report outcomes |

- a suitable (perhaps radical) range of development alternatives are identified and assessed in terms of their ability to meet the sustainability objectives
- stakeholders, statutory and non-statutory, are engaged from the outset and throughout the process of project development and decision-making
- selected solutions are resilient and adaptable to future environmental and social change
- progress of the chosen solution in meeting the objectives is effectively measured and reported.

Strategic environmental assessment (SEA) is now widely used in Europe to enhance the sustainability of plans and programmes, in accordance with the EU SEA Directive (2001). A fundamental tenet of SEA is that it forms an integral part of the process of plan-making. In the UK, SEA is now being extended to an 'appraisal of sustainability' (AoS) in which environmental topics – such as biodiversity, soil, water, air and climatic factors – normally considered in an SEA are considered alongside social and economic factors identified by stakeholders and in other relevant policies and plans.

Of course, sustainability principles should be applied not only at a strategic level to meet a legislative requirement but routinely at all levels down to the individual development project. The typical stages of such an approach are shown in Figure 11.6.

In fact, the water industry economic regulator, Ofwat, which itself has a sustainable development duty, now requires water companies to include assessments of social and environmental impacts in their proposals for capital investment. The challenge for the water practitioner is to ensure that such assessments are carried out transparently and so as to deliver the best solutions.

Assessment criteria will differ from project to project, depending on local pressures and stakeholder concerns. Importantly, the basis for selecting indicators should focus on what should be measured as well as the practicality of what can be measured. Some typical headings and factors relevant to water supply projects are presented in Table 11.1.

Various different techniques can be used for assessing the sustainability of alternative interventions. For example:

- Cost–benefit analysis (CBA) is used to take account of social, environmental and economic effects alongside financial costs, by assigning monetary values to costs

Table 11.1 Typical factors that may be considered in sustainability assessment

Heading	Example factors
Environment	Effect of abstraction on the water environment Resource efficiency (carbon emissions, energy, water, sludge, waste) Other local impacts (e.g. biodiversity impacts)
Economic	Costs – initial, recurring (operational, capital maintenance) Revenue – e.g. renewable obligation certificates Affordability Community wealth generation, employment
Social	Level of service, supply interruptions Community acceptability – e.g. odour, noise, visual impact, traffic movements, health and safety Population health Amenity benefit
Risk	Performance risk, financial risk
Practicality	Technical feasibility Physical and planning constraints
Adaptability	Legislative change Climate change impacts – increased flood risk, changes in water demand, changes in receiving water quality, etc. Future growth and development plans, construction phasing

and benefits. Although increasingly used by the water industry, CBA has short-comings – for example, selecting a comprehensive set of benefits and costs can be difficult, and the values assigned to individual benefits and costs are often subject to stakeholder bias – and so is not always best suited to complex decisions involving multiple stakeholders (Ackermann, 2008).

■ In contrast, qualitative techniques using look-up tables, graduated scorecards and/or expert judgement are commonly used in SEA and AoS. Sustainability threshold analysis is a development of this approach in which alternatives are first assessed against criteria, which have been assigned a threshold value. Where a threshold is not met, this is used as a veto to prevent unsuitable interventions from being taken forward. Such techniques provide a high level of transparency, but some effort is required to combine quantitative and qualitative information in a clear overall assessment.

■ Multi-criteria analysis (MCA) is a third type of approach developed for complex problems involving multiple stakeholders. Defra uses a form of MCA for flood risk management, where it provides the basis for scoring and weighting of different economic, environmental and social impacts. MCA was incorporated into UKWIR's 'best practicable sustainable option' (BPSO) methodology (UKWIR, 2004).

The approach most suited to a particular water supply project may be a hybrid of these approaches, and, depending on the scale/importance of the project, developed to an appropriate level of detail. But whatever method is adopted, it is vital that decision-making does not become a mechanistic process but, rather, is an integral part of the overall appraisal framework. Ideally, the assessment process and timeline should incorporate the opportunity to revisit and represent appraisals of particular interventions, to take account of the outcome of stakeholder discussions and wider consultation.

Simultaneous improvement in all three of the usual pillars of sustainability – the economy, the environment and society – is not generally possible through the lens of one stakeholder, never mind several. And even within a single sustainability pillar there can be conflicting outcomes, as outlined in the following sections.

11.4.1 The environmental dimension

An example of the conflict between water objectives and other objectives, described as the environmental 'Catch-22' is shown in Figure 11.7 and discussed below.

The progressive tightening of drinking water quality standards and emission standards required to comply with the Water Framework Directive (and its daughters/precursors) across the UK is still driving the need to re-examine conventional measures to deliver water quality improvements. The repeating 5 year cycle of water industry investment

Figure 11.7 The environmental 'Catch-22'

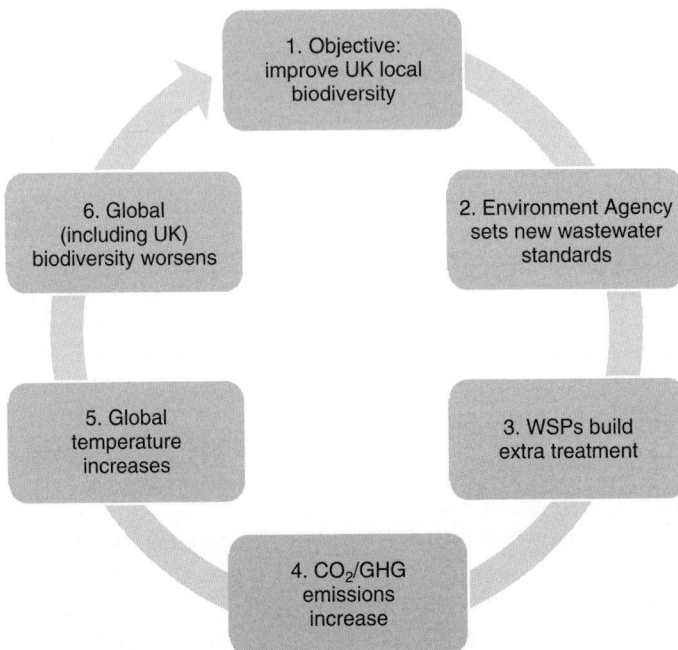

in ever-more energy-intensive processes to deliver incremental improvements in water quality – in drinking water and wastewater discharges – has now led the industry to question whether this is appropriate, as evidence mounts that resources are diminishing and global temperatures rising as a result of raised atmospheric CO_2 levels.

The concept of 'sustainable regulation' suggests there are alternatives to current solution paradigms which can minimise the whole impact on the environment of a particular activity, in this case providing a cost-effective water supply and wastewater treatment system while ensuring the highest customer service. The challenge is to identify alternatives that maintain the primary ethos of maintaining or improving environmental quality while also taking external environmental and societal impacts into account.

Catchment planning should recognise the multiple pressures within each catchment and look for innovative approaches to provide the best service outcome for the whole catchment. This would require a shift in the current regulatory framework. Basic tenets of environmental protection in the UK such as the concept of 'no deterioration' would require investigation to include all emissions, not just those to the aquatic environment. We also need to understand how liability is defined proportional to the source of the environmental emission load.

We need to apply a balance between H_2O and CO_2. This may include development of new models and extending current understanding of the science to establish the benefits of this approach.

As articulated by the Aldersgate Group (2010), 'Resource efficiency will be one of the key determinants of economic success and human well-being in the 21st century'. To whatever degree that anthropogenic greenhouse gas emissions are considered as contributing to climate change, it is clear that resources are diminishing. This is a key sustainability challenge. Energy costs from fossil fuels will not lessen (quite the opposite). The UK is predicted to be the most densely populated country in the EU by 2050, and the southeast of the UK has less available water per capita than some North African countries. At the same time, the significant tightening of emission standards required to comply with the EU Water Framework Directive (2000), Urban Wastewater Treatment Directive (1991), Habitats (1992) and Birds (2009) Directives, Bathing Water Directive (2006), Priority Substances Directive (2008), Freshwater Fish Directive (2006) and Groundwater Directives (2006) across the UK is resulting in significant increases in energy use within the water industry. Asset life means those assets will continue to operate at high levels of energy use unless we can produce alternative operating regimes, cost-effective retro-fitting or an economic model which incentivises rebuilding with less energy-intensive processes. Again, this raises some interesting issues – energy-intensive short-term compliance based on net present costs versus sustainability based on long-term values and which include proper carbon costing.

Similar conflicts can arise with rural water supplies, where the only alternative to upgrading small treatment systems is to employ advanced, energy-intensive treatment systems, often consolidating several sources into a larger one to achieve economies of scale in terms of treatment costs but with additional costs of distribution. But these are not the only effects.

Selecting the treatment process itself is not straightforward. In upland locations, the treatment process might also have to cope with rapidly varying water quality from peaty soils. In particular, colour and organics removal may be required to limit the formation of trihalomethanes, carcinogenic by-products of the disinfection process. The choice of process would also impact on the abstraction requirements, depending on the quantity of wastewater produced. If the process rejects a large quantity of water, then more must be abstracted initially.

For example, two typical options for small-scale water treatment in upland locations are nanofiltration membrane plants and sand filtration with granular activated carbon (GAC) contactors. The membrane plants offer a higher degree of security of achieving the required final water quality when the source waters have particularly high colours, but produce relatively large volumes of wastewater (up to 35% of the raw water abstracted). Sand filtration and GAC have lower wastewater requirements but may not provide the required colour and organics removal for the poorer-quality sources. In these processes the volume of wastewater produced is as low as 2% of the raw water.

Where water resources are scarce, minimising the volume of wastewater from the process could be an important factor. If the water course is sensitive to low flows and/or the receiving waters for the wastewater are vulnerable, the sand filters might be the preferred option, but if the treatment reliability is paramount, the membrane plant could be chosen, despite its higher energy consumption, higher abstraction requirements and higher volumes of wastewater having to be returned to the environment.

11.4.2 The social dimension

The social dimension of water supply and distribution encompasses a wide range of issues, including:

- access to and affordability of a sufficient quantity of wholesome water to individuals and communities across society
- consumer attitudes about the value of water and the wider environment and how this affects their behaviour in terms of the consumption and disposal of water.

These in turn inform issues related to how water should be produced, how it should priced and what measures should be put in place to fairly distribute water, manage demand and eliminate waste. These issues were explicitly linked to the environmental agenda through the Dublin Principles (see Box 11.2; World Meteorological Organisation, 1992), which came about because of escalating concerns about the sustainability of freshwater resources arising from continued population increase, overconsumption and pollution around the world.

The production and distribution of a wholesome water supply can be considered both a basic human right and an economic good. Both aspects are embodied in the Dublin

---

Box 11.2 The Dublin Principles

1. Fresh water is a finite and vulnerable resource, essential to sustain life, development and the environment.
2. Water development and management should be based on a participatory approach, involving users, planners and policy-makers at all levels.
3. Women play a central role in the provision, management and safeguarding of water.
4. Water has an economic value in all its competing uses, and should be recognised as an economic good.

Reproduced from World Meteorological Organisation (1992)

Principles (World Meteorological Organisation, 1992). However, the extent to which water is considered to be one or the other may differ significantly between individuals, communities and societies.

Herrington (2007) suggested two potentially competing models for household water services: water as an economic good and water as a social service. In the former, an economic value is placed on water in proportion to the amount used according to appraisal of the costs of production and benefits to water users. This approach:

- stresses the commercial nature of the water industry's output
- means water services are supplied in response to 'customer' demand
- produces sympathy towards domestic metering, volumetric charging and marginal cost pricing
- is financed by the industry via customer charges
- deals with affordability problems via the tax and social security system.

In the latter, water is considered a fundamental social service, with costs being distributed among users, irrespective of how much is being used. This approach:

- stresses consumer 'rights' to the service
- uses general taxation to finance the industry
- leads to scepticism over domestic metering and concern over volumetric charging systems (metering) lest universal access be compromised
- places an emphasis on public health benefits.

Environmental objectives can theoretically be built into the economic good model by incorporating environmental costs and benefits into the valuation. However, this can potentially conflict with the 'social service' model because the resulting tariff changes, involving charging on a volumetric basis, will tend to adversely affect large households on low incomes.

So which model should be employed? The problem can perhaps be overcome by classifying (household) water use in two ways: (1) essential ('basic needs') use for drinking, cooking, washing and basic sanitation; (2) discretionary or 'luxury' use for garden watering, leisure and luxury appliances. This enables the use of increasing block tariffs, in which the base price is set according to the size of the household and price at the margin increases the more a household consumes. This approach will serve to both progressively reduce higher levels of consumption with the consequential environmental benefit and ensure fair access. There is a key issue here, beyond the engineering canon, which is the extent to which social issues of such as affordability are acknowledged within the regulatory system – or simply that water services are regarded as an economic activity and that social issues are left to government.

11.4.3   The economic dimension

Investment in water infrastructure is increasingly based on delivering maximum benefits for the least whole-life costs. The move from decisions based solely on capital cost/expenditure (CAPEX) (with little consideration of operational cost/expenditure (OPEX)) is vital, but the concept of whole-life costing is too often now expressed in terms of the economic concept of 'net present value'.

The Age of Enlightenment spawned the growth of economic theory and the nature of capitalism, through the works of such as Adam Smith (1723–1790), Jeremy Bentham (1747–1832), David Ricardo (1772–1823), John Stuart Mill (1806–73) and Karl Marx (1818–1883).

Economics is essentially rooted in the concept that the human tendency is to maximise economic efficiency – i.e. 'the optimal allocation of resources'. It says nothing about whether this optimality results in an equitable, a better, a more-effective or a more-desirable state of affairs. And neither can the 'laws' of economics transcend the laws of thermodynamics and principles of social justice – perpetual growth in world GDP has its price.

As mentioned earlier, in a pre-echo of Brundtland (WECD, 1987), it was Bentham (1789) who coined the imperative 'the greatest good for the greatest number' – and with it the dilemma of satisfying two competing objectives simultaneously. Occam's razor[a] suggests using the simplest models available to understand the world. Engineers and economists alike have tended to do so wherever possible. In engineering, there is always a tendency to see if a linear model between cause and effect will adequately model reality. It often does, as with the use of Hooke's law to relate stress and strain (it does not work so well in water engineering!). Perhaps for economists the equivalent is the discount rate, and with it the notion of net present value, to represent the time value of money and reflect the sense of future risks and present-day alternative investment opportunities.

---

[a] Occam's razor, also called the law of parsimony: a principle stated by William of Ockham (1285–1347/49), *'Pluralitas non est ponenda sine necessitate'* – 'Plurality should not be posited without necessity'. The principle gives precedence to simplicity: of two competing theories, the simplest explanation of an entity is to be preferred. (From the *Encyclopaedia Britannica*.)

But net present value calculations do not necessarily lead to acceptable decision outcomes for those problems of global proportions. As the economist Heal (1996) has observed:

> If one discounts present world Gross National Product over 200 years at 5%, it is worth only a few hundred thousand dollars, the price of a good apartment. Discounted at 10%, it is equivalent to a used car. On the basis of such valuations, it is irrational to be concerned about global warming, nuclear waste, species extinction and other long run phenomena. Yet societies are worried about these issues, and are actively considering devoting very substantial resources to them.

By definition, discount theory reduces the value of future benefits and costs – in water engineering, this tends to favour low-CAPEX and high-OPEX solutions. Besides financial impact, high OPEX often means high-energy solutions and thus high CO_2 emissions. This is the trend we see in Figure 11.4 above. Worse, the value of the environment, if it features at all in the calculation, becomes less valuable year on year. 'Net present value' is thus an economic decision-making tool that is particularly inappropriate for major decisions which have long-term consequences, and there are hardly any bigger issues than water and water supply. In short, it is far too simplistic. It applies the same discount rate to all elements of the decision, despite the fact that the future value of some aspects of the decision denies the values and costs to future generations.

11.5   Water security

At a global scale, this stems from four major sources:

- Political failure – where the problem lies in international law/treaties, etc., and where the governing factors are not necessarily hydrologically driven.
- Development/financial failure – a failure to provide adequate infrastructure.
- Hydrological failure – including the long-term but inexorable closing of the gap between water availability and demand (drought/loss of headroom, etc.).
- Hydraulic failure – viewed as the catastrophic breakdown of the collection/treatment/distribution systems from random and non-random causes, and resulting in loss of power/key plant, etc., which causes sudden outages and sources of civil strife (e.g. post-Hurricane Katrina in New Orleans, USA).

The water industry – at least in the developed world – is capital-intensive, highly regulated (financially, environmentally) and often highly competitive. It has high customer/stakeholder expectations – a '24/7' business with the potential for catastrophic failure/loss of service. At the same time, it is still often reliant on many aged assets, from the point of abstraction, treatment, distribution, wastewater conveyance and wastewater treatment.

11.6   Systems – time to rethink

From the above discussion, it is clear that the WSP engaged in delivering, operating and maintaining water supply infrastructure is now faced with addressing not just technological issues but various environmental, social and economic issues too, to be sure of delivering sustainable outcomes. Given the scale of the challenge, a new approach is needed.

Figure 11.8 Sustainability pressures demand strategic innovation rather than reactive compliance
Reproduced with permission from Barker and Johnson (2010)

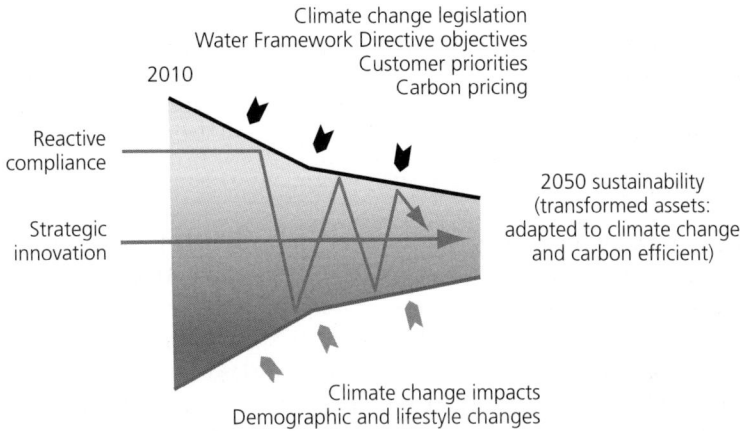

The range of pressures affecting the water industry now and into the future are summarised in Figure 11.8. To existing legislative and economic pressures can be added new pressures, such as the UK Climate Change Act (2008), which sets stretch targets for carbon reduction. The industry also has to respond to a range of other environmental and social pressures – climate change impacts but also demographic and lifestyle change, and technological change. The funnel indicates that, over time, these pressures will restrict the ability for the water sector to respond. How do we get from where we are now to sustainability in 2050?

A reactive, compliance-based approach is to wait for government regulation, for the price of carbon to increase, and respond to customer needs only as they arise – this is the harder and potentially higher-cost route. Conversely, a strategic, innovative approach will take a long-term view of asset development, think in new ways about customer service, pursue step changes in technology and approach, and deliver sustainable outcomes. A strategic approach requires careful thinking and targeted investment, but ultimately will be easier and lower cost.

A systems approach, which acknowledges the complexity inherent in considering this multiplicity of factors, is therefore vital. This is becoming increasingly recognised, but needs to be pressed home.

REFERENCES

Ackermann, F. (2008) *Critique of Cost–benefit Analysis, and Alternative Approaches to Decision-Making. A Report to Friends of the Earth England, Wales and Northern Ireland.* Medford, MA: Global Development and Environment Institute, Tufts University. www.foe.co.uk/resource/reports/policy_appraisal.pdf [accessed 01.09.10].

Ainger, C. M. (2010) Inventing the future – a sustainability strategy for water. In: *AWA Ozwater '10 Conference Proceedings.* St Leonards: Australian Water Association.

Aldersgate Group (2010) *Beyond Carbon: Towards a Resource Efficient Future*. London: Aldersgate Group. www.aldersgategroup.org.uk/reports [accessed 01.09.10].

Barker, C. and Johnson, A. (2010) Catchment-level approaches will deliver the most cost- and carbon-efficient outcomes for the water environment. In: *Water Quality Standards or Carbon Reduction – Is There a Balance?* London: Aqua Enviro Conference.

Bentham, J. (1789) *Introduction to the Principles of Morals and Legislation*, 1907 reprint of 1823 edn. Oxford: Clarendon Press.

Defra (2005) *The UK Government Sustainable Development Strategy*. London: TSO. www.defra.gov.uk/sustainable/government/publications/uk-strategy/ [accessed 01.09.10].

Defra (2008) *Future Water: The Government's Water Strategy for England*. London: TSO. www.defra.gov.uk/environment/quality/water/strategy/index.htm [accessed 01.09.10].

Environment Agency (2009) *Water for People and Environment: Water Resources Strategy for England and Wales*. London: Environment Agency. www.environment-agency.gov.uk/research/library/publications/40731.aspx [accessed 01.09.10].

Gibson, R. B., Hassan, S., Holtz, S., Tansey, J. and Whitelaw, G. (2005) *Sustainability Assessment: Criteria and Processes*. London: Earthscan.

Hardin, G. (1968) The tragedy of the commons. *Science* 162(3859): 1243–1248.

Heal, G. (1996) *Interpreting Sustainability*. *PaineWebber Working Paper PW-95-24 in Money, Economics and Finance*. New York, NY: Columbia Business School.

Herrington, P. R. (2007) *Waste Not, Want Not – Sustainable Water Tariffs*. Godalming: WWF-UK. www.wwf.org.uk/filelibrary/pdf/water_tariffs_report01.pdf [accessed 01.09.10].

Intergovernmental Panel on Climate Change (IPCC) (2007) *Climate Change 2007: Synthesis Report – Contribution of Working Groups I, II and III to the Fourth Assessment Report of the Intergovernmental Panel on Climate Change*. Geneva: IPCC.

Jenkins, G. J., Murphy, J. M., Sexton, D. M. H., Lowe, J. A., Jones, P. and Kilsby, C. G. (2009) *UK Climate Projections: UKCP09 Briefing Report*. Exeter: Met Office Hadley Centre. http://ukclimateprojections.defra.gov.uk/content/view/826/519/ [accessed 01.09.10].

MacKay, D. J. C. (2008) *Sustainable Energy – Without the Hot Air*. Cambridge: UIT Cambridge. www.withouthotair.com

Meadows, D. H., Meadows, D. L., Randers, J. and Behrens III, W. W. (1972) *The Limits to Growth*. New York: Universe Books.

Millennium Ecosystem Assessment (2005) *Ecosystems and Human Well-being: Synthesis*. Washington, DC: Island Press.

Pinchot, G. (1947) *Breaking New Ground*. New York: Harcourt, Brace.

Stern, N. (2007) *The Economics of Climate Change: The Stern Review*. Cambridge: Cambridge University Press. http://webarchive.nationalarchives.gov.uk/+/http://www.hm-treasury.gov.uk/sternreview_index.htm [accessed 01.09.10].

UKWIR (2004) *Sustainable WWTW for Small Communities*. Vol I, *Sustainability and the Water Industry*. Vol II, *BPSO Methodology Handbook*. London: UK Water Industry Research.

UKWIR (2007) *Climate Change, the Aquatic Environment and Water Framework Directive*. London: UK Water Industry Research.

Water UK and MWH (2007) *A Climate Change Adaptation Approach to Asset Management Planning*. London: Water.

World Commission on Environment and Development (1987) *Our Common Future*. Oxford: Oxford University Press.

World Meteorological Organisation (1992) *The Dublin Statement On Water And Sustainable Development.* Geneva: World Meteorological Organisation.

REFERENCED LEGISLATION

The Bathing Water Directive. Directive 2006/7/EC of the European Parliament and of the Council of 15 February 2006 concerning the management of bathing water quality and repealing Directive 76/160/EEC. *Official Journal of the European Communities* L064: 0037–0051.

The Birds Directive. Directive 2009/147/EC of the European Parliament and of the Council of 30 November 2009 on the conservation of wild birds. *Official Journal of the European Communities* L20: 7–25.

The Climate Change Act 2008 (chapter 27). London: TSO.

The Freshwater Fish Directive. Directive 2006/44/EC of the European Parliament and of the Council of 6 September 2006 on the quality of fresh waters needing protection or improvement in order to support fish life. *Official Journal of the European Communities* L264: 20–29.

The Groundwater Directive. Directive 2006/118/EC of the European Parliament and of the Council of 12 December 2006 on the protection of groundwater against pollution and deterioration. *Official Journal of the European Communities* L372: 19–30.

The Habitats Directive. Council Directive 92/43/EEC of 21 May 1992 on the conservation of natural habitats and of wild fauna and flora. *Official Journal of the European Communities* L206: 7–50.

The Priority Substances Directive. Directive 2008/105/EC of the European Parliament and of the Council of 16 December 2008 on environmental quality standards in the field of water policy, amending and subsequently repealing Council Directives 82/176/EEC, 83/513/EEC, 84/156/EEC, 84/491/EEC, 86/280/EEC and amending Directive 2000/60/EC of the European Parliament and of the Council. *Official Journal of the European Communities* L348: 84–97.

The SEA Directive. Directive 2001/42/EC of the European Parliament and of the Council of 27 June 2001 on the assessment of the effects of certain plans and programmes on the environment. *Official Journal of the European Communities* L197: 30–37.

The Urban Waste Water Directive. Council Directive 91/271/EEC of 21 May 1991 concerning urban waste-water treatment. *Official Journal of the European Communities* L135: 40–52.

The Water Framework Directive. Directive 2000/60/EC of the European Parliament and of the Council of 23 October 2000 establishing a framework for Community action in the field of water policy. *Official Journal of the European Communities* L327: 1–73.

Water Distribution Systems
ISBN: 978-0-7277-4112-7

ICE Publishing: All rights reserved
doi: 10.1680/wds.41127.331

ice

Institution of Civil Engineers

publishing

Index

Page numbers in *italics* refer to figures not on same page as text.